Process Systems Engineering Tools for the Design of Ionic Liquids and Integrated Biorefineries

Process Systems Engineering Tools for the Design of Ionic Liquids and Integrated Biorefineries

Editors

Nishanth G. Chemmangattuvalappil
Denny K. S. Ng

MDPI • Basel • Beijing • Wuhan • Barcelcna • Belgrade • Manchester • Tokyo • Cluj • Tianjin

Editors

Nishanth G. Chemmangattuvalappil
Chemical & Environmental Engineering
University of Nottingham Malaysia
Semenyih
Malaysia

Denny K. S. Ng
School of Engineering and Technology
Sunway University
Petaling Jaya
Malaysia

Editorial Office
MDPI
St. Alban-Anlage 66
4052 Basel, Switzerland

This is a reprint of articles from the Special Issue published online in the open access journal *Processes* (ISSN 2227-9717) (available at: https://www.mdpi.com/journal/processes/special_issues/PSE_biorefinery).

For citation purposes, cite each article independently as indicated on the article page online and as indicated below:

LastName, A.A.; LastName, B.B.; LastName, C.C. Article Title. *Journal Name* **Year**, *Volume Number*, Page Range.

ISBN 978-3-0365-7735-7 (Hbk)
ISBN 978-3-0365-7734-0 (PDF)

Contents

About the Editors

Nishanth G. Chemmangattuvalappil

Dr. Nishanth G. Chemmangattuvalappil is a Professor of Chemical Engineering in the department of Chemical and Environmental Engineering at University of Nottingham Malaysia. He received his Bachelor's degree in Chemical Engineering from the University of Calicut in 2001 and PhD in Chemical Engineering from Auburn University, AL, USA in 2010. His main areas of expertise include product and molecular design, application of machine learning tools on product design, carbon capture and storage using ionic liquids, and integration of molecular design techniques into the design of biorefineries.

Denny K. S. Ng

Professor Ir. Denny K. S. Ng is Dean, School of Engineering and Technology, Sunway University Malaysia. His areas of specialisation include development of netzero strategies, sustainable production and consumption strategies, optimisation of sustainable value chain of palm oil industry, energy management, resource conservation via process integration techniques, synthesis and analysis of biomass processing and integrated bio refineries. He is active in various professional bodies, which include IChemE, Young Scientist Network-Academy of Sciences Malaysia (YSN-ASM), etc. He is a fellow of IChemE, UK and The Higher Education Academy, UK; Chartered Engineer, Engineering Council, UK; Professional Engineer, Board of Engineers Malaysia; ASEAN Professional Engineer; Vice Chair of YSN-ASM, Chair of Academic Research on Palm Oil Sustainability, etc. With his excellent contributions, he received international and national recognition.

processes

Editorial

A Review of Process Systems Engineering (PSE) Tools for the Design of Ionic Liquids and Integrated Biorefineries

Nishanth G. Chemmangattuvalappil [1,*] and Denny K. S. Ng [2]

1 Department of Chemical & Environmental Engineering, University of Nottingham Malaysia, Jalan Broga, Semenyih 43500, Malaysia
2 School of Engineering and Physical Sciences, Heriot-Watt University Malaysia, No. 1, Jalan Venna P5/2, Precinct 5, Putrajaya 62200, Malaysia
* Correspondence: nishanth.c@nottingham.edu.my

Citation: Nishanth G. Chemmangattuvalappil and Denny K. S. Ng A Review of Process Systems Engineering (PSE) Tools for the Design of Ionic Liquids and Integrated Biorefineries. *Processes* **2022**, *10*, 2244. https://doi.org/10.3390/pr10112244

Received: 10 June 2022
Accepted: 13 July 2022
Published: 1 November 2022

Over the years, the global process industry is continually improving in product development and process performances. One of the areas is to identify more sustainable products and processes to manufacture products. Process Systems Engineering (PSE) tools have been instrumental in developing optimal processes and identifying better products based on different functions. With the development of novel tools, new sources of chemicals and alternative processing routes for producing products have been identified in the last few decades. Two of the most promising areas that possess the potential to provide optimal products through non-traditional processes are the integrated bio-refineries and ionic liquids. While the potential of these areas are obvious, commercial applications of these areas are still not completely explored, and PSE tools have been instrumental in the transition from the laboratory scale to the industrial scale [1]. Therefore, this Special Issue "Process Systems Engineering (PSE) Tools for the Design of Ionic Liquids and Integrated Biorefineries" makes an effort towards introducing the latest developments in these research areas to the scientific community by inviting research contributions from the top researchers in this field

Most recently, climate change due to greenhouse gas emissions, especially carbon dioxide (CO_2), has been a major concern of the society. Therefore, various research works have been developed to reduce CO_2. One of the topics of focus is carbon capture, storage, and utilization. To reduce CO_2 from emissions, the number of conventional solvents is currently being used in the absorption systems of CO_2. However, most of these processes suffer from issues such as solvent loss, environmental pollutions, and high energy regeneration, which also contributes to environmental issues. To address this issue, naturally occurring deep eutectic solvent (NADES) constituents have been studied using both experimental and Density Functional Theory [2]. NADES is found to possess most of the desirable attributes needed for CO_2 capture solvents. This work also has the potential to be extended into novel solvent design via NADES for different gases.

On the other hand, it has been recognized that ionic liquids can replace several conventional solvents in the absorption of CO_2 due to their tunability and environmentally benign nature. However, the lack of availability of experimental data and the high cost of ionic liquids are still major challenges to their application in the commercial scale[1]. In addition, the toxicity of ionic liquid is also an important factor to be considered to justify their selection over other conventional solvents. The work by Wang et al. [3] proposed an integrated approach between machine learning and computer-aided molecular design for identifying suitable ionic liquids for various applications. In the proposed approach, machine learning tools such as feedforward neural network (FNN) and support vector machine (SVM) algorithms were used to develop toxicity predictive models. It was found that the FNN-based models have a higher predictive power than SVM-based models. These models can also be incorporated into computer-aided molecular design to identify ionic liquids with desirable properties and less toxicity. We highlight that this approach can also illustrate the growing potential and role of machine learning applications in PSE.

Other than focusing on the removal of CO_2 from the end of pipe treatment in stack gas, etc., the replacement of fossil fuels with renewable sources such as biomass can also reduce CO_2 significantly. Biomass, in this work, refers to the organic materials that can be produced from agriculture products (i.e., energy crops) and wastes (i.e., rice husk, palm kernel shell, etc.), and it has been widely recognized as one of the most important renewable energy resources that reduces greenhouse gas emissions. An integrated bio-refinery is a processing facility that can convert biomass into multiple value-added products. It is extremely important to explore the potential of various biomass feedstocks for producing various products via integrated technologies. Said et al. [4] presented the potential of Napier grass to produce syngas through gasification. As reported in Said et al. [4], it was found that this energy crop has comparably high heating values and very high volatile matter, which has high potential for further investigation. Therefore, further exploring the possibility of scale up the gasification technology for practical applications is recommended.

The number of conceptual studies on the potential to produce value-added products in integrated biorefineries has also been explored recently and some of these works covered in this Special Issue. Lee et al. [5] developed a two-stage optimization approach that can identify the optimal product that can be produced from an integrated bio-refinery through a green pathway. A computer-aided molecular design approach is adapted to identify the optimal product, whereas a multi-objective optimization approach considers the economic, environmental, energy, and yield targets for production. It is noted that the bi-oil produced from biomass typically possesses poor properties such as low heating values, high viscosity, and non-homogeneity. To upgrade the bio-oil, one of the cheaper practical approaches is to blend the bio-oil with a solvent. However, the selection of solvent that can be used to upgrade bio-fuel must be based on both the functionalities and also on the cost. There are a number of solvents that have been developed for these purposes in the past. However, such specialty chemicals limit their practical applicability. To render the bio-oil solvent blend suitable for practical applications, a cost model has been integrated within a computer-aided molecular design framework [6]. The identified blends have the best economic potential to be used as fuel for real applications. We note that the final blend must be a homogeneous mixture, and a phase stability analysis was also performed in this work.

Another typical challenge encountered in the management of biomass is to synthesize an optimal supply chain of biomass resources to powerplants that can generate energy from biomass. In this Special Issue, a graphical method has been developed based on an established insight-based approach (Pinch Analysis) for the allocation of biomass to meet the demand of each powerplant [7]. Even though this approach is developed for palm-based biomass, the methodology is also applicable for other types of biomasses. A more complex scenario where the design and comparison of centralized versus decentralized bio-refining options has been performed [8]. This is a critical decision because the bio-refineries are generally small in size and the centralized options are not always the most profitable. The formulation of this optimization problem considerers several objectives such as economic, environmental, and safety aspects. This has been formulated as a multi-objective optimization problem to compare different scenarios for different performance targets. This work also proposes various future research directions that consider a more thorough environmental impact analysis, operational challenges, and uncertainty.

The effect of process intensification is another interesting area to be explored for higher profitability and a reduction in environmental impacts. The integration of both centralized and decentralized bio-refineries with other infrastructures can be considered for additional benefits by applying the concept of industrial symbiosis [8]. It has also been observed that the hydrothermal liquefaction process integrated with kraft mills can significantly mitigate CO_2 emissions [9]. The reduction is due to the replacement of fossil fuels with biofuels and produce biochar for sequestration. With such strategies, significant CO_2 emission reductions can be achieved when compared with the conventional process. An in-depth analysis of the processing conditions may further increase the benefits in these processes. In this issue, a new application of synthesis of integrated flower waste biorefinery is presented [10]. A superstructure approach is adapted to screen process alternatives to valorize flower waste into value added products. Multi-objective optimization approach (fuzzy optimization) is used to trade-off different objectives (maximize economic potential with minimum environmental impacts [10].

Based on the publications in this Special Issue, it can be seen that several promising technologies and novel products have been proposed by the PSE research community. The current work in this area focusses on incorporating more parameters related to cost, environmental impacts, uncertainty, operability, and scaled-up considerations. Considering the enormous interest in this field in the recent years, there is a very high potential for a higher industrial involvement based on the research output in these fields.

Author Contributions: Conceptualization, D.K.S.N. and N.G.C.; methodology, D.K.S.N. and N.G.C.; software, D.K.S.N. and N.G.C.; validation, D.K.S.N.; formal analysis, N.G.C.; investigation, D.K.S.N. and N.G.C.; resources, D.K.S.N. and N.G.C.; data curation, D.K.S.N. and N.G.C.; writing—original draft preparation, N.G.C.; writing—review and editing, D.K.S.N.; visualization, D.K.S.N. and N.G.C.; supervision, D.K.S.N. and N.G.C.; project administration, D.K.S.N. and N.G.C.; funding acquisition, D.K.S.N. and N.G.C. All authors have read and agreed to the published version of the manuscript.

Funding: This research received no external funding.

Conflicts of Interest: The authors declare no conflict of interest.

References

1. Chemmangattuvalappil, N.; Ng, D.; Ng, L.; Ooi, J.; Chong, J.; Eden, M. A Review of Process Systems Engineering (PSE) Tools for the Design of Ionic Liquids and Integrated Biorefineries. *Processes* **2020**, *8*, 1678. [CrossRef]
2. Altamash, T.; Amhamed, A.; Aparicio, S.; Atilhan, M. Effect of Hydrogen Bond Donors and Acceptors on CO_2 Absorption by Deep Eutectic Solvents. *Processes* **2020**, *8*, 1533. [CrossRef]
3. Wang, Z.; Song, Z.; Zhou, T. Machine Learning for Ionic Liquid Toxicity Prediction. *Processes* **2021**, *9*, 65. [CrossRef]
4. Md Said, M.; Wan Abdul Karim Ghani, W.; Hong Boon, T.; Hussain, S.; Ng, D. Thermochemical Conversion of Napier Grass for Production of Renewable Syngas. *Processes* **2019**, *7*, 705. [CrossRef]
5. Lee, J.W.Y.; Ng, L.Y.; Andiappan, V.; Chemmangattuvalappil, N.G.; Ng, D.K.S. Inverse Molecular Design Techniques for Green Chemical Design in Integrated Biorefineries. *Processes* **2021**, *9*, 1569. [CrossRef]
6. Chong, J.W.; Ng, L.Y.; Aboagwa, O.A.; Thangalazhy-Gopakumar, S.; Muthoosamy, K.; Chemmangattuvalappil, N.G. Computer-Aided Framework for the Design of Optimal Bio-Oil/Solvent Blend with Economic Considerations. *Processes* **2021**, *9*, 2159. [CrossRef]
7. Foo, D.C.Y. A Novel Graphical Targeting Technique for Optimal Allocation of Biomass Resources. *Processes* **2022**, *10*, 905. [CrossRef]
8. López-Molina, A.; Sengupta, D.; Shi, C.; Aldamigh, E.; Alandejani, M.; El-Halwagi, M. An Integrated Approach to the Design of Centralized and Decentralized Biorefineries with Environmental, Safety, and Economic Objectives. *Processes* **2020**, *8*, 1682. [CrossRef]
9. Ong, B.; Walmsley, T.; Atkins, M.; Walmsley, M. A Kraft Mill-Integrated Hydrothermal Liquefaction Process for Liquid Fuel Co-Production. *Processes* **2020**, *8*, 1216. [CrossRef]
10. Chong, E.H.Y.; Ng, D.K.S.; Andiappan, V.; Ng, L.Y.; Shivaprasad, P. Synthesis of Integrated Flower Waste Biorefinery: Multi-objective Optimisation with Economic and Environmental Consideration. *Processes* **2022**, *10*, 2240.

processes

MDPI

Review

A Review of Process Systems Engineering (PSE) Tools for the Design of Ionic Liquids and Integrated Biorefineries

Nishanth G. Chemmangattuvalappil [1], Denny K. S. Ng [2], Lik Yin Ng [2], Jecksin Ooi [3], Jia Wen Chong [1] and Mario R. Eden [4,*]

1 Department of Chemical and Environmental Engineering, University of Nottingham Malaysia, Jalan Broga, Semenyih 43500, Malaysia; Nishanth.C@nottingham.edu.my (N.G.C.); keby5cjw@nottingham.edu.my (J.W.C.)
2 School of Engineering and Physical Sciences, Heriot-Watt University Malaysia, No. 1, Jalan Venna P5/2, Precinct 5, Putrajaya 62200, Malaysia; denny.ng@hw.ac.uk (D.K.S.N.); l.ng@hw.ac.uk (L.Y.N.)
3 Department of Chemical & Petroleum Engineering, Faculty of Engineering, Technology and Built Environment, UCSI University Kuala Lumpur, No. 1 Jalan Menara Gading, UCSI Heights, Taman Connaught, Cheras 56000, Malaysia; jecksinooi@ucsiuniversity.edu.my
4 Department of Chemical Engineering, Auburn University, Auburn, AL 36849, USA
* Correspondence: edenmar@auburn.edu; Tel.: +1-334-844-2064

Received: 20 November 2020; Accepted: 16 December 2020; Published: 18 December 2020

Abstract: In this review paper, a brief overview of the increasing applicability of Process Systems Engineering (PSE) tools in two research areas, which are the design of ionic liquids and the design of integrated biorefineries, is presented. The development and advances of novel computational tools and optimization approaches in recent years have enabled these applications with practical results. A general introduction to ionic liquids and their various applications is presented followed by the major challenges in the design of optimal ionic liquids. Significant improvements in computational efficiency have made it possible to provide more reliable data for optimal system design, minimize the production cost of ionic liquids, and reduce the environmental impact caused by such solvents. Hence, the development of novel computational tools and optimization tools that contribute to the design of ionic liquids have been reviewed in detail. A detailed review of the recent developments in PSE applications in the field of integrated biorefineries is then presented. Various value-added products could be processed by the integrated biorefinery aided with applications of PSE tools with the aim of enhancing the sustainability performance in terms of economic, environmental, and social impacts. The application of molecular design tools in the design of integrated biorefineries is also highlighted. Major developments in the application of ionic liquids in integrated biorefineries have been emphasized. This paper is concluded by highlighting the major opportunities for further research in these two research areas and the areas for possible integration of these research fields.

Keywords: ionic liquids; integrated biorefineries; process systems engineering; process optimization; molecular design

1. Introduction

The competitive nature of the current chemical industry constantly demands improvements in process and product quality and efficiency. To maximize profitability and to succeed in the global competition, the chemical processes need to be operated optimally and also need to look for better products and ways to produce them. Process Systems Engineering (PSE) tools have been instrumental in developing optimal processes and identifying better products. PSE is defined by Grossmann and Westerberg [1] as "the field that is concerned with the improvement of decision-making processes for the creation and operation of the chemical supply chain. It deals with the discovery, design, manufacture, and distribution of chemical products in the context of many conflicting goals." The developed approaches towards chemical process design is a major shift from the conventional focus where the emphasis was to develop tools and procedures for the process design, control, and operation. In recent decades, the focus of research in PSE fields has shifted to the areas of development of new products, process networks and enterprise, supply chain optimization, and global life cycle assessment (LCA) [2].

The development of PSE tools and methodologies has enabled the applications of these tools in various research fields. While the focus of PSE tools was more process-oriented and towards specific industrial problems in the twentieth century, the effect of globalization has prompted researchers to extend the PSE tools into new areas, e.g., development of new products, process intensification, and to address sustainability challenges. Significant improvements in computational efficiency have contributed to the application of PSE tools in several areas where historically, size-related problems had made the solutions impractical. In this paper, the application of novel PSE approaches in the design of ionic liquids and the synthesis of integrated biorefineries are highlighted. Since these are relatively recent concepts, specific PSE tools had to be developed to address the specific/novel challenges raised by these problems.

It has been recognized in recent years that ionic liquids may be a suitable replacement for several traditional solvents that pose environmental risks. The major potential areas for their application as a solvent include carbon capture, extraction, as an entrainer in extractive distillation, and also in various chemical, biochemical, electrochemical, and pharmaceutical industries [3]. Ionic liquids have unique properties that make them suitable for these industrial applications. In addition, it is possible to alter the functional groups to meet the desired attributes for various applications. The two major challenges for the design of suitable ionic liquids and the application in industries are the unavailability of reliable data and the high production cost of ionic liquids. Because of that, the ionic liquids used for various applications and the conditions at which the systems are operated may not be optimal. In order to make an optimal selection of ionic liquids, various PSE approaches (both insight-based and mathematical optimization approaches) have been developed in the past decades. The development of novel group contribution (GC) models [3], computer-aided molecular design (CAMD) tools [4], and the development of quantum mechanical (QM) tools that can be integrated with CAMD approaches [5] had been the significant breakthroughs that allow the application of PSE tools in the ionic liquid design and selection.

In addition to the design and selection of ionic liquids, the development and application of PSE tools in the field of integrated biorefineries have been immense. In recent years, biomass utilization has shown promising results in addressing society's dependence on non-renewable energy resources and climate change caused by fossil fuel exploitation. Other than being utilized for heat generation through direct combustion, the nature of biomass enables it to be converted into other value-added products ranging from biomaterials, biofuel, bio-chemicals, biopharmaceuticals, etc. Through different biomass conversion technologies (physical, mechanical, chemical, and biological), biomass can be converted into other forms of energy products such as transportation fuels, or value-added products like commodities and specialty chemicals. In the past decades, the development of single biomass conversation systems such as gasification, fermentation, digestion, etc., has been established. However, due to the complexity and diversity of biomass in nature, a single biomass conversion system is typically not able to fully recover the potential of the biomass. In view of this, integrated biorefineries is a fast-developing research area that integrates a variety of technologies to convert biomass into the abovementioned value-added products [6]. As integrated biorefinery is able to integrate multiple technologies as a single integrated system, such a system provides more flexibility in product generation and generates sufficient energy to support the entire operation and reduce the overall energy consumption compared to the processes that operate independently. In addition, with the integration of multiple technologies, the waste/by-products can be used as feedstock for another process, therefore, the material recovery can be maximized as shown in Figure 1. Various research works [6–10] have been presented to optimize the synthesis and design of integrated biorefineries, ranging from insight-based approaches to complex mathematical optimization models, which is discussed in Section 3.

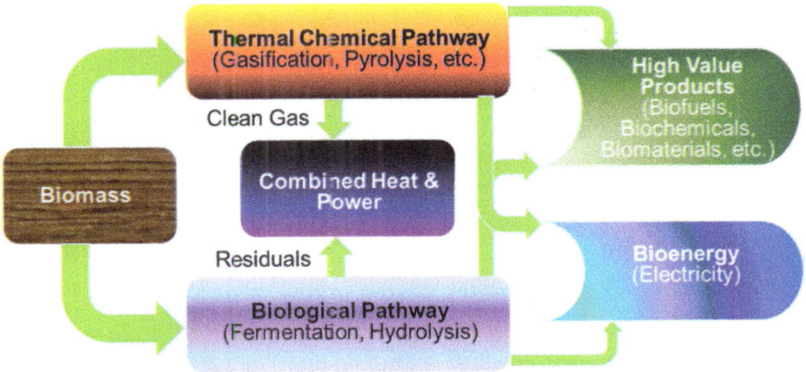

Figure 1. The integrated biorefinery concept.

In summary, advances in computing, data science, and optimization have allowed researchers to extend the application of PSE tools in several novel areas. In this paper, the application of PSE tools in the synthesis and design of ionic liquids, as well as integrated biorefineries, are to be reviewed. A general overview of these two research areas is provided followed by a detailed review of the recent development and application of PSE tools in the area. The major research challenges and the approaches used by the PSE community in addressing the identified challenges are highlighted in this paper. This review also focuses on the integration of these research fields where the role of ionic liquids as solvents in integrated biorefineries has been analyzed. Finally, the paper identifies some of the research scopes for further research and the way forward.

2. Applications of PSE in the Development of New and Green Chemicals

In this section, an introduction to ionic liquids and their potential applications is presented. A comprehensive review of the applications of ionic liquids in various industries is also provided. Furthermore, the latest developments and contributions in the field of ionic liquid design are reviewed. In addition, the challenges and limitations encountered in the design of optimal ionic liquids are discussed. Lastly, the recent applications of CAMD in the ionic liquid design are reviewed in detail.

2.1. Ionic Liquids

Ionic liquids are organic salts that comprise organic cations with inorganic or organic anions and melt at temperatures below 100 °C. They were first introduced by Paul Walden in 1914 when he successfully synthesized ethyl ammonium nitrate which melts at room temperature [11]. The introduction of ionic liquids has attracted considerable attention from researchers, especially in replacing traditional solvents with ionic liquids due to their extraordinary characteristics. For example, ionic liquids are known to exhibit non-flammability, as well as high thermal and chemical stabilities owing to the strong ionic bonds [12]. Besides, ionic liquids also possess the attributes of "green" solvents since their vapor pressure is significantly lower than the conventional solvents. The high volatility of conventional organic solvents can cause air pollution and human health problems if they leak from process equipment.

Ionic liquids are widely recognized as "designer" solvents because of the flexibility in turning their properties by altering the cations. For instance, their melting point can be reduced by having large asymmetric organic cations [13]. By selecting various combinations of anions and cations, ionic liquids with an attractive set of physicochemical properties can be synthesized. In addition, ionic liquids' capabilities, which include a wide range of intermolecular interactions such as hydrogen bonding, ionic and covalent interactions as well as π-stacking, have also contributed to their tunable properties [14]. For these reasons, they have huge potential to reduce the environmental impact caused by organic solvents. Nonetheless, the major drawback of applying ionic liquids in various chemical processes is their high purification cost. As ionic liquids consist of highly charged ions, the ionic liquid recovery and purification process becomes difficult and cost-intensive [15]. In spite of the high recovery and purification cost, the unique properties of ionic liquids are still important and attractive from an industrial point of view. This can, at least in part, be seen by an increasing number of patent filings, especially related to the applications of ionic liquids [16]. Moreover, there is also an exponential increase in Science Citation Index (SCI) papers published on ionic liquids in the last few decades [17]. For these reasons, it is feasible that ionic liquids can become viable options for various applications and industrial processes.

2.2. Potential Applications of Ionic Liquids

In recent years, there is a surge in the applications of ionic liquids in various industries. Table 1 shows some potential applications of ionic liquids that will be discussed in detail in this section.

Table 1. Summary of potential applications of ionic liquids discussed in this review.

Types of Ionic Liquids	Applications
1-methylimidazolium chloride	Biphasic acid scavenging [18]
1-butylpyridinium tetrafluoroborate	Geothermal fluid in organic Rankine cycle [19]
1-ethyl-3-methylimidazolium bis(trifluoromethylsulfonyl)imide	Hydrogen compressor [20]
1-Ethyl-3-methylimidazolium acetate	Sugarcane bagasse pre-treatment [21]
1-butyl-3-methylimidazolium chloride and 1-butyl-3-methylimidazolium acetate	Rubber woods pre-treatment [22]
1-butylimidazolium hydrogen sulfate (VI)	Kraft lignin activator [23]
1,3-dibutyl-2-methylimidazoliumbromide	Flavonoids extraction [24]
1-hydroxyethyl-3-methylimidazoliumbis (trifluoromethanesulfonyl)amide	Keratin extraction [25]
1-Ethyl-3-methylimidazolium methylsulfate	Ethyl acetate and ethanol azeotropic distillation [26]
1-butyl-4-methylpyridinium tricyanomethanide	Cyclohexane and benzene azeotropic distillation [27]
1-ethyl-3-methylimmidazolium-ethylsulphate	Carbon dioxide and hydrogen sulfide separation [28]
1-butyl-3-propylamineimidazolium tetrafluoroborate	Carbon dioxide capture [29]
1-butyl-3-methylimidazolium hexafluorophosphate	Carbon dioxide capture [30]
Tetraglyme-sodium salt ionic liquids	Sulfur dioxide separation [31]
1-butyl imidazolium bis(trifluoromethylsulfonyl)imide	Ammonia separation [32]
Bis(1-butyl-3-methyl imidazolium) copper tetrachloride salt and bis(1-butyl-3-methyl imidazolium) stannum tetrachloride salt	Ammonia separation [33]

The first industrial application of ionic liquids, called the BASIL process, was developed by BASF [18]. This process is mainly used for producing generic photo initiator precursor alkoxy phenyl phosphines. It is the first commercial publicly announced process that uses an ionic liquid (named 1-methylimidazolium chloride) to scavenge acid from reaction mixtures. The extraordinary properties of ionic liquids are also advantageous for several reactions such as bio-catalyzed reaction, dimerization, isomerization, hydrogenation, hydroformylations, and alkylation [34]. Higher reaction rates and selectivity can be achieved when organic solvents are replaced with ionic liquid solvents. For instance, the hydrolysis rate of casein by lumbrokinase is enhanced when ionic liquids are added to the reaction [35]. The existence of ionic liquids at low concentrations improves the hydrophilicity of casein by forming an ionic liquid-casein complex.

Ionic liquids have also been reported as excellent candidates for heat transfer fluids owing to their low flammability, high heat stability, and low volatility characteristics. When ionic liquids were used as heat transfer fluids in shell and tube heat exchangers, a comparable or larger heat transfer area can be achieved compared to other heat transfer fluids [36]. The ionic liquid, 1-butyl pyridinium tetrafluoroborate ($C_9H_{14}NBF_4$) was used as a geothermal fluid in an organic Rankine cycle was explored by Kazemi et al. [19]. Optimization and simulation work was carried out for a basic organic Rankine cycle using $C_9H_{14}NBF_4$ and water as the geothermal fluids, respectively. In their study, economic, thermodynamic, and thermo-economic evaluations were chosen as objective functions for minimizing specific investment costs and maximizing exergy efficiency. The results indicated that the basic organic Rankine cycle performance was improved thermodynamically and economically when $C_9H_{14}NBF_4$ was used as a geothermal fluid.

Moreover, ionic liquids have a high tendency to wet inorganic, metal, and polymeric surfaces [34]. With their wide liquidus range and high thermal stability, ionic liquids are ideal candidates to be used as lubricants in low pressure and/or high-temperature applications. Other than the aforementioned properties, ionic liquids are also applied as lubricating media in oxygen compressors due to their good lubricity and high chemical inertness [37]. Based on the long-term test conducted with pure oxygen, the result is both impressive and promising as the hydrocarbon concentrations in all tests were below 10 ppmv methane equivalents, which is beyond the threshold value of 1000 ppmv. In another recent study, it was proposed that the solid piston in reciprocating hydrogen compressors may be replaced by ionic liquids [20]. By replacing the solid piston in reciprocating compressors with a suitable liquid, efficiency can be improved along with significant cost reduction.

Apart from being suitable candidates for lubricants and thermal transfer fluids, ionic liquids have also been introduced in biomass conversion applications, such as lignocellulose pretreatment and fractionation. Most existing molecular solvents are not able to dissolve and hydrolyze cellulose into fermentable sugars as the cellulose chains are interconnected by strong hydrogen bonding in flat sheets [38]. Nonetheless, various ionic liquids have demonstrated their capabilities in dissolving and fractionating lignocellulose into its main constituent compounds. For example, cellulose isolated from sugarcane bagasse was pretreated with an ionic liquid called 1-Ethyl-3-methylimidazolium acetate ([Emim]Ac) at 90 °C prior to hydrolysis by cellulase [21]. After the pretreatment process, the glucose content is increased whereas the crystallinity index and degree of polymerization are reduced. The cellulose crystal structure has also transformed from cellulose I to cellulose II. These changes have enhanced enzymatic hydrolysis rate and glucose production yield. Furthermore, the effect of pretreating rubber woods with ionic liquids on thermodynamic properties and pyrolysis kinetics was explored by Khan et al. [22]. The pretreatment of rubber woods with both ionic liquids, 1-butyl-3-methylimidazolium chloride ([BMim][Cl]) and 1-butyl-3-methylimidazolium acetate ([BMIM][OAc]) successfully reduced the activation energy required for the pyrolysis reaction. Moreover, ionic liquids have also shown their potential in kraft lignin activation processes. After activating kraft lignin with ionic liquids, a better electrochemical performance was observed due to an increase in carbonyl groups, which are responsible for transporting protons and electrons in electrochemical applications [23].

Considering stringent environmental regulations as well as increased emphasis on green and clean manufacturing practices, ionic liquids have gained popularity in extraction processes. This is because most of the common organic solvents used in extraction processes are toxic and flammable, which leads to detrimental environmental issues when released into the atmosphere. The use of ionic liquids in extraction processes typically leads to shorter processing time and most importantly, the process can often be carried out at ambient temperature [39]. For this reason, the application of ionic liquid is widely found in the extraction and purification of bioactive compounds such as proteins, lipids, amino acids, and pharmaceuticals. Novel dual-chain ionic liquids were synthesized and applied to recover flavonoids from *Pinus massoniana* Lamb [24]. The results showed that ionic liquids are more efficient in extracting flavonoids than water and ethanol. In another study, a hydrophobic ionic liquid was shown to be capable of extracting keratin from chicken feathers [25]. The water extraction method can be used to isolate keratin from the ionic liquid as hydrophobic ionic liquid is insoluble in water while keratin is soluble in water.

A classic challenge in the process has been to choose a suitable entrainer in separating azeotropic mixtures. Ionic liquids have been suggested as potential entrainers in an azeotropic system since their selectivity, capacity and thermal stability can be easily altered [40]. For example, Zhu et al. [26] performed a process simulation study of ionic liquids applied in an extractive distillation process to separate an ethyl acetate and ethanol mixture, which have close boiling points. The results showed that 1-Ethyl-3-methylimidazolium methyl sulfate was the best solvent for this separation. In another contribution, 1-butyl-4-methyl pyridinium tricyanomethanide ([4bmpy][TCM]) has been proposed as an entrainer in the extractive distillation process to separate cyclohexane and benzene [27] as overall solvent consumption can be reduced. Another contribution evaluated the capability of protic ionic liquids as an entrainer to separate ethanol-water azeotropic mixtures [41]. In this process, the relative volatility between ethanol and water is increased and the complete azeotrope elimination was achieved. Conductor-like screening model for real solvents (COSMO-RS) was used to study the intermolecular interactions between the binary ethanol-water and the ternary ethanol-water-protic ionic liquids mixtures. The results were in agreement with the experimental values.

The possibility of using ionic liquids to separate liquids was explored by Fadeev and Meagher [42], where an ionic liquid was used to recover butyl alcohol from a fermentation broth. Ionic liquids are also reported to be efficient in gas separation. Due to their hygroscopic nature, ionic liquids can remove water vapor efficiently from gas mixtures [43]. Likewise, the solubility of gas can be altered by the proper selection of cation, anion, and substituents. For example, when certain anions are used and the alkyl chain is shortened, the Henry's Law constant of carbon dioxide is reduced to half [34]. Based on the experiments performed by Anthony et al. [44], gases were found to have different solubilities in ionic liquids, suggesting that there is a great potential for designing ionic liquids to be used in specific gas separations.

Carbon dioxide (CO_2) was first reported to be highly soluble in imidazolium-based ionic liquids like 1-butyl-3-methylimidazolium hexafluorophosphate by Blanchard et al. [30]. The finding reported that the dissolution of CO_2 in imidazolium-based ionic liquids is totally reversible meaning that pure ionic liquids can be easily recovered after the desorption process. Since then, there has been an increasing number of scientific studies investigating the capability of ionic liquid compounds in capturing CO_2. The efficiency of a new task-specific ionic liquid, which is an amino-functionalized ionic liquid named 1-butyl-3-propylamineimidazolium tetrafluoroborate ([NH_2p-bim][BF_4]), was comparable to conventional solvents in CO_2 absorption processes [12]. To maximize the interaction sites for CO_2 capture, a dual amino ionic liquid that contains a taurine anion and amino-functionalized imidazolium cation has been synthesized [45]. The result reported that 1 mole of the ionic liquid can absorb 0.9 mol of CO_2 at atmospheric pressure. Besides, the application of both an ionic liquid and a zeolitic imidazolate framework (ZIF) as a separating agent in adsorptive absorption has demonstrated its capability in capturing CO_2 [29]. The experimental and modeling results showed that CO_2 is more soluble in the ionic liquid and ZIF mixture compared to that of pure ionic liquid.

Ionic liquids have also been introduced for absorbing other environmentally polluting gases such as hydrogen sulfide, ammonia, and sulfur dioxide. A mixed matrix membrane with imidazolium-based ionic liquids blended with poly ether-block-amide elastomer was synthesized for CO_2 and H_2S separation [28]. This study showed that the chosen ionic liquid has higher H_2S selectivity than CO_2. In another contribution, tetraglyme-sodium salt ionic liquids which possess high thermal stability appeared to be potential candidates for SO_2 absorption [31]. The experimental results showed that SO_2 absorption capacity improved by 30% when tetraglyme–sodium salt ionic liquids were used instead of tetraglyme. Three different types of ionic liquid including traditional ionic liquids, protic ionic liquids, and Brønsted acidic ionic liquids were synthesized and investigated for NH_3 absorption [32]. Among the investigated ionic liquids, the protic ionic liquid (1-butyl imidazolium bis(trifluoromethylsulfonyl)imide ([Bim][NTf_2])) achieved the highest NH_3 absorption capacity. 1 mole of [Bim][NTf_2] was able to absorb 2.69 moles of NH_3 at ambient pressure. Another study synthesized a series of metal ionic liquids and explored their potential for NH_3 separation [33]. Bis(1-butyl-3-methyl imidazolium) copper tetrachloride salt ([Bmim]$_2$[$CuCl_4$]) and bis(1-butyl-3-methyl imidazolium) stannum tetrachloride salt ([Bmim]$_2$[$SnCl_4$]) showed great potential for NH_3 separation due to their high selectivity and absorption capacities.

The abovementioned works illustrate that ionic liquids have been extensively explored for various applications. Nonetheless, considering that an enormous number of ionic liquids can be synthesized by combining various organic cations and anions, it is difficult to select the optimum candidate(s) from this huge search space. Experimentally testing random ionic liquid combinations is both time consuming and costly. Thus, in order to minimize the time and expense required in performing experiments, various theoretical/computational methods have been proposed to guide the ionic liquids screening process preceding an experimental campaign. The following section presents the latest developments and challenges in the design of optimal ionic liquids.

2.3. Challenges in the Design of Optimal Ionic Liquids

In order to design ionic liquids that suit a certain industrial application, it is first important to know how the anion, cation, and side chains on the cation affect physicochemical properties. This can be done through two pathways, either via experimental work or simulation approaches. Considering a very huge number of possible ionic liquids, it would be a very tedious task to select an optimum candidate through experimentation. Because of this, researchers have been working on developing systematic computer simulation methods including CAMD and/or process design simulations to design optimal ionic liquids. Computer-aided methods provide alternative routes in determining potential ionic liquids as they can explore a larger number of options in a shorter time frame. However, it does not mean that computer-aided methods can replace experiments entirely as identified candidates should be further tested via experiments to verify their performance.

CAMD techniques are shown to be effective in designing various types of chemical, biochemical, and material products. It is able to design molecular structures that meet a predefined property target. The success of employing the CAMD technique to design ionic liquids depends greatly on the availability and reliability of the associated/underlying predictive models. Only when thermophysical properties such as density and viscosity (which influence mass transfer rates) are satisfactorily characterized, can optimal ionic liquids and/or the process in which they are used be designed [46]. Despite the availability of significant thermophysical property data for an extensive range of ionic liquids in the free ILThermo database [47], predictive models for most thermophysical properties with adequate accuracy are still being developed [48]. The availability of adequate predictive models is important in the design/development of new processes, improvement of operating conditions in a process, and reduction of energy consumption [48]. Various approaches have been proposed to develop property prediction models for thermophysical and transport properties of ionic liquids. These approaches can be categorized into two main classes which include both theoretical and empirical methods [49]. The property prediction models classified under theoretical approaches include molecular dynamics (MD) simulations, equations of state (EOS), Monte Carlo (MC) simulations as well as rough hard-sphere theory (RHST). On the other hand, machine learning (ML) tools, GC methods, and quantitative structure–property relationships (QSPR) are considered empirical approaches. These prediction models are all widely applied in predicting physicochemical properties of ionic liquids, which include density, viscosity, thermal conductivity, surface tension, heat capacity, etc. A comprehensive review of the current developments in prediction models for pure ionic liquids was presented by Hosseini et al. [48]. Moreover, Hosseini et al. [48] also reviewed the most extensively applied semi-classical EOSs in estimating thermodynamic and physical properties of pure ionic liquids. EOSs are typically developed from empirical, theoretical, and semi-theoretical approaches. Semi-classical EOSs are developed from a combination of molecular thermodynamic theories of fluid (theoretical EOSs) and classical thermodynamic approaches (empirical EOSs). Nonetheless, most of these existing prediction models for thermodynamic properties are limited to only a relatively small number of common families of ionic liquids.

To alleviate the limitations mentioned above, Chen et al. [3] presented newly developed GC models that are capable of estimating thermodynamic and physical properties for additional families of ionic liquids. The thermophysical properties include heat capacity, viscosity, surface tension, density as well as the melting point of the ionic liquids. New correlation equations with higher prediction accuracy that are easier have been proposed for the prediction of the aforementioned properties. The GC parameters were developed based on more than 13,300 experimental data points, including approximately 200 ionic liquids derived from 19 anions, 6 cations, and 4 substituents covering a large range of pressure and temperature conditions. These newly developed GC models offer the opportunity of applying them for CAMD to identify more potential ionic liquids solutions. Additionally, there is great potential for extending these methods to even more families of ionic liquids once experimental data become available.

With increasing awareness of environmental detriments, it is important to also consider environmental properties during the ionic liquid design stage. Since ionic liquids are normally highly soluble in an aqueous medium, they are readily discharged into the environment through wastewater [50]. Hence, the designed ionic liquids need to be environmentally friendly before they start accumulating in the environment. For this reason, various quantitative structure-activity response (QSAR) models were developed to assess the toxicity of ionic liquids. Recently, a comprehensive review of the expansion of QSAR models for toxicity prediction of ionic liquids was presented by Abramenko et al. [51]. The review concluded that the toxicity of ionic liquids is not solely reliant on the type of cation but increases with the length of the cation alkyl chain and the number of cation chain group branches. It was also found that the toxicity of ionic liquid is reduced with the presence of a polar side chain. Furthermore, the presence of polar side chains was found to increase the efficiency of biodegradation. However, most existing models for ionic liquid toxicity prediction only cover a small number of ionic liquids.

It is notable that pure ionic liquids may not always be the optimal candidates in satisfying the target properties, however, it is possible that ionic liquid mixtures could assist in addressing such problems. Unfortunately, property prediction models for ionic liquid mixtures are limited as this research area is still in its infancy. In a recent contribution, equations of state and artificial neural networks (ANN) were combined to predict volumetric properties of a limited number of pure and mixtures of amino acid ionic liquids [52]. Another contribution explored the behavior of some properties such as viscosity, diffusion coefficient, and electrical conductivity for binary mixtures of protic-protic ionic liquids as well as protic-aprotic ionic liquids [53]. The results presented that the electrical conductivity of ionic liquid mixtures is higher than that of the pure ionic liquids. Efforts have been devoted to developing property prediction models for ionic liquid mixtures, but the models are limited to only certain classes of ionic liquids. Future research should focus on developing reliable property prediction models for different families of ionic liquid mixtures.

2.4. CAMD for Ionic Liquid Design

The development of CAMD tools for the ionic liquid design has been limited due to the lack of accurate models in estimating ionic liquid properties. Most of the earlier work in the modeling of ionic liquids was based on molecular dynamics tools and quantum chemistry study. While these tools show reasonably accurate and reliable results, the computational time needed to complete the simulation of each candidate molecule is large. So, the applicability of these tools is more appropriate towards the final selection from a limited number of promising molecules. An early application of CAMD in ionic liquid design was the design of a suitable ionic liquid to be used in a refrigerant gas separation system. This work used molecular connectivity index based QSPR models to link the structure of the ionic liquids to target properties like diffusivity, Henry's law constant, and solubility [54]. A CAMD problem is formulated to determine the molecular structures is then formulated as a MILP.

Another earlier application of CAMD tools for the design of ionic liquid was the extension of existing CAMD approaches based on GC models [4]. The original MINLP problem was decomposed into a set of sub-problems to reduce the complexity of the model. This approach solves one sub-problem each for structural constraints, physical property constraints, and mixture property constraints. Each of these sub-problems can eliminate several infeasible structures. The solutions obtained through the sub-problems are considered in a final sub-problem to solve for the optimum ionic liquid molecule based on the objective function and process models.

Figure 2 shows the general framework of ionic liquid design via CAMD. The procedure starts with an ionic liquid design problem definition where the product needs are identified. These requirements are then translated into targeted properties which will be validated using a model-based approach. Next, an extensive search of ionic liquids is conducted. The information collected from a vast number of literature sources and online databases is stored in the ionic liquid library. Various property models are also collected and stored in the property model library. In Step 4, UNIFAC and/or COSMO models are used to screen and predict ionic liquid with desirable properties. Lastly, the performance of the shortlisted ionic liquid candidates is validated either by experiments or simulations. This step is crucial to ensure that the ionic liquids identified are feasible and practical.

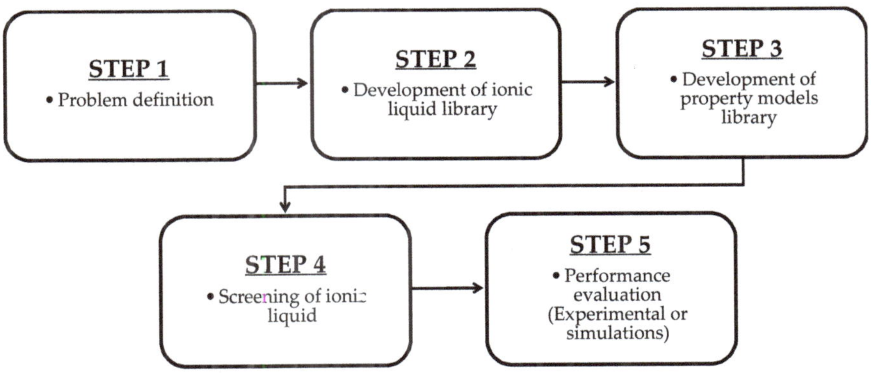

Figure 2. General framework of ionic liquid design via a computer-aided molecular design (CAMD) approach.

The application of GC-based models in CAMD has been well established and there are efficient algorithms for solving CAMD problems with different target properties. A comprehensive review of these works can be seen in the reviews of Austin et al. [55] and Chemmangattuvalappil [56]. In order to provide an accurate prediction of properties using GC models, the binary interaction parameters need to be obtained. QM calculations have been incorporated into several CAMD algorithms to address this issue in recent years [56]. One of the most popular advances in CAMD is the application of COSMO-based methods as the information of binary interaction parameters is not necessary [5]. COSMO-RS [57] and COSMO-SAC [58] are the two most popular methods applied in CAMD [5]. These methods only require the estimation of molecular volumes and sigma profiles for the thermodynamic calculations. This makes the COSMO based approaches extremely suitable for the design of ionic liquids.

A preliminary design strategy to screen a list of possible ionic liquids for CO_2 capture by using COSMO-RS was proposed by Farahipour et al. [59]. COSMO-RS was applied for activity coefficients estimation whereas viscosity and melting point constraints are modeled by employing GC models. The proposed methodology holds the capability to consider different trade-offs and optimal ionic liquids were identified by enumerating all possible cation and anion combinations. As a result, 10 ionic liquids with promising characteristics were identified out of a few thousand ionic liquid candidates considered. This method was improved with an extended model library for ionic liquid properties along with the consideration of several ionic liquid-based separation processes [60]. In their work, CAMD, property models based on GC, and process design simulations are combined and solved simultaneously to obtain the optimal ionic liquid separating agent. The developed method was then applied in case studies. For the first case study, CO_2 capture from natural gas was considered while the focus of the second case study was ethanol-water azeotrope separation.

In recent years, systematic approaches have been developed for ionic liquid screening as extraction solvents. In one study, a sequential approach combining COSMO-RS-based LLE prediction, GC-based physical property prediction, and process simulation using Aspen Plus was applied [61]. Firstly, COSMO-RS is utilized to select the cation-anion solutions which possess desirable properties with common organic solvents as a benchmark. In the next step, GC tools have been applied to screen ionic liquid with desirable properties. Finally, ionic liquids that give the optimal process performance are selected by evaluation through process simulation. This method was then further improved by extending the UNIFAC-IL model to describe the extractive desulfurization of fuel oils system with extensive experimental data [62]. Optimal ionic liquid molecules for the extractive distillation can be identified by integrating both process and molecular design [63]. UNIFAC-IL model was selected for its database. The database fitting the group binary interaction parameters in the UNIFAC-IL model agreed well with the characteristics of the compounds in the system. An MINLP model was formulated to represent the design problem by combining UNIFAC-IL and GC models. The potential of the shortlisted ionic liquid candidates was then evaluated by process simulation. Song et al. [64] employed a similar approach for the design of ionic liquids for an alkane/cycloalkane extractive distillation process. In this design, the UNIFAC-IL model had been extended based on the consideration of the proximity effect in alkanes/cycloalkanes as distinct groups. Optimal ionic liquid as an entrainer had been modeled by formulating an MINLP problem and then through process simulation and economic analysis.

A CAMD approach was also applied to design optimal ionic liquids for cellulose dissolution. In this work, the solubility of cellulose in ionic liquids was estimated using a QSAR model which was developed using GC and ANN methods [65]. The design problem was formulated as a MINLP and solved by a genetic algorithm with an objective function maximizing the cellulose solubility in ionic liquids. Experimental analysis and characterization of the identified ionic liquid candidates were also performed. A simple method for CAMD in ionic liquids design as CO_2 absorbents was proposed by Firaha et al. [66]. The strength of the interaction between CO_2 and ionic liquid anions corresponded to chemical and physical absorption, respectively. The type of absorption was predicted according to geometry optimization accompanied by a solvation model. Solvated geometries provide Gibbs free energies which are analogous to the experimental values and correlate well with the experimentally observable absorption capacity.

An extensive GC model for infinite dilution activity coefficient (IDAC) of molecular solutes has been introduced to design ionic liquids for extractive desulfurization of gasoline [67]. In this work, ionic liquids are generated by applying purely empirical correlations of GC-based IDAC data and the least-squares support vector machine (LSSVM) method. The LSSVM-based model was capable of predicting IDACs based on the GCs of ions and Abraham's solvation descriptors of solutes. The reliability of the developed CAMD tool was verified by LLE simulation of ternary systems.

Another significant contribution is the ionic liquid design approach Karunanithi et al. [68] in which GC models and the COSMO-SAC thermodynamic model are combined to predict the surface charge density of ionic liquid structures predicted by density functional theory (DFT). The CAMD model can be solved via either deterministic or stochastic methods. This approach was tested on the design of ionic liquids for ibuprofen dissolution and ionic liquids with high electrical conductivity. The properties of the identified ionic liquid were verified with the data on the properties of the predicted structure. In another contribution, a multi-scale simulation approach was introduced to design ionic liquid solvents for separation processes [69]. An extended GC-COSMO approach was developed to estimate the σ-profiles and cavity volumes of ionic liquids. In this work, activity coefficients were predicted by applying the COSMO-SAC model while the rest of the properties were predicted using semi-empirical models. A CAMD problem was formulated using an MINLP model. The MINLP was then solved using the branch and bound approach in order to obtain the optimal ionic liquid solvents. A hybrid process design approach has been developed by Chen et al. [70]. This approach combined ionic liquid design and process simulation to obtain the best ionic liquid for hybrid process schemes. Hybrid schemes that satisfy the desired demands were generated. The ionic liquid design problem was solved based on the hybrid schemes generated, structural constraints, along with the physical properties. Shortlisted ionic liquid candidates were evaluated by conducting the simulation of the process.

The advances in the ML tools have also been utilized in the ionic liquid design. In a recent contribution, a huge data library of ionic liquids has been used to train ML models [71]. ML models were developed using three learning methods simultaneously, including the random forests (RF), cubist, and gradient boosted regression (GBM). The ML models thus developed (trained) were applied to design ionic liquids by identifying the cation-anion pairs from a large database that covers varied chemical scaffolds. The performance of the proposed method was then validated theoretically and experimentally. Over 2600 promising ionic liquids were found to be potential ionic liquids for gas separation and cellulose dissolution applications.

3. Applications of PSE in Integrated Biorefineries

In this section, an introduction to the field of integrated biorefineries and the types of integrated biorefineries is presented. The different types of processes common in an integrated biorefinery are explained. Following this discussion, various design approaches and major contributions made in each of these approaches are reviewed in detail. Finally, the research gaps are identified and major current challenges in the design and current work on these challenging areas are presented.

3.1. Introduction to Integrated Biorefineries/Types of Integrated Biorefineries

Over the past few decades, in order to enhance the sustainability of chemical and energy production, there has been a shift from petroleum-based feedstock to biomass-based feedstock. Besides, the societal realization of limited non-renewable resources, concerning environmental issues, technological advancements, and the discovery of additional renewable energy resources have also contributed to this shift [72].

Early on, biomass was mainly used as renewable resources for energy production. Later, biomass has also been used as feedstock for the production of value-added products via different processing pathways (physical, biological, and thermal, etc.). Various independent processes (e.g., gasification, pyrolysis, fermentation, hydrolysis, palletization, etc.) for the production of bio-chemicals, biomaterials, and bio-specialty chemicals from biomass have been developed. The utilization of biomass as feedstock for the production of multiple products through a *biorefinery* has also gained attention from both industry and the scientific community in the past decades [6]. A biorefinery was first defined as *a complex system of sustainable, environmental, and resources-friendly technologies for the comprehensive utilization and the exploitation of biological raw materials (biomass)* by Kamm et al. [73]. Similar to a petroleum refinery, a biorefinery utilizes different biomass resources as feedstock, combines a diversified collection of conversion pathways to produce a wide range of value-added products for example bioenergy, bulk chemicals, and fine chemicals. According to Frost and Draths [74], the application of the biorefinery concept plays a crucial role in driving the change in the chemical industry to shift from using petroleum-based feedstock to biomass-based feedstock. As biomass consists of a wide range of organic constituents, it comes in a variety of forms with different properties and characteristics. Therefore, various processing technologies can be applied to convert biomass into higher-value market products. A range of pre-treatment systems (such as size reduction, drying system, acid, and base hydrolysis) are needed to standardize biomass into a form that can be further converted into products. In order to maximize the quality and performance of the biorefinery, process integration and optimization shall be applied to synthesize a biorefinery comprised of multiple processing systems. This realization provides the foundation for the concept of integrated biorefinery which integrates various biomass conversion platforms [6].

According to Gravitis et al. [75], an integrated biorefinery represents a processing facility consisting of multiple technologies including feedstock handling, pretreatment processes, and different biomass conversion/upgrading processes. This allows the by-products and waste to be minimized while recovering the energy generated within the biorefinery. Therefore, the integrated biorefinery concept provides an opportunity to create a variety of value-added products while enhancing the sustainability performance in terms of the economic, environmental, and social impacts. A general representation of the integrated biorefinery concept is illustrated in Figure 1 as provided in the Introduction section.

As shown in Figure 1, an integrated biorefinery consists of different conversion pathways that convert different types of biomass feedstock into heat, power, and value-added products through depolymerizing and deoxygenating biomass components [76]. According to the U.S. Department of Energy/National Renewable Energy Laboratory (NREL), existing conversion pathways and technologies can be generally categorized into five different platforms based on the products produced [77]. These five platforms are the sugar platform, the thermochemical/syngas platform, the biogas platform, the carbon-rich chains platform, and the plant products platform. Table 2 summarizes the foci of these five biomass processing platforms as classified by NREL:

Table 2. Comparison of different biomass conversion platforms.

Platform	Focus	Main Products
Sugar	Fermentation of sugars obtained via extraction of biomass feedstocks	Ethanol and other building block chemicals
Thermochemical syngas	Gasification of biomass feedstocks	Gaseous and liquid fuels
Biogas	Decomposition of biomass feedstocks	Cooking gas
Carbon-rich chains	Transesterification of vegetable oil or animal fat	Biodiesel (fatty acid methyl esters)
Plant products	Selective breeding and genetic engineering of biological plant	Plant strains that can be used as feedstock for further conversion into chemicals and compounds that are difficult to obtain from plant naturally

Other than classifying the biomass conversion processes into different conversion platforms, these conversion processes and technologies are more commonly categorized according to the nature of the processes. Based on the method and nature of the processes, these conversion processes and technologies can be divided into four main groups of physical/mechanical, thermochemical, chemical, and biochemical/biological processes [76]. A detailed discussion of each group of technologies is covered in the following sub-sections.

3.1.1. Physical/Mechanical Processes

Physical/mechanical processes refer to processes that change the physical properties of the biomass, such as reduction of the size of the biomass material or separating biomass feedstock components without changing the state/composition. These processes are usually applied as pre-treatments in the early stages of a biorefinery to process the biomass feedstock into appropriate size ranges for subsequent conversion processes. Note that size reduction technologies refer to the physical treatment of biomass feedstock that includes cutting or commuting processes that result in changing of particle size, shape, and bulk density of the biomass. On the other hand, separation processes are used to separate the bulk biomass feedstock into its constituent components. Extraction methods can also be utilized as physical/mechanical processes to isolate valuable compounds from a bulk and inhomogeneous substrate [78].

3.1.2. Thermochemical Processes

Thermochemical processes are generally referred to as the decomposition of biomass into smaller molecules via thermal energy (i.e., gasification and pyrolysis, etc.). Gasification is a process that partially oxidizes biomass into syngas by keeping the biomass at a high temperature (above 700 °C) [79]. Syngas is a gaseous mixture primarily consisting of carbon dioxide, carbon monoxide, hydrogen, methane, and water. It can be used directly as a stationary fuel or further converted into important intermediates such as methanol, ammonia, and oxyalcohols. Pyrolysis is a process that converts biomass into bio-oil, solid charcoal, and light gases similar to syngas in the absence of oxygen [80] and can generally be categorized as fast or slow pyrolysis. Fast pyrolysis normally operates at high heating rates with short vapor residence time (ranging from a few seconds to a few minutes) in the absence of oxygen. The typical operating temperature for fast pyrolysis is around 500 °C. Meanwhile, slow pyrolysis operates at a lower heating rate, longer vapor residence time, and a lower operating temperature at around 400 °C. According to Brownsort [81], bio-oil and charcoal are the main products of fast pyrolysis and slow pyrolysis, respectively. Additionally, direct combustion is another type of thermochemical conversion process. Direct combustion of biomass involves the burning of biomass in an oxygen-rich environment to produce heat [82]. According to Demirbas [83], liquefaction of biomass into heavy oil is also considered a thermochemical process. In this process, biomass is reacted with carbon monoxide/hydrogen in the presence of sodium carbonate to form heavy oil.

3.1.3. Chemical Processes

Chemical processes refer to processes that change the chemical structure of biomass molecules at high temperature and pressure. Most of the time these chemical processes occur in the presence of a catalyst. Although chemical processes include a wide class of chemical reactions, the most common chemical processes to convert biomass into value-added products are hydrolysis and transesterification [76]. The hydrolysis process utilizes acids, alkalis, or enzymes to decompose molecules with complex molecular structures such as polysaccharides and proteins into their component sugars or derivative chemicals [84]. For example, hydrolysis depolymerizes cellulose into glucose while the derivative compound, levulinic acid can then be obtained from the resulting glucose. Transesterification is a chemical process that converts vegetable oils into biodiesel such as methyl esters and ethyl esters of fatty acids [85]. Fischer-Tropsch (FT) synthesis, steam reforming, and methanization are other common chemical processes for converting biomass into bio-chemicals.

3.1.4. Biochemical/Biological Processes

Biochemical/biological processes utilize biological enzymes or living organisms as biological catalysts to transform biomass into a commodity and specialty bio-chemicals. Compared to thermochemical and chemical processes which involve high temperatures and pressures, biochemical processes normally occur at lower temperatures and thus results in lower reaction rates [76]. Some established biochemical processes are anaerobic digestion and fermentation. Anaerobic digestion utilizes bacteria for the breakdown of biodegradable organic material in the absence of oxygen at a mild temperature (30–65 °C). The main product of anaerobic digestion is biogas, which is a mixture of methane, carbon dioxide, hydrogen sulfide, water, and other impurities. According to Romano and Zhang [86], the product biogas can be purified into more than 97% methane via different technologies (e.g., pressure swing adsorption, membrane, etc.), which can serve as a natural gas substitute. On the other hand, in the fermentation processes, the fermentable substrate is converted into different products such as alcohols or organic acids via enzymes or microorganisms [87].

3.2. Synthesis and Design of Integrated Biorefineries

To synthesize a sustainable integrated biorefinery with maximum performance and minimum environmental impact, it is essential to integrate different conversion technologies in a systematic and efficient manner. Process synthesis methods developed for conventional chemical processes can be extended to enable the synthesis of integrated biorefineries. Process synthesis is defined as the activity to identify the optimal interconnection of unit operations involved in the overall process and the optimal design and configuration of the units within the process [88]. According to Douglas [89], process synthesis can generally be classified into seven categories, i.e., synthesis of reaction pathway, synthesis of reactor network, synthesis of separation network, synthesis of mass exchange network, synthesis of material, synthesis of heat exchanger network, and synthesis of the complete flowsheet.

According to Kokossis and Yang [90], PSE approaches have the potential to support process synthesis and design, which can be applied and utilized in the design of integrated biorefineries. PSE is the field that covers the actions and activities involved in the engineering of systems consist of physical, chemical, and/or biological processing operations [91]. Throughout the years, various PSE approaches have been developed for the synthesis and design of integrated biorefineries [7,8,92], each with its own advantages and disadvantages. Some of the commonly used approaches are presented below.

3.2.1. Hierarchical Approaches

According to Douglas [93], hierarchical approaches use a series of decision-making processes and short-cut models at different hierarchical levels/stages for solving process synthesis problems. These approaches emphasize the decomposition of the problem and the screening of alternatives to identify the solution. According to Li and Kraslawski [94], hierarchical approaches generally offer a quick solution for process synthesis problems, however, they are not able to guarantee an optimal solution as they employ a sequential decomposition strategy. Ng et al. [95] proposed a hierarchical approach to synthesis and analysis of integrated biorefineries based on the Forward-Reverse Synthesis Tree concept. The proposed approach is a sequence of interlinked activities that utilize an appropriate level of information for the initial design. Later, Conde-Mejía et al. [96] applied a hierarchical design approach for the selection of processing routes in biorefinery facilities. Most recently, Tey et al. [97] further extended the developed hierarchical approach (commonly known as "Douglas' hierarchical approach") to the synthesis of an integrated biorefinery with consideration of the pretreatment system.

3.2.2. Heuristic Searches

According to Stephanopoulos and Westerberg [98], heuristic searches utilize engineering expertise to generate solutions for process synthesis problems. The first step is to identify a suitable base case design. This is followed by subsequent modification and fine-tuning to enhance the overall process performance. While heuristic searches are effective in providing feasible solutions for process synthesis problems, it is difficult for such methods to guarantee an optimal solution [99]. A heuristic framework for debottlenecking of a palm oil-based integrated biorefinery was presented by Kasivisvanathan et al. [100].

3.2.3. Insight-Based Approaches

Insight-based approaches such as pinch analysis, ternary diagrams, distillation residue curve maps, etc. have been developed for various applications in process flow sheeting. For example, Tay et al. [8] presented a graphical synthesis approach based on a carbon-hydrogen-oxygen ternary diagram, which enables tracking the change of atomic components within the process. Patel [101] presented a thermodynamic targeting approach based on Van Krevelen diagrams and energy balances for screening, evaluating, and comparing various biomass conversion processes. According to Voll [10], insight-based approaches enable visualization of process synthesis problems, however, employing such approaches in solving complicated process synthesis problems is prone to be restricted by the limited number of parameters and variables that can be considered.

3.2.4. Algorithmic Approaches

Examples of algorithmic approaches include the process graph (P-graph) method. According to Oppenheim [102], algorithmic approaches execute a sequential set of actions based on automated reasoning, calculation, and data processing to determine solutions to process synthesis problems. Benjamin et al. [103] utilized the P-graph approach to generate alternative processes for an integrated biorefinery. As these approaches involve search space reduction, they are powerful and reliable to solve process synthesis problems related to process network synthesis, and often provide quick alternative solutions. However, algorithmic approaches are less flexible when applied to complicated process synthesis problems compared to mathematical optimization approaches [104]. Most recently, Yeo et al. [105] applied a graph-theoretic method to synthesize a sustainable biorefinery for the palm oil industry with optimum resources, which include fertilizer, steam, and electricity generation.

3.2.5. Mathematical Optimization Approaches

According to Grossmann [106], various mathematical optimization models (i.e., linear programming, mixed-integer linear programming, nonlinear programming, and mixed-integer nonlinear programming models) can be formulated to solve process synthesis problems. The models are then solved using different optimization techniques and/or commercial optimization software. Depending on the required information and the complexity/nature of the process synthesis problem, a range of mathematical programming models can be developed and solved to identify optimal solutions for various objectives. Bao et al. [107] presented a superstructure-based shortcut approach to track the feedstock to final products via chemical species for the synthesis of integrated biorefineries. Pham and El-Halwagi [108] presented a mathematical optimization model for integrated biorefineries based on the backward and forward synthesis concept. Later, various mathematical optimization models based on superstructure approaches have been developed for different configurations of integrated biorefineries; e.g., palm-based biorefinery [109], fast pyrolysis-gasification for production of liquid fuels and propylene [110], microalgal biorefinery [111], cyanobacteria biorefinery [112], seaweed-based biorefinery [113], etc. To further enhance the process performance in an integrated biorefinery, efforts have been devoted to the consideration of simultaneous reactor modeling, heating/cooling and work [114], simultaneous process synthesis, heat and power integration [115], production of biodiesel, hydrogen, and synthesis gas integrated with CHP [116], total site analysis [117]. Some studies have focused on retrofitting existing processes and integrating biorefinery technologies with such production facilities (e.g., palm oil industry [118], sago industry [119,120], pulp, and paper industry [121], etc.).

Additionally, many other alternative mathematical optimization methods such as value analysis [122], decomposition approach [9], disjunctive programming [123], modular optimization approach [124], fuzzy optimization [8,125], robust optimization [126], Monte Carlo type sampling [127], flexibility optimization [128], multi-objective optimization based on a genetic algorithm [129], multi-objective target-oriented robust optimization [130], etc. have been applied in synthesis and analysis of integrated biorefineries. Martín and Grossmann [131] reviewed alternative mathematical programming techniques for the design of sustainable biorefineries for the production of biofuels from different raw materials. While mathematical optimization approaches are useful and effective in solving process synthesis/design problems, such approaches may require intensive computational effort when rigorous process modeling is required [132].

3.2.6. Hybrid Methods

Hybrid methods that combine insights-based and mathematical optimization approaches have been developed to take advantage of the specific strengths of both approaches. An automated targeting method was developed to integrate the cascade approach into a mathematical optimization environment [133–135] for the synthesis of resource conservation networks. Ng [136] further developed the automated targeting method for the synthesis of integrated biorefineries by tracking carbon content from raw material (biomass) to final products. Tay and Ng [137] further extended the approach into a multiple-cascade automated targeting method for the synthesis of gasification based integrated biorefineries, where syngas composition is used as the quality measure of the gasification process. Finally, Shabbir et al. [138] combined a superstructure optimization approach with the automated targeting approach for the synthesis of integrated biorefineries with maximum economic performance and minimal environmental impact.

3.3. Challenges in Designing Integrated Biorefineries

While integrated biorefineries provide advantages as discussed in earlier sections, it should be noted that the unique features of integrated biorefineries make them less straightforward compared to conventional chemical processes. Due to the complex structure and varying compositions of biomass, the important thermodynamic properties (e.g., enthalpy, entropy, heat capacity, etc.) of biomass are often not well established. Besides, the rate of reaction for biomass conversion technologies such as fermentation, hydrolysis, etc. are difficult to determine accurately most of the time. Therefore, most of the available contributions in process synthesis for chemical processes cannot be directly applied for the synthesis of integrated biorefineries. Therefore, it is essential to develop systematic procedures to address the abovementioned issues. In order to synthesize and design a sustainable integrated biorefinery, the following criteria shall be fulfilled [139,140]:

i Able to minimize energy consumption and potential environmental impact through material and energy integrations between different conversion platforms.
ii Able to accommodate the varying seasonal patterns on feedstock availability and quality through integration of different biomass conversion platforms.
iii Able to depolymerize biomass components to intermediate products that match the requirements of subsequent processing technologies.
iv Able to maximize the yield and quality of value-added products.

Identifying new potential product molecules from a given biomass feedstock can be achieved through reaction pathway approaches. Andiappan et al. [141] presented a reaction pathway synthesis approach for integrated biorefineries with consideration of economic, energy assessment, and environmental impact in terms of incremental burden. Most recently, Tey et al. [142] further extended the chemical reaction pathway map to synthesize a sustainable integrated biorefinery for the production of fine chemicals from palm-based biomass. Additionally, Madenoor Ramapriya et al. (2018) extended the usage of the superstructure approach for integrated biorefineries synthesis with consideration of complex reaction networks. Similarly, Filho et al. [143] presented an integrated biorefinery for bioethanol production with an analysis of the molecular transformations.

Design of integrated biorefineries requires the ability to handle process/market uncertainties while maintaining the process performance. Tang et al. [144] presented a superstructure optimization approach for the synthesis of integrated biorefineries with consideration of biomass feedstock, value-added products, trenda of market price, constraints of technology, and system uncertainties at multi-periods. Later, Tay et al. [145] presented a mathematical optimization model for the synthesis of integrated biorefineries with supply and demand uncertainties based on a robust optimization approach. Moreover, a framework that uses a superstructure-based process synthesis approach integrated with uncertainty analysis was developed to consider price effects on the design of an integrated biorefinery [146]. To address the effect of changing biomass type and composition in a multi-product biorefinery, Giuliano et al. [147] presented a MINLP to synthesize an integrated biorefinery that converts lignocellulosic biomass into levulinic acid, succinic acid, and ethanol. Meanwhile, Čuček et al. [148] developed a multi-period optimization approach for a heat-integrated biorefinery's supply network with maximum economic performance. Additionally, the biomass supply chain also impacts the overall performance of the integrated biorefinery. In order to address such uncertainties, a two-stage optimization approach (macro- and micro-stage) has been presented.

The micro-stage involves waste (biomass, industrial waste, etc.) optimization and allocation, as well as the design of the integrated biorefinery, while the macro-stage focuses on supply chain network synthesis and optimization. Various review papers on biomass supply chains covering different aspects (e.g., uncertainties and sustainability [149]; value chain optimization [150]; economic, social and environmental perspectives [151]; design and management [152]; modeling and optimization [153]; techno-economic modeling and optimization [154] have been presented in the literature. Stewart and El-Halwagi [155] provided a detailed review and analysis of the integration of biorefinery innovations into existing processes. This book also covers the application of PSE tools in various stages of the design of such processes, biorefinery products, supply chains, policies, and environmental impacts. Most recently, Lo et al. [156] provided a state-of-the-art review of techno-economic analysis for biomass supply chains.

In addition to the synthesis of independent integrated biorefineries and supply chain networks, the synthesis of industrial symbiosis for bioenergy networks has also received much interest from the research community. Ng et al. [157] developed a superstructure model for the synthesis of palm-based biorefineries, which considers multiple owners in an industrial symbiosis network. Later, Ng et al. [158] presented a superstructure-based disjunctive fuzzy optimization approach for planning and synthesis of industrial symbiosis in bioenergy systems. More recently, Barla et al. [159] applied circular economy concepts in the design of a textile waste biorefinery.

4. Application of Molecular Design within the Context of Integrated Biorefineries

As presented in the previous section, the products produced from integrated biorefineries are generally classified into two major categories of chemical/material products and energy products [76]. Examples of energy products include biogas, bio-oil, biodiesel, etc. which are products that are utilized based on their energy content. These products are commonly used for the generation of heat, electricity, and energy for different electrification and transportation purposes. On the other hand, chemical/material products are often utilized based on different functionalities given by their physical and chemical properties. Carbohydrates are identified as one of the commonly utilized biomass feedstocks for the production of commodity chemicals and specialty chemicals [72]. According to Lichtenthaler and Mondel [160], the primary form of carbohydrates is in the form of polysaccharides (which are also known as starch). Traditionally, starch (polysaccharide) and its derivative D-glucose (monosaccharide) have been greatly utilized as raw material to produce commodity chemicals and polymers by chemical industries [161]. As starch is an important food resource, utilization of starch to produce chemical products has raised a lot of concerns from the community on the competition of starch utilization between food and industrial applications. For this reason, the utilization of lignocellulosic biomass to produce biochemical products has been receiving focus and attraction as lignocellulosic biomass is mostly plant-based waste in different forms [72]. According to Wyman [162], lignocellulosic biomass is made up of primarily cellulose, hemicellulose, and lignin. Lignocellulosic biomass can be converted into different value-added products via different conversion technologies. Moreover, lignocellulosic biomass can be converted into different intermediate products with a high content of hydroxyl groups through the liquefaction process [163].

In view of the possibility of lignocellulosic biomass conversion into different value-added chemical products, extensive effort has been invested by industrial and research sectors to identify these value-added chemical products. Elliott [164] conducted a study on potential chemical products that can be produced from biomass, based on different biomass conversion. According to Elliot [164], the potential chemical products can be commonly classified into fermentation products, gasification products, catalytic/bioprocessing products, derivatives of carbohydrates, and derivatives of plants. Later, findings on the identification of twelve chemical building blocks that are possible to be converted from starch through different processing technologies are presented by Werpy and Peterson [165]. There exists a potential for these twelve building blocks to be further converted into new families of useful chemical molecules. Hence, these twelve building blocks provide the potential to be utilized to produce different bio-based chemical products that fulfill market needs. In addition, Holladay et al. [166] presented their study on the potential of lignin recovery to produce macromolecules, aromatics, and miscellaneous monomers. These potential products include different fuel additives, adhesives, carbon fiber, etc. Skibar et al. [167] emphasized the future of biomass through their discussion on the efforts of the chemical industry in utilizing biomass to produce value-added products. In addition to the traditional usage of biomass to produce polymers, the future of biomass lies in the production of specialty products as biomass has the potential to be utilized in producing specialty chemicals such as pharmaceutical, beauty, and personal care products [167]. From the works discussed above, it is obvious that the future of biomass utilization covers the production of energy products, bulk chemical products, and new and novel products such as fine and specialty chemical products. In most cases, such products are designed to satisfy product needs [168]. Thus, the product design aspects have to be taken into account while synthesizing an optimally integrated biorefinery to make sure that the products produced by integrated biorefinery meet the required product needs. This goal can be fulfilled by integrating the synthesis of integrated biorefineries with chemical product design.

Chemical product design problems are suitable to be explained as a process to search for chemical products that fulfill preferred product needs. Traditionally, the product designers first hypothesize a target molecule that has the preferred product needs while designing a chemical product. The subsequent step would be the product synthesis step, which is followed soon by the product testing step to test for the preferred product needs. Target molecule revision and redesign are required repeated if the target molecule does not satisfy the preferred product needs. Hence, it can be seen that the traditional approaches of chemical product design require an iterative process, which leads them to being inefficient, laborious, and low in cost-effectiveness [169]. In light of these limitations, CAMD techniques provide efficient and effective options for the design of chemical products.

In the past decades, CAMD techniques have been gaining attraction and appeared as reliable techniques for their capability in identifying feasible molecules that possess the desired set of product needs [170]. This includes the integration of CAMD techniques with process synthesis for the consideration of product design aspects while synthesizing integrated biorefineries. Hechinger et al. (2010) [171] prosed an integrated approach to the design of biofuels by combining product and process design aspects. The presented study designs biofuels in terms of their molecular structure by using CAMD techniques and identify alternative conversion pathways for the production of biofuels by applying reaction network flux analysis. Ng et al. b [172] proposed an integrated optimization approach for the design of optimal chemical products and synthesis of optimal biomass processing pathways in integrated biorefineries. The optimal biochemical products are designed using CAMD techniques while optimum biomass processing pathways are identified through the developed mathematical superstructure optimization approach. Later, Ng et al. b [173] extended the integrated optimization approach for the design of optimal mixture and synthesis of optimal processing pathways in an integrated biorefinery. Later, Bertran et al. [174] presented two special tools which include a database that employs specially developed knowledge to represent the processing system, and Super-O, which is a software with a practical user interface to guide users for the synthesis of molecular product and integrated biorefinery.

4.1. Integrated Tools for Ionic Liquid Design within Integrated Biorefineries

The availability of huge data libraries and property prediction models along with the general improvements in computational tools has enabled the application of ionic liquids in the chemical process industry. One of the early applications of the CAMD approach for an integrated process is the design of ionic liquid via the CAMD approach for the purification of bioethanol [175]. CAMD is used to design ionic liquids that can be used to produce 99% pure ethanol from bioethanol with an ethanol concentration of 85%. Various GC models have been used to predict the properties and activity coefficients required to estimate/predict vapor-liquid equilibria.

An extended CAMD approach for carbon capture was developed by Chong et al. [176] to model task-specific ionic liquids. In this work, optimal ionic liquid candidates and conditions for the carbon capture process were determined simultaneously by considering the effect of process conditions as part of the developed approach. To reduce the search space, continuous variables of operating conditions are discretized by introducing disjunctive programming. The thermophysical properties of ionic liquids are predicted using an extended UNIFAC model. To ensure the feasibility of the synthesized ionic liquids, appropriate structural constraints, including cation-anion pairing constraints are included in the design.

In another effort, optimal ionic liquid candidates were screened using the COSMO-RS model, absorption mechanisms, and experimental data for the decarbonization of shale gas [177]. In addition to the Henry's constant, viscosity and toxicity of the ionic liquids are considered in the screening of candidates. Thermodynamic property models, physicochemical properties, and phase equilibria related to the decarbonization process were established. The new ionic liquid-based decarbonization technology was then evaluated by process simulation. Simulations confirmed that the use of ionic liquid requires less energy than the conventional methyl diethanolamine (MDEA) process for CO_2 capture.

Chong et al. [178] presented a comprehensive study, where the ionic liquid design was integrated with the entire bioenergy production system that produces multiple energy products from biomass. In this work, ionic liquid is used to remove CO_2 from a bioenergy production plant. This process is a bioenergy with carbon capture and storage (BECCS) system. Since the production of bioenergy is considered to be carbon neutral, the CO_2 removal and storage make this a negative emission process. One of the challenges in this design was identifying the optimal conditions for bioenergy production and for CO_2 removal. Therefore, the bioenergy production system was retrofitted to provide utilities for the carbon capture system.

Liu et al. [179] applied a CAMD approach to screen and predict ionic liquid candidates for shale gas separation. In this work, an extensive ionic liquid-database was established for various gas-ionic liquid systems which consisted of the solubility data of gases in several ionic liquids. Henry's law constants have also been established for several combinations of gases and ionic liquids. A corrected COSMO-RS model was developed to generate pseudo-experimental data after validation to model the systems for which no experimental data were available. A UNIFAC-IL method was established to predict the solubility of shale gas components in ionic liquids. The model showed good agreement between the predictions and experimental solubility data.

Ionic liquids have also been introduced as suitable solvents for the recovery of bioisoprene, which is a promising alternative to petroleum-derived isoprene [180]. In this work, CAILD and UNIFAC-IL models are applied for the thermodynamic calculations and GC based methods are used for the prediction of physical properties. The thermodynamic behavior of the bioisoprene systems involving ionic liquids was described using the extended UNIFAC-IL model that combines gas-organic chemicals using extensive experimental data from literature and pseudo-experimental data from COSMO-RS calculations. The final verification of the process improvements with the designed ionic liquid has been performed via a detailed simulation of the process.

To design ionic liquid for absorbents, the Absorption-Selectivity-Desorption index (ASDI) has been integrated with CAMD tools for ionic liquid design [181]. Thermodynamic properties of ionic liquids are predicted by the COSMO-GC-IL integrated with the COSMO-SAC model. The ionic liquid design was formulated as an optimization problem where the rest of the physical properties were modeled using GC models. A case study has been performed on sour gas absorption in a syngas production process. The integrated ASDI was shown to be capable of determining promising ionic liquid absorbents that can meet the thermodynamic property targets while maintaining low energy consumption for the process.

4.2. Opportunities for Further Research

4.2.1. Expanding the Optimization Scope/Parameters for Integrated Biorefinery Design

The above-detailed review of existing synthesis/optimization approaches for integrated biorefineries has discussed various approaches ranging from insights-based approaches to complex mathematical optimization models. The developed approaches can address steady-state processes, batch processes, as well as process and market uncertainties. In order to incorporate a sustainable integrated biorefinery, global life cycle optimization ought to be applied during the synthesis stage. The Life Cycle Optimization approach [182] can be extended for concurrent synthesis and optimization of sustainable integrated biorefineries with consideration of economic, environmental, and societal factors/impacts. It is noted that most of the above-mentioned methods focus on technical feasibility, environmental sustainability, market uncertainties, etc. However, there is limited consideration of societal impacts during the design/development of the integrated biorefinery. Therefore, society impacts should be taken into consideration in future efforts.

It should also be noted that a lot of research and development activities are moving towards human-centered design approaches; which is an approach to problem-solving that incorporates the human perspective in all steps. Human-centered design [183] is generally applied in design and management frameworks to solve problems with developed solutions. In order to consider human perspectives in the synthesis and analysis of integrated biorefineries, integration of the human-centered design methodology into PSE approaches would need to be developed.

4.2.2. Design of Novel Ionic Liquids

As discussed in earlier sections, most property prediction models are only applicable for commonly studied ionic liquids such as pyrrolidinium-, ammonium-, imidazolium-, phosphonium- and pyridinium-based ionic liquids, which hinder the discovery of potential new ionic liquids using CAMD tools. Future directions should focus on developing property prediction models that are able to estimate properties for a wider range of pure ionic liquids and/or ionic liquid mixtures. This shortcoming may be addressed by predicting properties of ionic liquids based on functional groups rather than ionic liquid constituents. Accurate correlation equations are key to ensuring that the results generated by process/product design methods/tools are reliable. In addition, there are several works conducted on the effect of solvents on reaction rates. However, no such studies have been conducted on the influence of ionic liquid solvents on reactions. Therefore, it is worth exploring the application of quantum mechanical calculations to predict the rate constants of reactions in various ionic liquid media. Another new class of solvents called deep eutectic solvents (DESs) have received much attention in recent years as they are relatively cheap, simple to synthesize, and have good biodegradability [184]. DESs are solvents that comprise both hydrogen bond acceptors and donors. Some researchers have validated their feasibility as entrainers in the separation of azeotropic mixtures through experiments. With the use of DESs as entrainers in the ethanol-water azeotropic system, the relative volatility of the ethanol-water system increased 4.7 times compared to the DES-free system [185]. Despite the potential of these materials, studies on modeling extractive distillation processes using DESs as entrainers remain limited. This may be due to the fact that there is very limited thermophysical property data available, as this particular research field is still relatively new. A general model that requires only critical pressure, temperature, and one reference viscosity point has been developed to predict the viscosities of DESs [186]. However, there is still a need to improve the model accuracy and develop more property prediction models for DESs to facilitate the computational design and/or screening of solvents.

5. Conclusions

Recent advances in the development of computational PSE tools have enabled the design of ionic liquids and the design of integrated biorefineries. It is imperative that ionic liquids may serve as potential replacements for current organic solvents when considering stringent environmental regulations and the implementation of a clean manufacturing practice. Nevertheless, the process of identifying an optimal ionic liquid through experimentation is a very tedious task as a very large number of possible ionic liquid combinations exist. Various PSE approaches have been developed in order to make an optimal selection of ionic liquids prior to experimental testing. Similarly, numerous research works have been performed to optimize the synthesis and design of sustainable integrated biorefineries with maximum performance and minimal environmental impact. The application of existing PSE tools to these problems had been extremely challenging because of the difficulty in incorporating accurate property prediction tools in the design of ionic liquids and the computational challenges in the design of such a large and complex processing facility. Decomposition-based approaches developed by different researchers have been instrumental in reducing the complexity of these design problems.

Author Contributions: Conceptualization, N.G.C. and D.K.S.N.; Methodology, L.Y.N. and J.O.; Software, validation, formal analysis, J.W.C.; Investigation, resources, data curation; N.G.C. and D.K.S.N.; writing- original draft preparation, L.Y.N. and J.O.; Writing—review and editing, M.R.E.; visualization, J.W.C.; supervision, N.G.C. and D.K.S.N.; project administration—N.G.C. and M.R.E.; funding acquisition, N.G.C. All authors have read and agreed to the published version of the manuscript.

Funding: This research was funded by the Ministry of Higher Education, Malaysia, Grant number FRGS/1/2019/TK02/UNIM/02/1.

Conflicts of Interest: The authors declare no conflict of interest.

Nomenclature

[4bmpy][TCM]	1-butyl-4-methylpyridinium tricyanomethanide
[Bim][NTf_2]	1-butyl imidazolium bis(trifluoromethylsulfonyl)imide
[BMim][Cl]	1-butyl-3-methylimidazolium chloride
[BMIM][OAc]	1-butyl-3-methylimidazolium acetate
$[Bmim]_2[CuCl_4]$	Bis(1-butyl-3-methyl imidazolium) copper tetrachloride salt
$[Bmim]_2[SnCl_4]$	Bis(1-butyl-3-methyl imidazolium) stannum tetrachloride salt
[EMIM][MS]	1-Ethyl-3-methylimidazolium methyl sulfate
[Emim]Ac	1-Ethyl-3-methylimidazolium acetate
[EMIM]Cl	1-methylimidazolium chloride
[NH2p-bim][BF_4]	1-butyl-3-propylamineimidazolium tetrafluoroborate
BMIM-BF_4	1-butyl-3-methylimidazolium hexafluorophosphate
$C_9H_{14}NBF_4$	1-butylpyridinium tetrafluoroborate
CO_2	Carbon dioxide
EAN	Ethyl ammonium nitrate
H_2S	Hydrogen sulfide
NH_3	Ammonia
SO_2	Sulfur dioxide

Abbreviations

ANN	Artificial neural networks
ASDI	Absorption selectivity desorption index
CAILD	Computer-aided ionic liquid design
CAMD	Computer-aided molecular design
CHP	Combined heat and power
COSMO	Conductor-like screening model
COSMO-RS	COSMO for real solvents
COSMO-SAC	COSMO for segment activity coefficient
DES	Deep eutectic solvent
DFT	Density functional theory
EOS	Equations of state

FT	Fischer-Tropsch
GBM	Gradient boosted regression
GC	Group contribution
GC-COSMO	Group contribution-based COSMO
IDAC	Infinite dilution activity coefficient
IL	Ionic liquid
LCA	Life cycle assessment
LLE	Liquid-liquid extraction
LSSVM	Least-squared support vector machine
MC	Monte Carlo simulations
MD	Molecular dynamics simulations
MILP	Mixed-integer linear program
MINLP	Mixed-integer nonlinear program
ML	Machine learning
PPMV	Parts per million volume
PSE	Process systems engineering
QM	Quantum mechanics
QSAR	Quantitative structure-activity relationship
QSPR	Quantitative structure-property relationship
RF	Random forests
RHST	Rough hard-sphere theory
UNIFAC	Universal quasichemical functional-group activity coefficients
ZIF	Zeolitic imidazolate framework

References

1. Grossmann, I.E.; Westerberg, A.W. Research challenges in process systems engineering. *AIChE J.* **2000**, *46*, 1700–1703. [CrossRef]
2. Grossmann, I.E. Challenges in the new millennium: Product discovery and design, enterprise and supply chain optimization, global life cycle assessment. *Comput. Aided Chem. Eng.* **2004**, *29*, 29–39. [CrossRef]
3. Chen, Y.; Kontogeorgis, G.M.; Woodley, J.M. Group Contribution Based Estimation Method for Properties of Ionic Liquids. *Ind. Eng. Chem. Res.* **2019**, *58*, 4277–4292. [CrossRef]
4. Karunanithi, A.T.; Mehrkesh, A. Computer-aided design of tailor-made ionic liquids. *AIChE J.* **2013**, *59*, 4627–4640. [CrossRef]
5. Austin, N.D.; Sahinidis, N.V.; Trahan, D.W. A COSMO-based approach to computer-aided mixture design. *Chem. Eng. Sci.* **2017**, *159*, 93–105. [CrossRef]
6. Fernando, S.; Adhikari, S.; Chandrapal, C.; Murali, N. Biorefineries: Current Status, Challenges, and Future Direction. *Energy Fuels* **2006**, *20*, 1727–1737. [CrossRef]
7. Sammons, N.E.; Yuan, W.; Eden, M.R.; Aksoy, B.; Cullinan, H.T. Optimal biorefinery product allocation by combining process and economic modeling. *Chem. Eng. Res. Des.* **2008**, *86*, 800–808. [CrossRef]
8. Tay, D.H.S.; Ng, D.K.; Sammons, N.E.; Eden, M.R. Fuzzy Optimization Approach for the Synthesis of a Sustainable Integrated Biorefinery. *Ind. Eng. Chem. Res.* **2011**, *50*, 1652–1665. [CrossRef]
9. Kokossis, A.; Tsakalova, M.; Pyrgakis, K. Design of integrated biorefineries. *Comput. Chem. Eng.* **2015**, *81*, 40–56. [CrossRef]
10. Voll, P. *Automated Optimization-Based Synthesis of Distributed Energy Supply Systems*; RWTH Aachen University: Aachen, Germany, 2013.
11. Walden, P. Molecular weights and electrical conductivity of several fused salts. *Bull. Acad. Imp. Sci.* **1914**, *1800*, 405–422.
12. Bates, E.D.; Mayton, R.D.; Ntai, I.; Davis, J.H. CO_2 Capture by a Task-Specific Ionic Liquid. *J. Am. Chem. Soc.* **2002**, *124*, 926–927. [CrossRef] [PubMed]
13. Marsh, K.N.; Boxall, J.A.; Lichtenthaler, R. Room temperature ionic liquids and their mixtures—A review. *Fluid Phase Equilib.* **2004**, *219*, 93–98. [CrossRef]
14. Olivier-Bourbigou, H.; Magna, L.; Morvan, D. Ionic liquids and catalysis: Recent progress from knowledge to applications. *Appl. Catal. A Gen.* **2010**, *373*, 1–56. [CrossRef]

15. Eftekhari, A. Supercapacitors utilising ionic liquids. *Energy Storage Mater.* **2017**, *9*, 47–69. [CrossRef]
16. Morton, M.D.; Hamer, C.K. Ionic liquids—The beginning of the end or the end of the beginning? A look at the life of ionic liquids through patent claims. *Sep. Purif. Technol.* **2018**, *196*, 3–9. [CrossRef]
17. Lei, Z.; Chen, B.; Koo, Y.-M.; Macfarlane, D.R. Introduction: Ionic Liquids. *Chem. Rev.* **2017**, *117*, 6633–6635. [CrossRef]
18. Freemantle, M. Chemistry Basf's Smart Ionic Liquid. *Chem. Eng. News Arch.* **2003**, *81*, 9. [CrossRef]
19. Kazemi, S.; Nor, M.I.M.; Teoh, W.H. Thermodynamic and economic investigation of an ionic liquid as a new proposed geothermal fluid in different organic Rankine cycles for energy production. *Energy* **2020**, *193*, 116722. [CrossRef]
20. Kermani, N.A.; Petrushina, I.; Rokni, M. Evaluation of ionic liquids as replacements for the solid piston in conventional hydrogen reciprocating compressors: A review. *Int. J. Hydrogen Energy* **2020**, *45*, 16337–16354. [CrossRef]
21. Bian, J.; Peng, F.; Peng, X.-P.; Xiao, X.; Peng, P.; Xu, F.; Sun, R.-C. Effect of [Emim]Ac pretreatment on the structure and enzymatic hydrolysis of sugarcane bagasse cellulose. *Carbohydr. Polym.* **2014**, *100*, 211–217. [CrossRef]
22. Khan, A.S.; Man, Z.; Bustam, M.A.; Kait, C.F.; Ullah, Z.; Nasrullah, A.; Khan, M.I.; Gonfa, G.; Ahmad, P.; Muhammad, N. Kinetics and thermodynamic parameters of ionic liquid pretreated rubber wood biomass. *J. Mol. Liq.* **2016**, *223*, 754–762. [CrossRef]
23. Szalaty, T.J.; Klapiszewski, Ł.; Kurc, B.; Skrzypczak, A.; Jesionowski, T. A comparison of protic and aprotic ionic liquids as effective activating agents of kraft lignin. Developing functional MnO_2/lignin hybrid materials. *J. Mol. Liq.* **2018**, *261*, 456–467. [CrossRef]
24. Zuo, L.; Ao, X.; Guo, Y. Study on the synthesis of dual-chain ionic liquids and their application in the extraction of flavonoids. *J. Chromatogr. A* **2020**, *1628*, 461446. [CrossRef]
25. Wang, Y.-X.; Cao, X.-J. Extracting keratin from chicken feathers by using a hydrophobic ionic liquid. *Process. Biochem.* **2012**, *47*, 896–899. [CrossRef]
26. Zhu, Z.; Ri, Y.; Jia, H.; Li, X.; Wang, Y.; Wang, Y. Process evaluation on the separation of ethyl acetate and ethanol using extractive distillation with ionic liquid. *Sep. Purif. Technol.* **2017**, *181*, 44–52. [CrossRef]
27. Ayuso, M.; Cañada-Barcala, A.; Larriba, M.; Navarro, P.; Delgado-Mellado, N.; García, J.; Rodríguez, F. Enhanced separation of benzene and cyclohexane by homogeneous extractive distillation using ionic liquids as entrainers. *Sep. Purif. Technol.* **2020**, *240*, 116583. [CrossRef]
28. Bhattacharya, M.; Mandal, M.K. Synthesis and characterization of ionic liquid based mixed matrix membrane for acid gas separation. *J. Clean. Prod.* **2017**, *156*, 174–183. [CrossRef]
29. Lei, Z.; Dai, C.; Song, W. Adsorptive absorption: A preliminary experimental and modeling study on CO_2 solubility. *Chem. Eng. Sci.* **2015**, *127*, 260–268. [CrossRef]
30. Blanchard, L.A.; Hancu, D.; Beckman, E.J.; Brennecke, J.F. Green processing using ionic liquids and CO_2. *Nature* **1999**, *399*, 28–29. [CrossRef]
31. Xu, Q.; Jiang, W.; Xiao, J.; Wei, X.-H. Absorption of Sulfur Dioxide by Tetraglyme–Sodium Salt Ionic Liquid. *Mol.* **2019**, *24*, 436. [CrossRef]
32. Shang, D.; Zhang, S.; Zeng, S.; Jiang, K.; Gao, H.; Dong, H.; Yang, Q.; Zhang, S. Protic ionic liquid [Bim][NTf2] with strong hydrogen bond donating ability for highly efficient ammonia absorption. *Green Chem.* **2017**, *19*, 937–945. [CrossRef]
33. Wang, J.; Zeng, S.; Huo, F.; Shang, D.; He, H.; Bai, L.; Zhang, X.; Li, J. Metal chloride anion-based ionic liquids for efficient separation of NH_3. *J. Clean. Prod.* **2019**, *206*, 661–669. [CrossRef]
34. Brennecke, J.F.; Maginn, E.J. Ionic liquids: Innovative fluids for chemical processing. *AIChE J.* **2001**, *47*, 2384–2389. [CrossRef]
35. Fan, Y.; Dong, X.; Zhong, Y.; Li, J.; Miao, J.; Hua, S.; Li, Y.; Cheng, B.; Chen, W. Effects of ionic liquids on the hydrolysis of casein by lumbrokinase. *Biochem. Eng. J.* **2016**, *109*, 35–42. [CrossRef]
36. Franca, J.M.P.; De Castro, C.A.N.; Lopes, M.M.; Nunes, V.M.B. Influence of Thermophysical Properties of Ionic Liquids in Chemical Process Design. *J. Chem. Eng. Data* **2009**, *54*, 2569–2575.
37. Predel, T.; Schlücker, E. Ionic Liquids in Oxygen Compression. *Chem. Eng. Technol.* **2009**, *32*, 1183–1188. [CrossRef]
38. Nasirpour, N.; Mohammadpourfard, M.; Heris, S. Ionic liquids: Promising compounds for sustainable chemical processes and applications. *Chem. Eng. Res. Des.* **2020**, *160*, 264–300. [CrossRef]

39. Bogdanov, M.G.; Svinyarov, I. Ionic liquid-supported solid–liquid extraction of bioactive alkaloids. II. Kinetics, modeling and mechanism of glaucine extraction from Glaucium flavum Cr. (Papaveraceae). *Sep. Purif. Technol.* **2013**, *103*, 279–288. [CrossRef]
40. Seiler, M.; Jork, C.; Kavarnou, A.; Arlt, W.; Hirsch, R. Separation of azeotropic mixtures using hyperbranched polymers or ionic liquids. *AIChE J.* **2004**, *50*, 2439–2454. [CrossRef]
41. Boli, E.; Voutsas, E. Ionic liquids as entrainers for the separation of azeotropic mixtures: Experimental measurements and COSMO-RS predictions. *Chem. Eng. Sci.* **2020**, *219*, 115579. [CrossRef]
42. Fadeev, A.G.; Meagher, M.M. Opportunities for ionic liquids in recovery of biofuels. *Chem. Commun.* **2001**, *3*, 295–296. [CrossRef]
43. Anthony, J.L.; Maginn, A.E.J.; Brennecke, J.F. Solution Thermodynamics of Imidazolium-Based Ionic Liquids and Water. *J. Phys. Chem. B* **2001**, *105*, 10942–10949. [CrossRef]
44. Anthony, J.L.; Maginn, E.J.; Brennecke, J.F. Gas Solubilities in 1-n-Butyl-3-methylimidazolium Hexafluorophosphate. *ACS Symp. Ser.* **2002**, *818*, 260–269.
45. Xue, Z.; Zhang, Z.; Han, J.; Chen, Y.; Mu, T. Carbon dioxide capture by a dual amino ionic liquid with amino-functionalized imidazolium cation and taurine anion. *Int. J. Greenh. Control.* **2011**, *5*, 628–633. [CrossRef]
46. Chen, H.; He, Y.; Zhu, J.; Alias, H.; Ding, Y.; Nancarrow, P.; Hardacre, C.; Rooney, D.W.; Tan, C. Rheological and heat transfer behaviour of the ionic liquid, [C4mim] [NTf2]. *Int. J. Heat Fluid Flow* **2008**, *29*, 149–155. [CrossRef]
47. Dong, Q.; Muzny, C.D.; Kazakov, A.; Diky, V.; Magee, J.W.; Widegren, J.A.; Chirico, R.D.; Marsh, K.N.; Frenkel, M. ILThermo: A Free-Access Web Database for Thermodynamic Properties of Ionic Liquids. *J. Chem. Eng. Data* **2007**, *52*, 1151–1159. [CrossRef]
48. Hosseini, S.M.; Mulero, A.; Alavianmehr, M.M. Predictive methods and semi-classical Equations of State for pure ionic liquids: A review. *J. Chem. Thermodyn.* **2019**, *130*, 47–94. [CrossRef]
49. Coutinho, J.A.P.; Carvalho, P.J.; Oliveira, N.M. Predictive methods for the estimation of thermophysical properties of ionic liquids. *RSC Adv.* **2012**, *2*, 7322–7346. [CrossRef]
50. Torrecilla, J.S.; García, J.; Rojo, E.; Rodríguez, F. Estimation of toxicity of ionic liquids in Leukemia Rat Cell Line and Acetylcholinesterase enzyme by principal component analysis, neural networks and multiple lineal regressions. *J. Hazard. Mater.* **2009**, *164*, 182–194. [CrossRef]
51. Abramenko, N.; Kustov, L.M.; Metelytsia, L.; Kovalishyn, V.; Tetko, I.; Peijnenburg, W. A review of recent advances towards the development of QSAR models for toxicity assessment of ionic liquids. *J. Hazard. Mater.* **2020**, *384*, 121429. [CrossRef]
52. Taghizadehfard, M.; Hosseini, S.M.; Pierantozzi, M.; Alavianmehr, M.M. Predicting the volumetric properties of pure and mixture of amino acid-based ionic liquids. *J. Mol. Liq.* **2019**, *294*, 111604. [CrossRef]
53. Thawarkar, S.; Khupse, N.; Shinde, D.R.; Kumar, A. Understanding the behavior of mixtures of protic-aprotic and protic-protic ionic liquids: Conductivity, viscosity, diffusion coefficient and ionicity. *J. Mol. Liq.* **2019**, *276*, 986–994. [CrossRef]
54. McLeese, S.E.; Eslick, J.C.; Hoffmann, N.J.; Scurto, A.M.; Camarda, K.V. Design of Ionic Liquids via Computational Molecular Design. *Comput. Chem. Eng.* **2010**, *34*, 1476–1480. [CrossRef]
55. Austin, N.D.; Sahinidis, N.V.; Trahan, D.W. Computer-aided molecular design: An introduction and review of tools, applications, and solution techniques. *Chem. Eng. Res. Des.* **2016**, *116*, 2–26. [CrossRef]
56. Chemmangattuvalappil, N.G. Development of solvent design methodologies using computer-aided molecular design tools. *Curr. Opin. Chem. Eng.* **2020**, *27*, 51–59. [CrossRef]
57. Klamt, A. Conductor-like Screening Model for Real Solvents: A New Approach to the Quantitative Calculation of Solvation Phenomena. *J. Phys. Chem.* **1995**, *99*, 2224–2235. [CrossRef]
58. Lin, S.-T.; Sandler, S.I. A Priori Phase Equilibrium Prediction from a Segment Contribution Solvation Model. *Ind. Eng. Chem. Res.* **2002**, *41*, 899–913. [CrossRef]
59. Farahipour, R.; Mehrkesh, A.; Karunanithi, A.T. A systematic screening methodology towards exploration of ionic liquids for CO_2 capture processes. *Chem. Eng. Sci.* **2016**, *145*, 126–132. [CrossRef]
60. Chen, Y.; Woodley, J.M.; Kontogeorgis, G.; Gani, R. Integrated Ionic Liquid and Process Design involving Hybrid Separation Schemes. *Comput.-Aided Chem. Eng.* **2018**, *44*, 1045–1050.
61. Song, Z.; Zhou, T.; Qi, Z.; Sundmacher, K. Systematic Method for Screening Ionic Liquids as Extraction Solvents Exemplified by an Extractive Desulfurization Process. *ACS Sustain. Chem. Eng.* **2017**, *5*, 3382–3389. [CrossRef]

62. Song, Z.; Zhang, C.; Qi, Z.; Zhou, T.; Sundmacher, K. Computer-aided design of ionic liquids as solvents for extractive desulfurization. *AIChE J.* **2018**, *64*, 1013–1025. [CrossRef]
63. Chao, H.; Song, Z.; Cheng, H.; Chen, L.; Qi, Z. Computer-aided design and process evaluation of ionic liquids for n-hexane-methylcyclopentane extractive distillation. *Sep. Purif. Technol.* **2018**, *196*, 157–165. [CrossRef]
64. Hessel, V.; Li, X.; Chao, H.; Mo, F.; Zhou, T.; Cheng, H.; Chen, L.; Qi, Z. Computer-aided ionic liquid design for alkane/cycloalkane extractive distillation process. *Green Energy Environ.* **2019**, *4*, 154–165.
65. Mai, N.L.; Koo, Y.-M. Computer-Aided Design of Ionic Liquids for High Cellulose Dissolution. *ACS Sustain. Chem. Eng.* **2016**, *4*, 541–547. [CrossRef]
66. Firaha, D.S.; Hollóczki, O.; Kirchner, B. Computer-Aided Design of Ionic Liquids as CO_2 Absorbents. *Angew. Chem. Int. Ed.* **2015**, *54*, 7805–7809. [CrossRef] [PubMed]
67. Paduszyński, K.; Królikowski, M.; Zawadzki, M.; Orzeł, P. Computer-Aided Molecular Design of New Task-Specific Ionic Liquids for Extractive Desulfurization of Gasoline. *ACS Sustain. Chem. Eng.* **2017**, *5*, 9032–9042. [CrossRef]
68. Karunanithi, A.T.; Farahipour, R.; Dilmurat, K. Ionic Liquids: Applications by Computational Design. In *Encyclopedia of Inorganic and Bioinorganic Chemistry*; Wiley: Chichester, UK, 2016; pp. 1–13.
69. Peng, D.; Zhang, J.; Cheng, H.; Chen, L.; Qi, Z. Computer-aided ionic liquid design for separation processes based on group contribution method and COSMO-SAC model. *Chem. Eng. Sci.* **2017**, *159*, 58–68. [CrossRef]
70. Chen, Y.; Koumaditi, E.; Gani, R.; Kontogeorgis, G.M.; Woodley, J.M. Computer-aided design of ionic liquids for hybrid process schemes. *Comput. Chem. Eng.* **2019**, *130*, 106556. [CrossRef]
71. Venkatraman, V.; Evjen, S.; Lethesh, K.C.; Raj, J.J.; Knuutila, H.K.; Fiksdahl, A. Rapid, comprehensive screening of ionic liquids towards sustainable applications. *Sustain. Energy Fuels* **2019**, *3*, 2798–2808. [CrossRef]
72. Xu, Y.; Hanna, M.A.; Isom, L. "Green" Chemicals from Renewable Agricultural Biomass—A Mini Review. *Open Agric. J.* **2008**, *2*, 54–61. [CrossRef]
73. Kamm, B.; Kamm, M.; Soyez, K. The Green Biorefinery, Concept of Technology. In *Proceeding of the first International Symposium on Green Biorefinery, Neuruppin, Germany, October 1997*; Society of Ecological Technology and System Analysis: Berlin, Germany, 1998.
74. Frost, J.W.; Draths, K.M. Biocatalytic Syntheses of Aromatics from D-Glucose: Renewable Microbial Sources of Aromatic Compounds. *Annu. Rev. Microbiol.* **1995**, *49*, 557–579. [CrossRef] [PubMed]
75. Gravitis, J.; Zandersons, J.; Vedernikov, N.; Kruma, I.; Ozols-Kalnins, V. Clustering of bio-products technologies for zero emissions and eco-efficiency. *Ind. Crop. Prod.* **2004**, *20*, 169–180. [CrossRef]
76. Cherubini, F. The biorefinery concept: Using biomass instead of oil for producing energy and chemicals. *Energy Convers. Manag.* **2010**, *51*, 1412–1421. [CrossRef]
77. NREL. National Renewable Energy Laboratory [Internet]. In *Biomass Research.*; 2009. Available online: http://www.nrel.gov/biomass/biorefinery.html (accessed on 13 October 2020).
78. Huang, H.-J.; Ramaswamy, S.; Tschirner, U.W.; RamaRao, B.V. A review of separation technologies in current and future biorefineries. *Sep. Purif. Technol.* **2008**, *62*, 1–21. [CrossRef]
79. Spath, P.L.; Dayton, D.C. *Preliminary Screening—Technical and Economic Assessment of Synthesis Gas to Fuels and Chemicals with Emphasis on the Potential for Biomass-Derived Syngas*; National Renewable Energy Laboratory: Golden, CO, USA, 2003.
80. Bridgwater, A. Fast pyrolysis processes for biomass. *Renew. Sustain. Energy Rev.* **2000**, *4*, 1–73. [CrossRef]
81. Brownsort, P.A. *Biomass Pyrolysis Processess: Performance Parameters and Their Influence on Biochar System Benefits*; The University of Edinburgh: Edinburgh, UK, 2009.
82. Senneca, O. Kinetics of pyrolysis, combustion and gasification of three biomass fuels. *Fuel Process. Technol.* **2007**, *88*, 87–97. [CrossRef]
83. Demirbas, M.F. Biorefineries for biofuel upgrading: A critical review. *Appl. Energy* **2009**, *86*, S151–S161. [CrossRef]
84. Sun, Y.; Cheng, J. Hydrolysis of lignocellulosic materials for ethanol production: A review. *Bioresour. Technol.* **2002**, *83*, 1–11. [CrossRef]
85. Demirbaş, A. Biodiesel fuels from vegetable oils via catalytic and non-catalytic supercritical alcohol transesterifications and other methods: A survey. *Energy Convers. Manag.* **2003**, *44*, 2093–2109. [CrossRef]
86. Romano, R.T.; Zhang, R. Co-digestion of onion juice and wastewater sludge using an anaerobic mixed biofilm reactor. *Bioresour. Technol.* **2008**, *99*, 631–637. [CrossRef]

87. Hamelinck, C.N.; van Hooijdonk, G.; Faaij, A.P. Ethanol from lignocellulosic biomass: Techno-economic performance in short-, middle- and long-term. *Biomass Bioenergy* **2005**, *28*, 384–410. [CrossRef]
88. Nishida, N.; Stephanopoulos, G.; Westerberg, A.W. A review of process synthesis. *AIChE J.* **1981**, *27*, 321–351. [CrossRef]
89. Douglas, J.M. Process synthesis for waste minimization. *Ind. Eng. Chem. Res.* **1992**, *31*, 238–243. [CrossRef]
90. Kokossis, A.C.; Yang, A. On the use of systems technologies and a systematic approach for the synthesis and the design of future biorefineries. *Comput. Chem. Eng.* **2010**, *34*, 1397–1405. [CrossRef]
91. Stephanopoulos, G.; Reklaitis, G.V. Process systems engineering: From Solvay to modern bio- and nanotechnology. *Chem. Eng. Sci.* **2011**, *66*, 4272–4306. [CrossRef]
92. Ng, D.K.; Ng, R.T.L. Applications of process system engineering in palm-based biomass processing industry. *Curr. Opin. Chem. Eng.* **2013**, *2*, 448–454. [CrossRef]
93. Douglas, J.M. A hierarchical decision procedure for process synthesis. *AIChE J.* **1985**, *31*, 353–362. [CrossRef]
94. Li, X.; Kraslawski, A. Conceptual process synthesis: Past and current trends. *Chem. Eng. Process. Process. Intensif.* **2004**, *43*, 583–594. [CrossRef]
95. Ng, D.; Pham, V.; Jiménez-Gutiérrez, A.; Spriggs, H. A Hierarchical Approach to the Synthesis and Analysis of Integrated Biorefineries. In *Design Energy and the Environment, Proceedings of the Seventh International Conference on the Foundations of Computer-Aided Process Design*; CRC Press: Boca Raton, FL, USA, 2009; pp. 425–432.
96. Conde-Mejía, C.; Jiménez-Gutiérrez, A.; El-Halwagi, M.M. Application of a Hierarchical Approach for the Synthesis of Biorefineries. In *Process Design Strategies for Biomass Conversion Systems*; Wiley: Chichester, UK, 2015; pp. 39–61.
97. Tey, S.-Y.; Wong, S.S.; Lam, J.A.; Ong, N.Q.; Foo, D.C.; Ng, D.K. Extended hierarchical decomposition approach for the synthesis of biorefinery processes. *Chem. Eng. Res. Des.* **2020**, *166*, 40–54. [CrossRef]
98. Stephanopoulos, G.; Westerberg, A.W. Studies in process synthesis—II. *Chem. Eng. Sci.* **1976**, *31*, 195–204. [CrossRef]
99. Frangopoulos, C.A.; von Spakovsky, M.R.; Sciubba, E. A brief review of methods for the design and synthesis optimization of energy systems. *Int. J. Appl. Thermodyn.* **2002**, *4*, 151–160.
100. Kasivisvanathan, H.; Tan, R.R.; Ng, D.K.; Aziz, M.K.A.; Foo, D.C. Heuristic framework for the debottlenecking of a palm oil-based integrated biorefinery. *Chem. Eng. Res. Des.* **2014**, *92*, 2071–2082. [CrossRef]
101. Patel, B. A Thermodynamic Targeting Approach for the Synthesis of Sustainable Biorefineries. *Comput.-Aided Chem. Eng.* **2015**, *37*, 1283–1288.
102. Oppenheim, A.V. *Study Guide for Discrete-Time Signal Processing*; Prentice-Hall: Upper Saddle River, NJ, USA, 2010.
103. Benjamin, M.F.D.; Cayamanda, C.D.; Tan, R.R.; Razon, L.F. P-graph approach to criticality analysis in integrated bioenergy systems. *Clean Technol. Environ. Policy* **2017**, *19*, 1841–1854. [CrossRef]
104. Lam, H.L.; Klemeš, J.J.; Varbanov, P.S.; Kravanja, Z. P-Graph Synthesis of Open-Structure Biomass Networks. *Ind. Eng. Chem. Res.* **2013**, *52*, 172–180. [CrossRef]
105. Yeo, J.Y.J.; How, B.S.; Teng, S.Y.; Leong, W.D.; Ng, W.P.; Lim, C.H.; Ngan, S.L.; Sunarso, J.; Lam, H.L. Synthesis of Sustainable Circular Economy in Palm Oil Industry Using Graph-Theoretic Method. *Sustainability* **2020**, *12*, 8081. [CrossRef]
106. Grossmann, I.E. Review of Nonlinear Mixed-Integer and Disjunctive Programming Techniques. *Optim. Eng.* **2002**, *3*, 227–252. [CrossRef]
107. Bao, B.; Ng, D.K.; Tay, D.H.; Jiménez-Gutiérrez, A.; El-Halwagi, M.M. A shortcut method for the preliminary synthesis of process-technology pathways: An optimization approach and application for the conceptual design of integrated biorefineries. *Comput. Chem. Eng.* **2011**, *35*, 1374–1383. [CrossRef]
108. Pham, V.; El-Halwagi, M. Process synthesis and optimization of biorefinery configurations. *AIChE J.* **2011**, *58*, 1212–1221. [CrossRef]
109. Ng, R.T.L.; Ng, D.K. Systematic Approach for Synthesis of Integrated Palm Oil Processing Complex. Part 1: Single Owner. *Ind. Eng. Chem. Res.* **2013**, *52*, 10206–10220. [CrossRef]
110. Yuan, Z.; Eden, M.R. Superstructure optimization of integrated fast pyrolysis-gasification for production of liquid fuels and propylene. *AIChE J.* **2016**, *62*, 3155–3176. [CrossRef]
111. Rizwan, M.; Almansoori, A.; Elkamel, A. An overview on synthesis and design of microalgal biorefinery configurations by employing superstructure-based optimization approach. *Energy Syst.* **2019**, *10*, 941–966. [CrossRef]

112. Fasahati, P.; Wu, W.; Maravelias, C.T. Process synthesis and economic analysis of cyanobacteria biorefineries: A superstructure-based approach. *Appl. Energy* **2019**, *253*, 113625. [CrossRef]
113. Dickson, R.; Liu, J.J. Optimization of seaweed-based biorefinery with zero carbon emissions potential. *Comput.-Aided Chem. Eng.* **2019**, *46*, 247–252.
114. Ramapriya, G.M.; Won, W.; Maravelias, C T. A superstructure optimization approach for process synthesis under complex reaction networks. *Chem. Eng. Res. Des.* **2018**, *137*, 589–608. [CrossRef]
115. Ng, R.T.L.; Tay, D.H.S.; Ng, D.K. Simultaneous Process Synthesis, Heat and Power Integration in a Sustainable Integrated Biorefinery. *Energy Fuels* **2012**, *26*, 7316–7330. [CrossRef]
116. Ayodele, B.V.; Cheng, C.K. Biorefinery for the Production of Biodiesel, Hydrogen and Synthesis Gas Integrated with CHP from Oil Palm in Malaysia. *Chem. Prod. Process Model.* **2018**, *11*, 305–314. [CrossRef]
117. Pyrgakis, K.; Kokossis, A.C. Total Site Analysis as a Synthesis Model to Select, Optimize and Integrate Processess in Multiple-Product Biorefineries. *Chem. Eng. Trans.* **2016**, *52*, 913–918.
118. Kasivisvanathan, H.; Ng, R.T.; Tay, D.H.; Ng, D.K. Fuzzy optimisation for retrofitting a palm oil mill into a sustainable palm oil-based integrated biorefinery. *Chem. Eng. J.* **2012**, *200*, 694–709.
119. Wan, Y.K.; Sadhukhan, J.; Ng, K.S.; Ng, D.K. Techno-economic evaluations for feasibility of sago-based biorefinery, Part 1: Alternative energy systems. *Chem. Eng. Res. Des.* **2016**, *107*, 263–279. [CrossRef]
120. Wan, Y.K.; Sadhukhan, J.; Ng, D.K. Techno-economic evaluations for feasibility of sago-based biorefinery, Part 2: Integrated bioethanol production and energy systems. *Chem. Eng. Res. Des.* **2016**, *107*, 102–116. [CrossRef]
121. Mongkhonsiri, G.; Charoensuppanimit, P.; Anantpinijwatna, A.; Gani, R.; Assabumrungrat, S. Process development of sustainable biorefinery system integrated into the existing pulping process. *J. Clean. Prod.* **2020**, *255*, 120278. [CrossRef]
122. Sadhukhan, J.; Mustafa, M.A.; Misailidis, N.; Mateos-Salvador, F.; Du, C.; Campbell, G.M. Value analysis tool for feasibility studies of biorefineries integrated with value added production. *Chem. Eng. Sci.* **2008**, *63*, 503–519. [CrossRef]
123. Ponce-Ortega, J.M.; Pham, V.; El-Halwagi, M.M.; El-Baz, A.A. A Disjunctive Programming Formulation for the Optimal Design of Biorefinery Configurations. *Ind. Eng. Chem. Res.* **2012**, *51*, 3381–3400. [CrossRef]
124. Tay, D.H.S.; Ng, R.T.; Ng, D.K. Modular Optimization Approach for Process Synthesis and Integration of an Integrated Biorefinery. *Comput. Aided Chem. Eng.* **2012**, *31*, 1045–1049.
125. Ng, R.T.L.; Hassim, M.H.; Ng, D.K. Process synthesis and optimization of a sustainable integrated biorefinery via fuzzy optimization. *AIChE J.* **2013**, *59*, 4212–4227. [CrossRef]
126. Kasivisvanathan, H.; Ubando, A.T.; Ng, D.K.; Tan, R.R. Robust Optimization for Process Synthesis and Design of Multifunctional Energy Systems with Uncertainties. *Ind. Eng. Chem. Res.* **2014**, *53*, 3196–3209. [CrossRef]
127. Kelloway, A.; Daoutidis, P. Process Synthesis of Biorefineries: Optimization of Biomass Conversion to Fuels and Chemicals. *Ind. Eng. Chem. Res.* **2014**, *53*, 5261–5273. [CrossRef]
128. Kasivisvanathan, H.; Ng, D.K.; Poplewski, G.; Tan, R.R. Flexibility Optimization for a Palm Oil-Based Integrated Biorefinery with Demand Uncertainties. *Ind. Eng. Chem. Res.* **2016**, *55*, 4035–4044. [CrossRef]
129. Albarelli, J.Q.; Onorati, S.; Caliandro, P.; Peduzzi, E.; Meireles, M.A.A.; Marechal, F.; Ensinas, A.V. Multi-objective optimization of a sugarcane biorefinery for integrated ethanol and methanol production. *Energy* **2017**, *138*, 1281–1290. [CrossRef]
130. Sy, C.L.; Ubando, A.T.; Aviso, K.B.; Tan, R.R. Multi-objective target oriented robust optimization for the design of an integrated biorefinery. *J. Clean. Prod.* **2018**, *170*, 496–509. [CrossRef]
131. Martin, M.; Grossmann, I.E. On the Systematic Synthesis of Sustainable Biorefineries. *Ind. Eng. Chem. Res.* **2013**, *52*, 3044–3064. [CrossRef]
132. Caballero, J.A.; Odjo, A.; Grossmann, I.E. Flowsheet optimization with complex cost and size functions using process simulators. *AIChE J.* **2007**, *53*, 2351–2366. [CrossRef]
133. Ng, D.K.; Foo, D.C.; Tan, R.R. Automated Targeting Technique for Single-Impurity Resource Conservation Networks. Part 1: Direct Reuse/Recycle. *Ind. Eng. Chem. Res.* **2009**, *48*, 7637–7646. [CrossRef]
134. Ng, D.K.; Foo, D.C.; Tan, R.R. Automated Targeting Technique for Single-Impurity Resource Conservation Networks. Part 2: Single-Pass and Partitioning Waste-Interception Systems. *Ind. Eng. Chem. Res.* **2009**, *48*, 7647–7661. [CrossRef]

135. Ng, D.K.; Foo, D.C.; Tan, R.R.; Pau, C.H.; Tan, Y.L. Automated targeting for conventional and bilateral property-based resource conservation network. *Chem. Eng. J.* **2009**, *149*, 87–101. [CrossRef]
136. Ng, D.K. Automated targeting for the synthesis of an integrated biorefinery. *Chem. Eng. J.* **2010**, *162*, 67–74. [CrossRef]
137. Tay, D.H.S.; Ng, D.K.S. Automated Targeting for the Synthesis of an Integrated Biorefinery. *J. Clean. Prod.* **2012**, *34*, 38–48. [CrossRef]
138. Shabbir, Z.; Tay, D.H.S.; Ng, D.K. A hybrid optimisation model for the synthesis of sustainable gasification-based integrated biorefinery. *Chem. Eng. Res. Des.* **2012**, *90*, 1568–1581. [CrossRef]
139. Koukios, E.; Koullas, D.; Koukios, I.D.; Avgerinos, E. Critical parameters for optimal biomass refineries: The case of biohydrogen. *Clean Technol. Environ. Policy* **2010**, *12*, 147–151. [CrossRef]
140. Fernando, A.L.; Duarte, M.P.; Almeida, J.; Boléo, S.; Mendes, B. Environmental impact assessment of energy crops cultivation in Europe. *Biofuels Bioprod. Biorefining* **2010**, *4*, 594–604. [CrossRef]
141. Andiappan, V.; Ko, A.S.Y.; Lau, V.W.S.; Ng, L.Y.; Ng, R.T.L.; Chemmangattuvalappil, N.G.; Ng, D.K. Synthesis of sustainable integrated biorefinery via reaction pathway synthesis: Economic, incremental enviromental burden and energy assessment with multiobjective optimization. *AIChE J.* **2015**, *61*, 132–146. [CrossRef]
142. Tey, T.O.; Chen, S.; Cheong, Z.X.; Choong, A.S.X.; Ng, L.Y.; Chemmangattuvalappil, N.G. Synthesis of a sustainable integrated biorefinery to produce value-added chemicals from palm-based biomass via mathematical optimisation. *Sustain. Prod. Consum.* **2020**, *26*, 288–315. [CrossRef]
143. Filho, J.F.S.D.C.; Romano, P.N.; De Almeida, J.M.A.R.; Sousa-Aguiar, E.F. Critical catalytic routes: From the conventional bioethanol production model toward the integrated biorefinery concept. *Curr. Opin. Green Sustain. Chem.* **2019**, *20*, 33–38. [CrossRef]
144. Tang, M.C.; Chin, M.W.S.; Lim, K.M.; Mun, Y.S.; Ng, R.T.L.; Tay, D.H.S.; Ng, D.K. Systematic approach for conceptual design of an integrated biorefinery with uncertainties. *Clean Technol. Environ. Policy* **2013**, *15*, 783–799. [CrossRef]
145. Tay, D.H.S.; Ng, D.K.; Tan, R.R. Robust optimization approach for synthesis of integrated biorefineries with supply and demand uncertainties. *Environ. Prog. Sustain. Energy* **2013**, *32*, 384–389. [CrossRef]
146. Cheali, P.; Quaglia, A.; Gernaey, K.V.; Sin, G. Effect of Market Price Uncertainties on the Design of Optimal Biorefinery Systems—A Systematic Approach. *Ind. Eng. Chem. Res.* **2014**, *53*, 6021–6032. [CrossRef]
147. Giuliano, A.; Poletto, M.; Barletta, D. Process optimization of a multi-product biorefinery: The effect of biomass seasonality. *Chem. Eng. Res. Des.* **2016**, *107*, 236–252. [CrossRef]
148. Čuček, L.; Martín, M.; Grossmann, I.E.; Kravanja, Z. Multi-period synthesis of optimally integrated biomass and bioenergy supply network. *Comput. Chem. Eng.* **2014**, *66*, 57–70. [CrossRef]
149. Awudu, I.; Zhang, J. Uncertainties and sustainability concepts in biofuel supply chain management: A review. *Renew. Sustain. Energy Rev.* **2012**, *16*, 1359–1368. [CrossRef]
150. Shabani, N.; Akhtari, S.; Sowlati, T. Value chain optimization of forest biomass for bioenergy production: A review. *Renew. Sustain. Energy Rev.* **2013**, *23*, 299–311. [CrossRef]
151. Cambero, C.; Sowlati, T. Assessment and optimization of forest biomass supply chains from economic, social and environmental perspectives—A review of literature. *Renew. Sustain. Energy Rev.* **2014**, *36*, 62–73. [CrossRef]
152. De Meyer, A.; Cattrysse, D.; Rasinmäki, J.; Van Orshoven, J. Methods to optimise the design and management of biomass-for-bioenergy supply chains: A review. *Renew. Sustain. Energy Rev.* **2014**, *31*, 657–670. [CrossRef]
153. Atashbar, N.Z.; Labadie, N.; Prins, C. Modelling and optimisation of biomass supply chains: A review. *Int. J. Prod. Res.* **2018**, *56*, 3482–3506. [CrossRef]
154. Mirkouei, A.; Haapala, K.R.; Sessions, J.; Murthy, G.S. A review and future directions in techno-economic modeling and optimization of upstream forest biomass to bio-oil supply chains. *Renew. Sustain. Energy Rev.* **2017**, *67*, 15–35. [CrossRef]
155. Stuart, P.R.; El-Halwagi, M.M. *Integrated Biorefineries: Design, Analysis, and Optimization*, 1st ed.; CRC Press: Boca Raton, FL, USA, 2013; p. 873.
156. Lo, S.L.Y.; How, B.S.; Leong, W.D.; Teng, S.Y.; Rhamdhani, M.A.; Sunarso, J. Techno-economic analysis for biomass supply chain: A state-of-the-art review. *Renew. Sustain. Energy Rev.* **2020**, *135*, 110164. [CrossRef]
157. Ng, R.T.L.; Ng, D.K.; Tan, R.R. Systematic Approach for Synthesis of Integrated Palm Oil Processing Complex. Part 2: Multiple Owners. *Ind. Eng. Chem. Res.* **2013**, *52*, 10221–10235. [CrossRef]

158. Ng, R.T.L.; Ng, D.K.; Tan, R.R.; El-Halwagi, M.M. Disjunctive fuzzy optimisation for planning and synthesis of bioenergy-based industrial symbiosis system. *J. Environ. Chem. Eng.* **2014**, *2*, 652–664. [CrossRef]
159. Barla, F.; Nikolakopoulos, A.; Kokossis, A.C. Design of Circular Economy Plants—The Case of the Textile Waste Biorefinery. *Comput. Aided Chem. Eng.* **2017**, 1933–1938.
160. Lichtenthaler, F.W.; Mondel, S. Perspectives in the use of low molecular weight carbohydrates as organic raw materials. *Pure Appl. Chem.* **1997**, *69*, 1853–1866. [CrossRef]
161. Wilpiszewska, K.; Spychaj, T. Chemical modification of starch with hexamethylene diisocyanate derivatives. *Carbohydr. Polym.* **2007**, *70*, 334–340. [CrossRef]
162. Wyman, C.E. Potential Synergies and Challenges in Refining Cellulosic Biomass to Fuels, Chemicals, and Power. *Biotechnol. Prog.* **2003**, *19*, 254–262. [CrossRef] [PubMed]
163. Liang, L.; Mao, Z.; Li, Y.; Wan, C.; Wang, T.; Zhang, L.; Zhang, L. Liquefaction of crop residues for polyol production. *Bioresource* **2006**, *1*, 248–256. [CrossRef]
164. Elliott, D.C. Biomass, Chemicals from. In *Encyclopedia of Energy*; Elsevier B.V.: Amsterdam, The Netherlands, 2004; pp. 163–174.
165. Werpy, T.; Petersen, G. *Top Value Added Chemicals from Biomass—Volume I—Results of Screening for Potential Candidates from Sugars and Synthesis Gas*; U.S Department of Energy: Washington, DC, USA, 2004.
166. Holladay, J.; Bozell, J.; White, J.; Johnson, D. *Top Value-Added Chemicals from Biomass—Volume II—Results of Screening for Potential Candidates from Biorefinery Lignin*; U.S. Department of Energy: Washington, DC, USA, 2007.
167. Skibar, W.; Grogan, G.; McDonald, J.; Pitts, M. *UK Expertise for Exploitation of Biomass-Based Platform Chemicals*; The FROPTOP Group: Runcorn, UK, 2009.
168. Achenie, L.E.K.; Gani, R.; Venkatasubramanian, V. *Computer Aided Molecular Design: Theory and Practice*, 1st ed.; Elsevier, B.V.: Amsterdam, The Netherlands, 2003; p. 404.
169. Venkatasubramanian, V.; Chan, K.; Caruthers, J.M. Computer-aided molecular design using genetic algorithms. *Comput. Chem. Eng.* **1994**, *18*, 833–844. [CrossRef]
170. Samudra, A.P.; Sahinidis, N.V. Optimization-based framework for computer-aided molecular design. *AIChE J.* **2013**, *59*, 3686–3701. [CrossRef]
171. Hechinger, M.; Voll, A.; Marquardt, W. Towards an integrated design of biofuels and their production pathways. *Comput. Chem. Eng.* **2010**, *34*, 1909–1918. [CrossRef]
172. Ng, L.Y.; Andiappan, V.; Chemmangattuvalappil, N.G.; Ng, D.K. Novel Methodology for the Synthesis of Optimal Biochemicals in Integrated Biorefineries via Inverse Design Techniques. *Ind. Eng. Chem. Res.* **2015**, *54*, 5722–5735. [CrossRef]
173. Ng, L.Y.; Andiappan, V.; Chemmangattuvalappil, N.G.; Ng, D.K. A systematic methodology for optimal mixture design in an integrated biorefinery. *Comput. Chem. Eng.* **2015**, *81*, 288–309. [CrossRef]
174. Bertran, M.-O.; Frauzem, R.; Sanchez-Arcilla, A.-S.; Zhang, L.; Woodley, J.M.; Gani, R. A generic methodology for processing route synthesis and design based on superstructure optimization. *Comput. Chem. Eng.* **2017**, *106*, 892–910. [CrossRef]
175. Chaávez-Islas, L.M.; Vasquez-Medrano, R.; Flores-Tlacuahuac, A. Optimal molecular design of ionic liquids for high-purity bioethanol production. *Ind. Eng. Chem. Res.* **2011**, *50*, 5153–5168.
176. Chong, F.K.; Foo, D.C.; Eljack, F.; Atilhan, M.; Chemmangattuvalappil, N.G. A systematic approach to design task-specific ionic liquids and their optimal operating conditions. *Mol. Syst. Des. Eng.* **2016**, *1*, 109–121. [CrossRef]
177. Liu, X.; Huang, Y.; Zhao, Y.; Gani, R.; Zhang, X.; Zhang, S. Ionic Liquid Design and Process Simulation for Decarbonization of Shale Gas. *Ind. Eng. Chem. Res.* **2016**, *55*, 5931–5944. [CrossRef]
178. Chong, F.K.; Andiappan, V.; Ng, D.K.; Foo, D.C.; Eljack, F.; Atilhan, M.; Chemmangattuvalappil, N.G. Design of Ionic Liquid as Carbon Capture Solvent for a Bioenergy System: Integration of Bioenergy and Carbon Capture Systems. *ACS Sustain. Chem. Eng.* **2017**, *5*, 5241–5252. [CrossRef]
179. Liu, X.; Zhou, T.; Zhang, X.; Zhang, S.; Liang, X.; Gani, R.; Kontogeorgis, G.M. Application of COSMO-RS and UNIFAC for ionic liquids based gas separation. *Chem. Eng. Sci.* **2018**, *192*, 816–828. [CrossRef]
180. Chen, Y.; Liu, X.; Kontogeorgis, G.M.; Woodley, J.M. Ionic-Liquid-Based Bioisoprene Recovery Process Design. *Ind. Eng. Chem. Res.* **2020**, *59*, 7355–7366. [CrossRef]
181. Wang, J.; Song, Z.; Cheng, H.; Chen, L.; Deng, L.; Qi, Z. Computer-Aided Design of Ionic Liquids as Absorbent for Gas Separation Exemplified by CO2 Capture Cases. *ACS Sustain. Chem. Eng.* **2018**, *6*, 12025–12035. [CrossRef]

182. Ramadhan, N.J.; Wan, Y.K.; Ng, R.T.; Ng, D.K.; Hassim, M.H.; Aviso, K.B.; Tan, R.R. Life cycle optimisation (LCO) of product systems with consideration of occupational fatalities. *Process. Saf. Environ. Prot.* **2014**, *92*, 390–405. [CrossRef]
183. Cooley, M. Human-centred Systems. In *Designing Human-centred Technology: A Cross-disciplinary Project in Computer-Aided Manufacturing*, 1st ed.; Rosenbrock, H., Ed.; Springer: London, UK, 1989; pp. 133–143.
184. Francisco, M.; van den Bruinhorst, A.; Kroon, M.C. Low-Transition-Temperature Mixtures (LTTMs): A New Generation of Designer Solvents. *Angew. Chem. Int. Engl.* **2013**, *52*, 3074–3085. [CrossRef]
185. Peng, Y.; Lu, X.; Liu, B.; Zhu, J. Separation of azeotropic mixtures (ethanol and water) enhanced by deep eutectic solvents. *Fluid Phase Equilibria* **2017**, *448*, 128–134. [CrossRef]
186. Bakhtyari, A.; Haghbakhsh, R.; Duarte, A.R.C.; Raeissi, S. A simple model for the viscosities of deep eutectic solvents. *Fluid Phase Equilibria* **2020**, *521*, 112662. [CrossRef]

processes MDPI

Article

Synthesis of Integrated Flower Waste Biorefinery: Multi-Objective Optimisation with Economic and Environmental Consideration

Emily Hau Yan Chong [1], Viknesh Andiappan [1,2], Lik Yin Ng [1,3], Parimala Shivaprasad [4] and Denny K. S. Ng [1,5,*]

1 School of Engineering and Physical Sciences, Heriot-Watt University Malaysia, Putrajaya 62200, Malaysia
2 Research Centre of Sustainable Technologies, Faculty of Engineering, Computing and Science, Swinburne University of Technology Sarawak, Kuching 93350, Malaysia
3 Faculty of Engineering, Technology & Built Environment, UCSI University, Puncak Menara Gading, Taman Connaught, Cheras, Kuala Lumpur 56000, Malaysia
4 Faculty of Engineering, University of Nottingham, Nottingham NG 7 2RD, UK
5 Centre of Hydrogen Energy (CHE), Universiti Teknologi Malaysia (UTM), Johor Bahru 81310, Malaysia
* Correspondence: denny.ng@hw.ac.uk; Tel.: +60-3-88943784

Abstract: The improper disposal of flower waste from cultural activities is one of the main challenges in certain countries such as India. If the flower waste is not managed properly, it causes a number of environmental issues. Therefore, various technologies have been developed to transform flower waste into value-added products. To integrate multiple technologies holistically to maximise the energy and material recovery, an integrated flower-waste biorefinery is required. Since there are a wide range of technologies available that can convert the waste into multiple products, there is a need to develop a systematic approach to evaluate all the technologies. This research proposes a systematic approach to synthesise an integrated flower-waste biorefinery based on different optimisation objectives, e.g., maximum economic performance and minimum environmental impact. Due to the conflicting nature between the two objectives, a fuzzy optimisation approach has been adapted to synthesise a sustainable integrated flower-waste biorefinery that satisfies both objectives at once. The efficacy of the proposed approach is demonstrated through a case study in India based on the optimised results with fuzzy optimisation—a synthesised flower-waste integrated biorefinery with economy performance of $400,932 and carbon emission of 46,209 kg CO_2/h.

Keywords: flower waste; integrated biorefinery; waste valorisation; value-added product; mathematical optimisation

Citation: Emily Hau Yan Chong, Viknesh Andiappan, Lik Yin Ng, Parimala Shivaprasad and Denny K. S. Ng Synthesis of Integrated Flower Waste Biorefinery: Multi-Objective Optimisation with Economic and Environmental Consideration. *Processes* **2022**, *10*, 2240. https://doi.org/10.3390/pr10112240

Academic Editor: Blaž Likozar

Received: 11 August 2022
Accepted: 17 October 2022
Published: 1 November 2022

1. Introduction

Disposal of large amounts of flower waste is one of the major concerns that leads to environmental issues. Cultural and heritage activities, especially in India, have produced the highest flower waste worldwide [1]. India has a population of religious devotees who commonly practice religious customs using flowers, as they are tokens of devotion in South Asian cultures [2]. Thus, large amounts of flowers are left as offerings and accumulate at religious sites like temples. According to Statista 2021, India's flower production reached about 3 million metric tonnes in the financial year 2021 [1]. Meanwhile, Varanasi is the holiest city in India, which generates around 3.5 to 4 tonnes of flower waste from temples daily. Other than religious activities, flower waste is also generated during different ceremonies, functions and festivals. Most of the flowers utilised in these activities are left unused, resulting in a common source of waste. Viewing the issues above, it is important to have an effective waste disposal policy for the recovery of such waste, with an aim towards reducing the environmental impacts. In current practice, most wilted and unused flowers in India are discarded into landfills or water bodies [3]. This causes water pollution as toxic pesticides and insecticides remaining in the flower waste would seep into the water bodies.

In addition, organic matter in flower waste will cause negative impacts on water quality by depleting oxygen levels in water, threatening the marine ecosystem.

To address the above-mentioned issues, flower waste can be recovered and converted into value-added products. Recently, various technologies have been developed to recover organic waste materials and convert them into value-added products. To maximise the material and energy recovery potential, an integrated flower-waste biorefinery that integrates multiple technologies will be introduced. Detailed discussions on the potential of flower-waste valorisation and synthesis of a sustainable integrated flower-waste biorefinery will be presented in the following sub-sections.

1.1. Valorisation of Flower Waste

Waste valorisation is an application of industrial ecology that allows closed-loop manufacturing facilities to gain the benefits of the cyclic pattern of the used material present in the consumption sectors [4]. The waste generated by using flowers for decorative or religious purposes can be converted into value-added products for different applications: agricultural, bioenergy, food, beverage, pharmaceutical, etc. For example, phenolic compounds present in flowers could provide a high antioxidant capacity compared to fruits and vegetables. It provides positive effects on oxidative stress-related diseases such as cardiovascular diseases, neurodegenerative diseases, cancer, diabetes, obesity, and liver diseases [5]. Hence, to conserve the medical constituents in flowers, a proper drying method is required. Drying can improve the edible nature of flowers and prolong their shelf life. The dry-flower industry has the great potential to provide employment opportunities to thousands of people, especially to housewives and rural women in India, as unlimited aesthetic value and decorative products can be created by using the dry-flower technology [6]. In addition to being utilised for food and pharmaceutical applications [7,8], flower waste can be processed through different technologies for a variety of purposes. Slavov et al. [9] and Dutta et al. [10] reviewed valorisation opportunities of flower waste to essential oil, recovery of valuable biologically active substances, biofuels production, activated carbon and their application to food industry or medicine. Research work on investigating various strategies for recovery of flower waste have been done over the past years. Table 1 summarises the possible technologies for the utilisation of flower waste.

Table 1. Possible technologies for utilisation of flower waste.

Application	Technology/Biorefinery	Description	Final Product
Food	Drying [7]	Removal of water from flower and retain the medical constituents	Edible flower and flower for garnishing purpose
Pharmaceutical	Extraction [8]	Secure biological active compound in flower	Flower for medical purpose
Agricultural application	Composting [9]	Utilised for soil replenishment after composting	Bio-fertilizers
Environmental remediation	Biosorption [9]	Heavy metals from wastewater are accumulated	Removal of dyes, wastewater treatment
Bioenergy application	Gasification and fermentation [9]	Turn waste into source of energy	Biogas and bioethanol

1.2. Synthesis of Sustainable Integrated Biorefinery

According to Kamm (1997), a biorefinery is defined as a complex system of sustainable, environmental and resources-friendly technologies for fully utilising and exploiting biological raw materials [11]. All technologies stated in Table 1 may be applied to convert biological raw materials into value-added products. Research based on different types of biological raw materials has been conducted in the past. To maximise the energy and material recovery, an integrated biorefinery was introduced [12]. An integrated biorefinery consists of multiple technologies including feedstock handling, pretreatment processes, and different biomass conversion (e.g., thermochemical, biological, physical, etc.) and upgrading processes to produce value-added products and energy [13].

Viewing the importance of sustainable production and consumption in the industry, systematic synthesis and optimisation of integration biorefinery have been developed since the last decade. A number of systematic tools have been developed, ranging from hierarchical approaches [14,15], algorithmic approaches [16,17], mathematical optimisation approaches and hybrid approaches. For example, Ng et al. [14] proposed a hierarchical approach known as the forward-reverse synthesis tree to synthesise and analyse integrated biorefineries. Most recently, Tey et al. [15] presented an extended hierarchical decomposition approach, which was developed for chemical process synthesis for the synthesis of biorefinery processes. Based on the benefit of algorithmic approaches to execute a sequential set of actions based on automated reasoning, calculation and data processing, P-graph was adapted to synthesise an integrated biorefinery [16] and integrated palm-based biorefinery based on the circular economy concept [17]. To address a complex synthesis problem with multi-criteria requirements, mathematical optimisation approaches have been developed. For example, a superstructure-based shortcut approach was developed by Bao et al. for the conceptual design of integrated biorefineries [18]. A superstructure optimisation model was then adapted in the synthesis of integrated biorefinery for different feedstock, such as a palm-based biorefinery [19], microalgal biorefinery [20], seaweed-based biorefinery [21], etc. A detailed review of the synthesis tools can be found in the recent review papers by Chemmangattuvalappil et al. [22].

Environmental Assessment on Integrated Flower-Waste Biorefinery

To synthesise a sustainable integrated biorefinery, an environmental assessment is one of the important factors. As shown by Romero-García et al. [23], various methods can be adopted to evaluate the environmental impact of an integrated biorefinery. These include life cycle assessment (LCA), potential environmental impact (PEI), greenhouse gases (GHG) emission, footprints (carbon, water and sustainability), etc. LCA is a systematic, scientific method for calculating and evaluating long-term environmental impact [24]. LCA has been applied in various integrated biorefineries such as cyclamen plants in greenhouse cultivation [25] and Ethiopian rose cultivation [26]. Russo et al. [25] applied LCA towards roses and cyclamens in greenhouse cultivation. According to Saraiva [27], there are a number of parameters that need to be considered when doing life-cycle assessment, as shown in Table 2.

Table 2. Parameters considered in measuring environmental impacts.

Parameters		Consideration
Direct environment impacts	Input-related emission	Greenhouse gases emissions from crop production
	Transportation of feedstock	Furl use for feedstock collection and transportation of residues and by-product
	Direct land-use change	Soil quality changes made by above-ground biomass in terms of carbon sequestration
Indirect environmental impacts	Indirect land-use change	Substitution of one feed into another feed will affect greenhouse gases emission (this section is not elaborated further in journal review)

Alig et al. [28] applied LCA to determine the environmental impacts of production of different types of cut roses from Holland, Kenya and Ecuador. In the previous work [28], the emission of gases from the packaging and transportation of cut roses are considered when performing LCA.

Limited research has been done on the synthesis of a flower-waste integrated biorefinery. Thus, this work aims to synthesise an integrated flower-waste biorefinery. To design a sustainable integrated flower-waste biorefinery, other than an environmental assessment, the economic performance of the synthesised biorefinery should also be considered. Therefore, such a design problem is required to be solved via multi-objective optimisation approaches. Note that a number of works have been developed to synthesise an integrated biorefinery with multi-objective functions, such as trade-off between environmental and economic performances [29–32], under uncertainty [33], chemical product design [34], inherent safety and health [35], etc.

According to Tay and Ng [29], fuzzy optimisation identifies the ideal alternatives in decision-making problems by solving an objective function on a set of alternatives given by fixed constraints; the preferable alternatives have the more desired maximum or minimum objective function values. Thus, a fuzzy optimisation is adapted in this work.

2. Problem Statement

Given a set of flower waste based on various compositions (i.e., petal, leaf, stem), $i \in I$ $\left(F_i^{\text{Biomass}}\right)$ that may be transformed into intermediates $k \in K$ and then further converted into products $k' \in K'$ via technologies j and j' based on fixed conversion factors, these conversions factors for intermediates k and final product k' are denoted as X_{ijk} and $X_{kj'k'}$. Based on the conversion factors, the flowrates F of each intermediate k and product k' can then be determined. A generic superstructure of the model is illustrated in Figure 1.

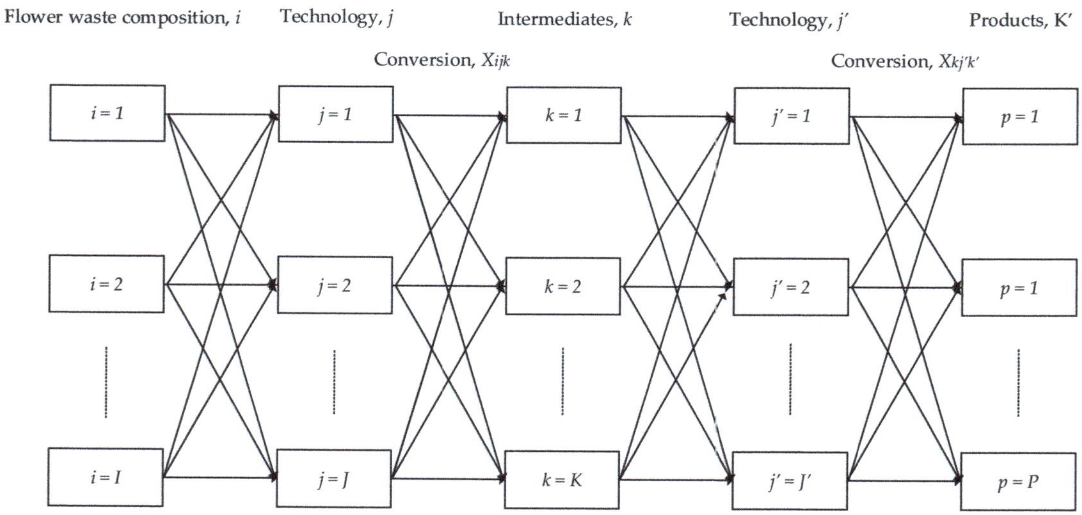

Figure 1. Generic superstructure for flower-waste biorefinery.

The objective of this work is to produce an integrated flower-waste biorefinery with economic and environmental considerations. Three scenarios are analysed in the case study: (i) maximum economic performance, (ii) minimum environmental impact and (iii) multi-objective optimisation with maximum economic performance and minimum environmental impact.

3. Mathematical Optimisation Model

Before formulating the mathematical optimisation model, data collection (e.g., conversion performance, emission factor, material and energy requirement, price of product and cost of materials, etc.) for all possible technologies and processes that valorise flower waste into value-added products were collected. Then, a superstructure model that includes all possible process alternatives was generated and a mathematical optimisation model was developed based on different optimisation objectives.

Model Formulation

Referring to Figure 1, the total amount of flower waste i is first sent to various biomass conversion technologies j to produce intermediates k. The intermediate then undergoes upgrading process j' to be further converted into products p or K'. The mass flow distribution within the integrated biorefinery is represented by Equation (1).

$$F_i^{\text{Biomass}} = \sum_j f_{ij} \; \forall \, i \tag{1}$$

where f_{ij} is the mass flowrate of the flower waste to technologies j.

$$F_k = \sum_j \sum_i f_{ij} X_{ijk} \; \forall \, k \tag{2}$$

where X_{ijk} is the conversion factor of flower waste i to intermediate k.

For intermediate k, which requires further processing in technologies j', the mass flowrate of final product k' is given as below:

$$F_k = \sum_{j'} f_{kj'} \; \forall \, k \tag{3}$$

where $f_{kj'}$ is the mass flowrate of flower waste intermediate to technologies j'.

$$F_{k'} = \sum_{j'} \sum_k f_{kj'} X_{kj'k'} \; \forall \, k' \tag{4}$$

where $X_{kj'k'}$ $X_{kj'k}$ is conversion factor of intermediate k to final product k'.

Note that the above mathematical model can be modified to suit different technologies j and j'. For example, in the drying process, water within the biomass conversion is performed to determine the water lost during the process and total mass changed. For the technologies that convert the biological material into bioenergy, the conversion factor X can be replaced by an energy conversion factor.

Other than tracking the final products, waste generated throughout the biorefinery is also determined. In this work, it is assumed that all the waste that cannot be converted into value-added products will be sent to a landfill. The equation below describes the mass flow of the collected waste to a landfill, from intermediates k and final product k' in technologies j and j'.

$$F^{\text{Landfill}} = \sum_j \sum_i f_{ij} \left(1 - X_{ijk}\right) + \sum_{j'} \sum_k f_{kj'} \left(1 - X_{kj'k'}\right) \tag{5}$$

The economic performance of the integrated biorefinery (EP), the revenue (REV) and landfill cost ($COST^{\text{Landfill}}$) are determined as below:

$$REV = \sum_k F_k \times P_k + \sum_{k'} F_{k'} \times P_{k'} \tag{6}$$

$$COST^{\text{Landfill}} = F^{\text{Landfill}} \times \text{TR} \tag{7}$$

$$EP = REV - COST^{\text{Landfill}} \tag{8}$$

where P_k and $P_{k'}$, TR are the selling price of intermediate k, final product k' and disposure cost, respectively.

Environmental impact is one of the crucial factors when developing a process. In this work, the parameters that affect the total environmental impact IMP^{Total} are the energy required for different technologies and carbon emissions based on one unit of electricity usage. The overall environmental impact in this work is determined via Equation (9):

$$IMP^{\text{Total}} = \sum_i \sum_j T_j \times CE_{ijk} + \sum_k \sum_{j'} T_{j'} \times CE_{kj'k'} \tag{9}$$

where T_j and $T_{j'}$ is the energy required for technologies in the biomass conversion process and upgrading process, while CE_{ijk} and $CE_{kj'k'}$ are the carbon emissions when one unit of electricity is utilised in technologies j and j'.

4. Case Study

A case study from India has been used to illustrate the proposed systematic optimisation approach. As mentioned previously, valorisation of flower waste in an integrated biorefinery involves various types of technologies to transform flower waste into valuable product. Based on a detailed literature review, all available technology pathways such as drying, extraction and production to generate bioenergy are summarised in Table 3. Note that the technologies can be widely divided into three categories, which are drying, extraction and generation of bioenergy.

Table 3. Summary of flower-waste biorefineries.

Technology/Biorefinery		Description
Drying	Press Drying [36]	Apply pressure while drying
	Sun Drying [8]	Natural drying with the aid of sun
	Shade Drying [37]	Natural drying without direct sunlight
	Embedded Drying [7]	Silica gel as desiccants for dying
	Freeze Drying [7]	Sublimation process in low temperature
	Glycerine Drying [7]	Glycerol solution as medium to absorb water
Extraction	Hydrodistillation [38]	Extraction process involving direct heat boiling
	Hydrolysis [38]	Chemical reaction to form acid and alcohol
	Supercritical Carbon Dioxide [38]	Carbon dioxide extraction in supercritical state
	Phytosol Extraction [38]	Cold extraction without heat
	Simultaneous Distillation [38]	Extraction with gas chromatography with two flasks
	Membrane Extraction [38]	Extraction involves partitioning in a membrane
	Fermentation [39]	Chemical breakdown of flower waste
Bioenergy Generation and Transesterification Reaction	Composting [40]	Utilised for soil replenishment after composting
	Biosorbent [41]	Heavy metals from wastewater are accumulated
	Gasification [42]	Convert flower waste into gas in high temperature
	Transesterification [43]	Convert into geranyl acetate and citronellyl acetate

4.1. Superstructure

Based on data collected as shown in Table 3, all possible technology pathways are collected and presented in a superstructure shown in Figure 2. For easier comprehension, the technologies are divided into different categories based on the function of the technologies, namely drying, extraction and bioenergy generation. Note that the superstructure can be divided into several parts: first pretreatment of flower waste, a biomass conversion process that converts the pretreated flower waste into intermediates, and the intermediates are then further converted into final products via upgrading processes.

In this work, the moisture content of flower waste is identified as the major limitation in the main process; therefore, drying processes are taken as pretreatment systems. Based on a literature review, the drying process allows for the reduction in moisture content in flower waste from 40 wt% to 10 wt% [36]. As a result, a two-stage drying process is necessary as one-stage drying is insufficient to meet the requirement. A thermal treatment technique is carried out in the pretreatment section when flower waste is exposed to sunlight before it is sent into the biomass conversion process.

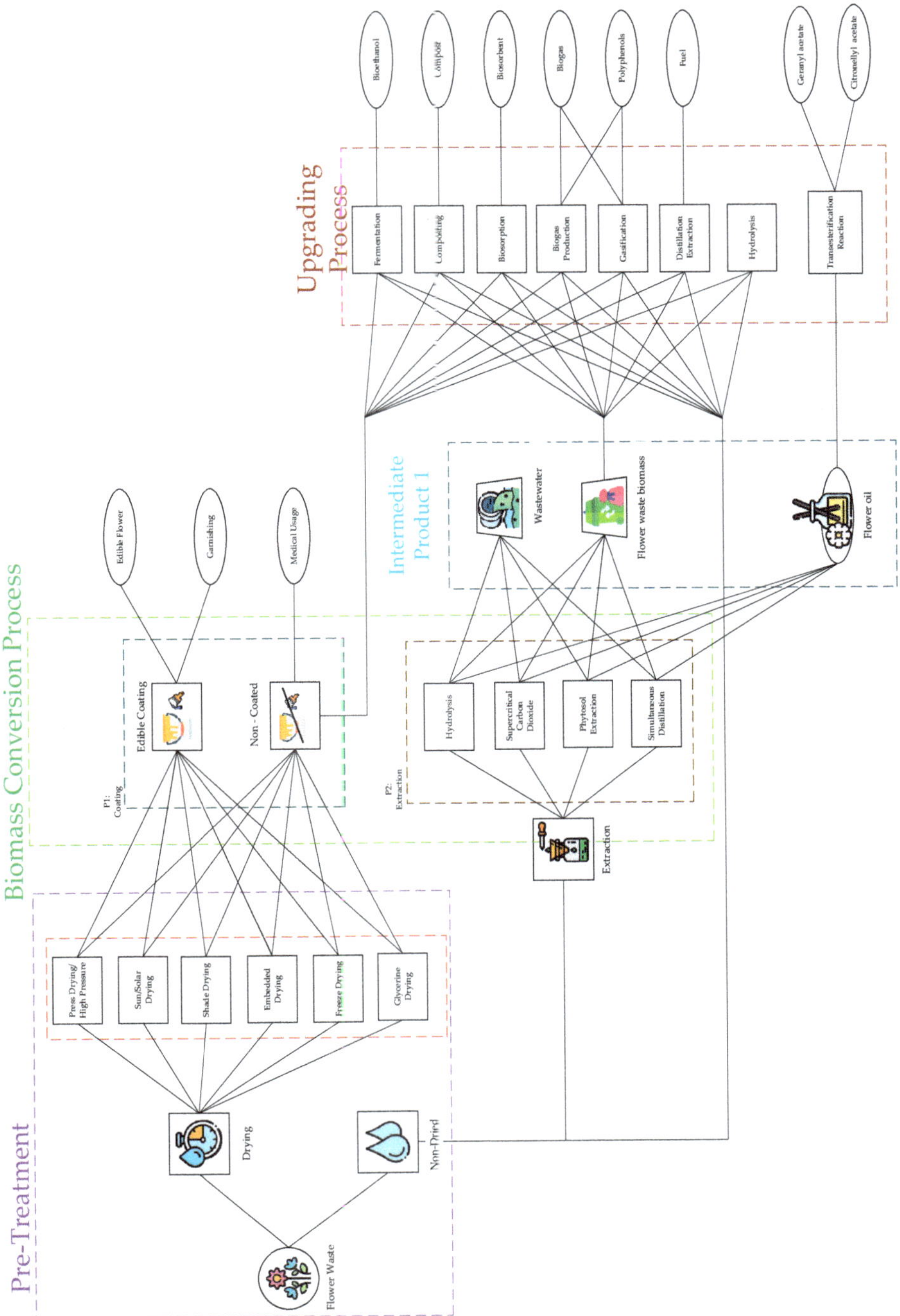

Figure 2. Overall superstructure for synthesis of integrated flower.

As shown in Figure 2, flower waste that undergoes pretreatment can then be sent into coating or extraction processes (biomass conversion process). The coating process is to preserve the edible flower with a longer shelf time and to provide mechanical support, preserving the original form of the flower [42]. In this process, retention of quality parameters (i.e., moisture content, colour, flavour, gaseous exchange and oxidation) are considered.

According to Report Linker [44], the U.S. essential oils market by revenue is predicted to grow at a CAGR of 9% by 2026 due to their value for medicinal and recreational purposes. As a result, this superstructure provides another pathway for extraction of flower essential oil. Intermediates like wastewater, biomass waste and flower essential oil are generated after the extraction process. Wastewater will be sent to waste treatment while biomass waste can go through intermediates process to form bioethanol, biogas, geranyl acetate, etc. Based on research work done by Shivaprasad [43], the flower essential oil produced can undergo a transesterification reaction to form geranyl acetate and citronellyl acetate.

According to Statista [1], the flower production reaches 3 million metric tonnes per year, which amounts to 8220 metric tonnes of flowers produced daily. According to Waghmode [45], 40% of India's entire flower production is sold, amounting to 3300 metric tonnes of flowers sold each day. In this case study, it is assumed that half of the flower waste sold is utilised for decorative purposes and for heritage activities; this will result in a total of 1,640,000 kg of waste produced per day, and roughly 68,500 kg every hour. The properties of flower waste, such as density, composition and weight distribution based on flower parts, are summarised in Table 4. Note that three main parts in flower waste are petals, leaves and stems. However, according to case study in India, flower waste left behind from religious offering is frequently incomplete, only consisting of flower petals and leaves. Thus, the stem parts are neglected in the flower-waste weight distribution.

Table 4. Properties of flower waste.

Flower Composition	Petals	Leaves
Density, kg/m^3 [46]	276	276
Weight distribution, % [47]	38	62

Performance of the drying processes and the conversion factor required for extraction technologies are summarised in Tables 5 and 6, respectively. Note that for the drying process, the final moisture content is used to compute the mass balance of the integrated flower-waste biorefinery.

Table 5. Performance of drying processes.

Technology/Biorefinery		Final Moisture Content, %	Intermediates/Product
Drying	Press Drying [36]	11.55	Edible flower/garnishing
	Sun Drying [36]	11.55	
	Shade Drying [36]	11.55	
	Embedded Drying [36]	11.55	
	Freeze Drying [37]	24.56	
	Glycerine Drying [37]	11.55	

Table 6. Conversion factor of extraction technologies.

Technology/Biorefinery		Conversion, %	Intermediates/Product
Extraction	Hydrolysis [43]	80	Wastewater, Biomass, and Flower Essential Oil
	Supercritical Carbon Dioxid [48]	95	
	Phytosol Extraction [49]	87	
	Simultaneous Distillation [50]	60	

Table 6. *Cont.*

Technology/Biorefinery		Conversion, %	Intermediates/Product
Bioenergy generation	Fermentation [51]	77	Bioethanol
	Composting [40]	77	Compost
	Biosorbent [52]	63	Biosorbent
	Biogas Production and Gasification [41]	65	Biogas/Polyphenols
	Distillation Extraction [53]	60	Fuel
	Hydrolysis [43]	70	Pectin Oligosaccharide
Flower oil further process	Transesterification Reaction [43]	95 and 70	Geranyl Acetate and Citronellyl Acetate

To evaluate the economic performance of the integrated flower-waste biorefinery, the selling price of all final products from different technologies are determined. According to Geitner and Block [54], an average operational hour of a production plant is taken as 8000 h per annum, which is used as the basis of operating hours in this work. To determine the number of batches per year ($n^{dry,annual}$), the total operating hours in a year (t^{annual}) can be divided with the time required for one batch of drying process ($t^{dry,batch}$) shown in Equation (10):

$$n^{dry,annual} = \frac{t^{annual}}{t^{dry,batch}} \tag{10}$$

The product cost can then be determined by the ratio of cost per batch with weight per batch for each technology. Table 7 shows the product cost for different drying processes. For processes that are not time-constrained, product cost can be determined by a literature review as listed in Table 8.

Table 7. Product cost for drying processes.

Drying Process	Time Required, h/Batch	Product Cost, USD/kg
Press drying/High pressure [55]	2	0.50
Sun/Solar drying [36]	48	0.02
Shade drying [36]	72	0.01
Embedded drying [56]	48	0.02
Freeze drying [7]	240	0.004
Glycerine drying [57]	4	0.25

Table 8. Product cost for upgrading processes.

Element	Price, USD/kg	Element	Price, USD/kg
Edible flower [58]	1.45	Polyphenols [59]	9.50
Flower for garnishing [60]	20.00	Fuel [61]	2.60
Medical flower [62]	85.30	Pectin oligosaccharide [63]	5.00
Bioethanol [64]	1.64	Flower oil [65]	35.00
Compost [66]	9.00	Geranyl acetate [67]	372.00
Biosorbent [63]	0.30	Citronellyl acetate [67]	311.00
Biogas [63]	0.33		

To evaluate the environmental impact of the integrated biorefinery, the energy consumption of each technology is required and has been summarised in Table 9.

Table 9. Energy required, and carbon emission for different technologies.

Biorefinery/Technology		Energy Required, kWh/kg	Carbon Emission, kg CO_2/kg Feed
Drying	Press drying [68]	3.70	1.89
	Sun drying	N/A	N/A
	Shade drying	N/A	N/A
	Embedded drying	N/A	N/A
	Freeze drying [69]	1.60	0.82
	Glycerine drying	N/A	N/A
Extraction	Hydrolysis [70]	1.90	0.97
	Supercritical CO_2 [71]	0.40	0.20
	Phytosol extraction [72]	0.60	0.30
	Simultaneous distillation [73]	1.50	0.002
Bioenergy generation	Fermentation [74]	0.43	0.001
	Composting [75]	13.14	6.70
	Biosorption [75]	0.43	0.22
	Biogas production [76]	0.32	0.0006
	Gasification [77]	0.30	0.15
	Distillation extraction [71]	1.50	0.003
	Hydrolysis [78]	44.70	22.8
Flower oil further process	Transesterification [79]	6.77	3.45

Note that most energy required for drying processes are labelled as N/A, as the drying processes do not require electricity nor power consumption. Sun and shade drying processes occur naturally with the aid of heat energy from the sun to reduce the moisture content in flower waste. Meanwhile, for embedded drying and glycerine drying, external material (i.e., silica gel and glycerol solution) is provided to cover the surface of flower waste, and water will be transferred into the silica body and glycerol solution. In this work, it is assumed 0.51 kg CO_2 is released when one unit of electricity is utilised [80].

4.2. Result Analysis

As mentioned previously, in this work, three scenarios are solved to illustrate the proposed model, which are:

(i) Maximum economic performance,
(ii) Minimum environmental impact,
(iii) Multi-objective optimisation with maximum economic performance and minimum environmental impact.

Scenario 1: Optimisation of flower waste biorefinery with maximum economic performance.

Solving Equation (11) subject to Equations (1)–(8) in a commercial optimisation model, LINGO version 18 with global solver.

$$\text{Maximise } EP \quad (11)$$

Figure 3 illustrates the optimised pathway for scenario 1. Based on the optimisation model, 68,493 kg/h of flower waste undergo a drying process in the pretreatment section to reduce the moisture content to 11.55 wt%. As a result, 23,445 kg/h of flower waste is formed and sent into the biomass conversion process.

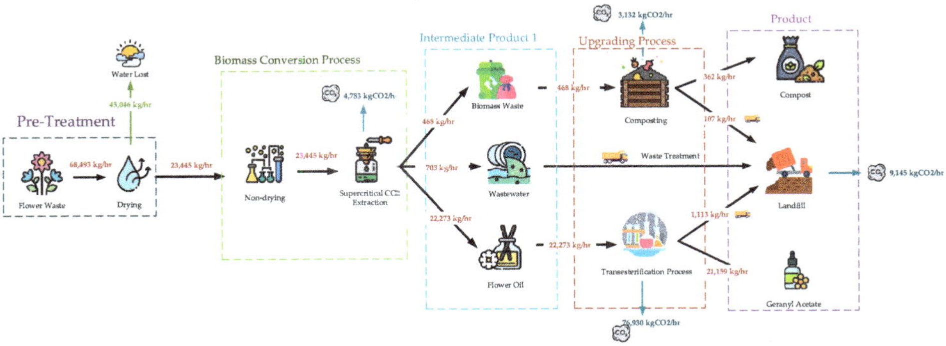

Figure 3. Optimisation based on maximum economic potential—scenario 1.

Next, flower waste is extracted with a supercritical carbon dioxide extraction method to be further converted into different intermediate products such as biomass waste, wastewater and flower essential oil. The amount of intermediate product is 468 kg/h, 703 kg/h and 22,273 kg/h, respectively.

Then, both biomass waste and wastewater are sent to the upgrading process and waste treatment process. The composting method is selected among the upgrading processes (e.g., fermentation, biogas production, biosorbent production, etc.) to form 362 kg/h of compost. Flower essential oil is converted into 21,159 kg/h of geranyl acetate via the transesterification process. The rest of the waste produced is sent to a landfill. The overall economic potential and total carbon emission in scenario 1 are summarised in Table 10.

Table 10. Economic potential and environmental impact in scenario 1.

Total Revenue	Total Carbon Emission
USD 7,874,455	93,999 kgCO_2/h

Scenario 1 shows that the pathways chosen to have significantly high revenue, but the total carbon dioxide released is extremely high as the environmental impact is being neglected. The carbon generated in scenario 1 is mostly from the transesterification process, which is 76,930 kg CO_2/h, as this process requires the use of machinery and electricity.

Table 11 shows the parameters that affect revenue in scenario 1. The waste generated in scenario 1 is relatively less, around 1924 kg/h. This is due to the biomass formed after extraction process being sent to the upgrading process to form other value-added products, which are compost and geranyl acetate, both with extremely high selling costs. Thus, the main revenue in scenario 1 only depends on the product selling price; treatment cost in this scenario does not have a big impact on the economic potential.

Table 11. Parameters that affect revenue.

Product Selling Price	Treatment Cost
USD 7,874,474	1924 $

Scenario 2: Optimisation of flower waste biorefinery with minimum environmental impact.

The synthesis of an integrated flower-waste biorefinery in scenario 2 is solved based on the optimisation objective listed in Equation (12). With this, the pathway of choosing the minimum environmental is determined.

$$\text{Miminise } TCE \tag{12}$$

Figure 4 shows the summarised refinery pathway based on the lowest environmental impact generated. First, 68,493 kg/h of flower waste will undergo a two-stage drying; first, sun drying to produce 23,445 kg/h of flower waste and, second, glycerine drying to form 15,904 kg/h of flower waste. Next, in the biomass conversion process, the flowers are coated to form the exact amount of flowers for decoration purposes. The overall economic potential and total carbon emission in scenario 2 is summarised in Table 12.

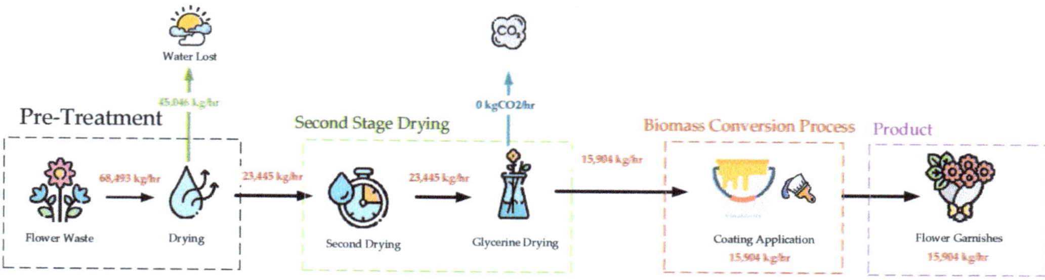

Figure 4. Optimisation based on minimum environmental impact—scenario 2.

Table 12. Economic potential and environmental impact in scenario 2.

Total Revenue	Total Carbon Emission
USD 11,927	0.01 kg CO_2/h

According to optimised results, the flower-waste mass flow is accumulated and sent to the glycerine drying and coating process, which requires no energy consumption nor release of carbon dioxide into the atmosphere. In this case, zero emissions can be demonstrated. Table 12 shows that the economic potential and environmental impact of optimised pathway in scenario 2. Table 13 shows carbon emission from landfill.

Table 13. Carbon emission from landfill.

Carbon Emission, kg CO_2/h
Landfill
0.01

Scenario 3: Optimisation of flower waste biorefinery with maximum economic potential and minimum environmental impact.

Scenario 1 and scenario 2 can only achieve one objective at a time. Thus, fuzzy optimisation is adopted for the synthesis of an integrated flower-waste biorefinery that considers both objectives simultaneously. The value of total revenue and total carbon emission from both scenario 1 and scenario 2 shown in Table 14 are taken as the upper and lower limits for fuzzy optimization equations shown in Equations (13) and (14), where X^U, X^L, Y^U and Y^L are the upper and lower limits of total revenue and total carbon emissions, respectively.

Table 14. Parameters required for fuzzy optimisation.

	Total Revenue	Parameters	Total Carbon Emission	Parameters
Scenario 1	USD 7,874,474	X^U	93,999 kg CO_2/h	Y^U
Scenario 2	USD 11,927	X^L	0.01 kg CO_2/h	Y^L

Equations (13) and (14) describe the basis of fuzzy:

$$\frac{TR - X^L}{X^U - X^L} \geq \lambda \tag{13}$$

$$\frac{Y^U - TCE}{Y^U - Y^L} \geq \lambda \tag{14}$$

with this, the pathway of choosing the minimum environmental is determined as:

$$\text{Minimise } \lambda \tag{15}$$

Figure 5 shows the optimised pathway for scenario 3. The pathways chosen by Lingo are similar to the combination of both scenarios 1 and 2, but with a different mass distribution and upgrading process. The amount of pre-dried flower waste in pretreatment is first divided into half and sent to two pathways. First, 11,521 kg/h of pre-dried flower undergo another stage of drying and coating process to form 7815 kg/h of flowers for decorative purposes. Second, another 11,923 kg/h of pre-dried flower is sent to a supercritical carbon dioxide extraction process to form 238 kg/h of biomass waste, 358 kg/h of wastewater and 11,327 kg/h of flower essential oil as intermediate products. Next, biomass waste undergoes fermentation to form 184 kg/h of bioethanol; flower essential oil goes through a transesterification process and eventually forms 10,761 kg/h of geranyl acetate.

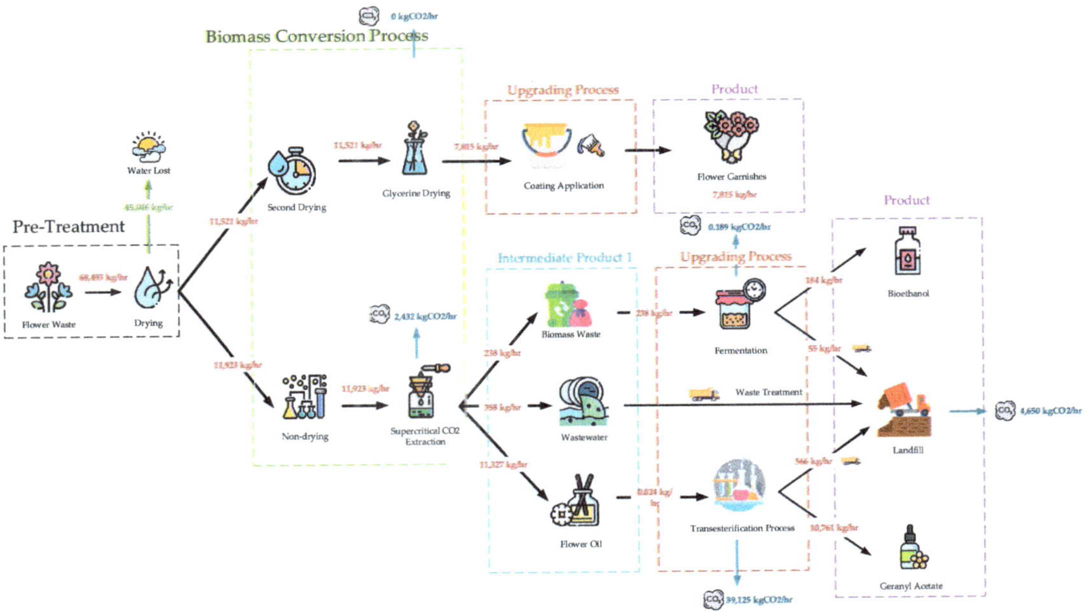

Figure 5. Optimisation based on maximum economic potential and minimum environmental impact—scenario 3.

The overall economic potential and total carbon emission in scenario 3 is summarised in Table 15. This tables shows that both revenue and environmental impact have achieved a balance between scenarios 1 and 2.

Table 15. Economic potential and environment impact in scenario 3.

Total Revenue	Total Environmental Impact
USD 4,009,320	46,209 kg CO_2/h

5. Conclusions

This paper presented a generic mathematical optimisation approach for synthesis of a sustainable integrated flower-waste biorefinery. An Indian case study was solved to illustrate the proposed optimisation model. Multi-objective optimisation (fuzzy optimisation) is adopted as a trade-off between the economic and environmental performances of the integrated biorefinery. Note that the proposed approach can be modified to develop an integrated flower-waste biorefinery based on different geographical constraints and availability of flower waste. This proposed approach demonstrates a preliminary feasibility study of integrated biorefineries based on different parameters, such as conversion ratio, electricity demand product unit cost, etc. Future work involves integrating sensitivity analyses to the above approach on selected parameters to understand their economic impact towards the selection of technologies and processes.

Author Contributions: Conceptualization, E.H.Y.C., V.A., L.Y.N., P.S. and D.K.S.N.; methodology, E.H.Y.C., V.A., L.Y.N. and D.K.S.N.; software, E.H.Y.C. and D.K.S.N.; validation, V.A., L.Y.N., P.S. and D.K.S.N.; formal analysis, E.H.Y.C.; investigation, E.H.Y.C.; resources, D.K.S.N.; data curation, E.H.Y.C.; writing—original draft preparation, E.H.Y.C.; writing—review and editing, E.H.Y.C., V.A., L.Y.N., P.S. and D.K.S.N.; visualization, E.H.Y.C.; supervision, V.A., L.Y.N., P.S. and D.K.S.N.; project administration, V.A., L.Y.N. and D.K.S.N.; funding acquisition, D.K.S.N. All authors have read and agreed to the published version of the manuscript.

Funding: This research received no external funding.

Acknowledgments: This research was funded by Royal Academy of Engineering under the Frontiers of Development scheme.

Conflicts of Interest: The authors declare no conflict of interest.

References

1. Statista Research Department. Production Volume of Flowers in India Between Financial Year 2016 and 2020, with an Estimate for 2021. 2021. Available online: https://www.statista.com/statistics/1036938/india-production-of-flowers/ (accessed on 30 November 2021).
2. Menon, S. Why Do We Offer Flowers to Gods? 2018. Available online: https://www.boldsky.com/yoga-spirituality/faith-mysticism/2018/why-do-we-offer-flowers-to-god-121808.html (accessed on 30 November 2021).
3. Frost, R. Waste from Worship: Solving India's Unique River Pollution Problem. 2020. Available online: https://www.euronews.com/green/2020/02/26/waste-from-worship-solving-india-s-unique-river-pollution-problem (accessed on 30 November 2021).
4. Kabongo, J.D. Waste Valorization. In *Encyclopedia of Corporate Social Responsibility*; Springer: Berlin/Heidelberg, Germany, 2013; pp. 2701–2706.
5. Rivas-Garcia, L.; Navarro-Hortal, M.; Romero- Marquez, J.M.; Forbes-Hernandez, T.Y.; Varela-Lopez, A.; Llopis, J.; Sanchez-Gonzalez, C.; Quiles, J.L. Edible Flowers as a Health Promoter: An Evidence-Based Review. *Trends Food Sci. Technol.* **2021**, *117*, 46–59. [CrossRef]
6. Vidhya, C.; Senthilkumar, S.; Manivannan, S. Recent Trends in Production of Dry Flowers and Foliages. *Pharma Innov. J.* **2021**, *SP-10*, 2135–2139.
7. Prasad, V.J.K.; Pal, R.K.; Voleti, S.R. Drying of Flowers: An Upcoming Industry. *Dry Flower Ind. Floric. Today* **1997**, *4*, 20–23.
8. Purohit, S.R.; Rana, S.S.; Idrishi, R.; Sharma, V.; Ghosh, P. A Review on Nutritional, Bioactive, Toxicological Properties and Preservation of Edible Flowers. *Future Foods* **2021**, *4*, 100078. [CrossRef]
9. Slavov, A.; Vasileva, I.; Stefanov, L.; Stoyanova, A. Valorization of Wastes from the Rose Oil Industry. *Rev. Environ. Sci. Biotechnol.* **2017**, *16*, 309–325. [CrossRef]
10. Dutta, S.; Kumar, M.S. Potential of Value-Added Chemicals Extracted from Floral Waste: A Review. *J. Clean. Prod.* **2021**, *294*, 126280. [CrossRef]
11. Kamm, B.; Kamm, M.; Soyez, K. The Green Biorefinery, Concept of Technology. In Proceedings of the First International Symposium on Green Biorefinery, Society of Ecological Technology and System Analysis, Berlin, Germany, 1997. Available online: https://www.handboekbodemenbemesting.nl/upload_mm/7/d/b/71d063bb-c719-4815-8f94-e67b4a76637c_12.%20Kamm%20Green%20Biorefinery.pdf (accessed on 11 August 2022).

12. Ng, D.K.S. Automated Targeting for Synthesis of An Integrated Biorefinery. *Chem. Eng. J.* **2010**, *162*, 67–74. [CrossRef]
13. Gravitis, J.; Zandersons, J.; Vedernikov, N.; Kurma, L.; Ozols-Kalnins, V. Clustering of Bioproducts Technologies for Zero Emissions and Eco-Efficiency. *Ind. Crops Prod.* **2004**, *20*, 169–180. [CrossRef]
14. Ng, D.; Pham, V.; Jiménez-Gutiérrez, A.; Spriggs, H. A Hierarchical Approach to the Synthesis and Analysis of Integrated Biorefineries. In *Design Energy and the Environment: Proceedings of the Seventh International Conference on the Foundations of Computer-Aided Process Design*; CRC Press: Boca Raton, FL, USA, 2009; pp. 425–432.
15. Tey, S.-Y.; Wong, S.S.; Lam, J.A.; Ong, N.Q.; Foo, D.C.; Ng, D.K. Extended hierarchical decomposition approach for the synthesis of biorefinery processes. *Chem. Eng. Res. Des.* **2020**, *166*, 40–54. [CrossRef]
16. Benjamin, M.F.D.; Cayamanda, C.D.; Tan, R.R.; Razon, L.F. P-graph approach to criticality analysis in integrated bioenergy systems. *Clean Technol. Environ. Policy* **2017**, *19*, 1841–1854. [CrossRef]
17. Yeo, J.Y.J.; How, B.S.; Teng, S.Y.; Leong, W.D.; Ng, W.P.; Lim, C.H.; Ngan, S.L.; Sunarso, J.; Lam, H.L. Synthesis of Sustainable Circular Economy in Palm Oil Industry Using Graph-Theoretic Method. *Sustainability* **2020**, *12*, 8081. [CrossRef]
18. Bao, B.; Ng, D.K.; Tay, D.H.; Jiménez-Gutiérrez, A.; El-Halwagi, M.M. A Shortcut Method for the Preliminary Synthesis of Process-Technology Pathways: An Optimization Approach and Application for The Conceptual Design of Integrated Biorefineries. *Comput. Chem. Eng.* **2011**, *35*, 1374–1383. [CrossRef]
19. Ng, R.T.L.; Ng, D.K. Systematic Approach for Synthesis of Integrated Palm Oil Processing Complex. Part 1: Single Owner. *Ind. Eng. Chem. Res.* **2013**, *52*, 10206–10220. [CrossRef]
20. Rizwan, M.; Almansoori, A.; Elkamel, A. An Overview on Synthesis and Design of Microalgal Biorefinery Configurations by Employing Superstructure-Based Optimization Approach. *Energy Syst.* **2019**, *10*, 941–966. [CrossRef]
21. Dickson, R.; Liu, J.J. Optimization of Seaweed-Based Biorefinery with Zero Carbon Emissions Potential. *Comput.-Aided Chem. Eng.* **2019**, *46*, 247–252.
22. Chemmangattuvalappil, N.G.; Ng, D.K.S.; Ng, L.Y.; Ooi, J.; Chong, J.W.; Eden, M.R. A Review of Process Systems Engineering (PSE) Tools for the Design of Ionic Liquids and Integrated Biorefineries. *Processes* **2020**, *8*, 1678. [CrossRef]
23. Romero-García, J.M.; Gutiérrez, C.D.B.; Toro, J.C.S.; Alzate, C.A.C.; Castro, E. Environmental Assessment of Biorefineries. In *Biosynthetic Technology and Environmental Challenges*; Varjani, S., Parameswaran, B., Kumar, S., Khare, S., Eds.; Energy, Environment, and Sustainability; Springer: Berlin/Heidelberg, Germany, 2017; pp. 377–401.
24. Muralikrishna, I.V.; Manickam, V. Life Cycle Assessment. In *Environmental Management*; Butterworth-Heinemann: Oxford, UK, 2017; pp. 57–75.
25. Russo, G.; Buttol, P.; Tarantini, M. LCA (Life Cycle Assessment) of Roses and Cyclamens in Greehouse Cultivation. *Acta Hortic.* **2008**, *801*, 359–366. [CrossRef]
26. Shale, A.; Potting, J. Environmental Life Cycle Assessment of Ethiopian Rose Cultivation. *Sci. Total Environ.* **2013**, *443*, 163–172. [CrossRef]
27. Saraiva, A.B. System Boundary Setting in Life Cycle Assessment of Biorefineries: A Review. *Int. J. Environ. Sci. Technol.* **2017**, *14*, 435–452. [CrossRef]
28. Alig, M.; Frischknecht, R. *Life Cycle Assessment Cut Roses*; Treeze Ltd.: Uster, Switzerland, 2018.
29. Tay, D.H.S.; Ng, D.K.S. Fuzzy Optimization Approach for the Synthesis of a Sustainable Integrated Biorefinery. *Ind. Eng. Chem. Res.* **2011**, *50*, 1652–1665. [CrossRef]
30. Bressanin, J.M.; Geraldo, V.C.; Assis Magalhães Gomes, F.; Klein, B.C.; Chagas, M.F.; Watanabe, M.D.B.; Bonomi, A.; Morais, E.R.d.; Cavalett, O. Multiobjective Optimization of Economic and Environmental Performance of Fischer-Tropsch Biofuels Production Integrated to Sugarcane Biorefineries. *Ind. Crops Prod.* **2021**, *170*, 113810. [CrossRef]
31. Solis, C.A.; Mayol, A.P.; San Juan, J.G.; Ubando, A.T.; Culaba, A.B. Multi-objective Optimal Synthesis of Algal Biorefineries Toward A Sustainable Circular Bioeconomy. *IOP Conf. Ser. Earth Environ. Sci.* **2020**, *463*, 012051. [CrossRef]
32. Andiappan, V.; Ko, A.S.Y.; Lau, V.W.S.; Ng, L.Y.; Ng, R.T.L.; Chemmangattuvalappil, N.G.; Ng, D.K.S. Synthesis of Sustainable Integrated Biorefinery via Reaction Pathway Synthesis: Economic, Incremental Burden and Energy Asssessments with Multi-Objective Optimisation. *AIChE J.* **2015**, *61*, 132–146. [CrossRef]
33. Geraili, A.; Romagnoli, J.A. A Multiobjective Optimization Framework for Design of Integrated Biorefineries Under Uncertainty. *AIChE J.* **2015**, *61*, 3208–3222. [CrossRef]
34. Ng, L.Y.; Andiappan, V.; Chemmangattuvalappil, N.G.; Ng, D.K.S. Novel Methodology for the Synthesis of Optimal Biochemicals in Integrated Biorefineries via Inverse Design Techniques. *Ind. Eng. Chem. Res.* **2015**, *54*, 5722–5735. [CrossRef]
35. Ng, R.T.L.; Ng, D.K.S.; Hassim, M.H. Process Synthesis and Optimization of a Sustainable Integrated Biorefinery via Fuzzy Optimization. *AIChE J.* **2013**, *59*, 4212–4227. [CrossRef]
36. Dilta, B.S.; Sharma, B.P.; Baweja, H.S.; Kashyap, B. Flower Drying Techniques—A Review. *Int. J. Arm Sci.* **2011**, *1*, 1–16.
37. Chen, C.; Gast, K.L.B.; Smithey, S. The Effect of Different Freeze-Drying Processes on the Moisture Content, Colour, and Physical Strength of Roses and Carnations. *Sci. Hortic.* **1999**, *84*, 321–332. [CrossRef]
38. Handa, S.S.; Khanuja, S.P.S.; Longo, G.; Rakesh, D.D. *Extraction Technologies for Medicinal and Aromatic Plants*; International Centre for Science and High Technology: Trieste, Italy, 2008.
39. Doda, S.; Sahu, O. Production of Bioethanol from Biomass (Marigold flower). *Mater. Today Proc.* **2021**, *48*, 932–937. [CrossRef]
40. Vazquez, M.A.; Soto, M. The Efficiency of Home Composting Programmes and Compost Quality. *Waste Manag.* **2017**, *64*, 39–50. [CrossRef]

41. Ryckebosch, E.; Drouillon, M.; Vervaeren, H. Techniques for Transformation of Biogas to Biomethane. *Biomass Bioenergy* **2011**, *35*, 1633–1645. [CrossRef]
42. Kamal, I. Edible Films and Coatings: Classification, Preparation, Functionality and Applications—A Review. *Arc. Org. Inorg. Chem. Sci.* **2019**, *4*, 501–510.
43. Shivaprasad, P. Methods of Extracting Essential Oils from Discarded Flower Petals in India. 2018. Available online: https://www.bath.ac.uk/announcements/student-start-up-to-tackle-indias-flower-waste/ (accessed on 31 March 2022).
44. ReportLinker. ReportLinker. In-Depth Analysis and Data-Driven Insights on the Impact of COVID-19 Included in This U.S. Essential Oils Market Report. 2020. Available online: https://www.globenewswire.com/news-release/2020/12/17/2146704/0/en/United-States-Essential-Oils-Market-Industry-Outlook-and-Forecast-2021-2026-with-In-depth-Analysis-and-Data-driven-Insights-on-the-Impact-of-COVID-19.html (accessed on 31 March 2022).
45. Waghmode, M.S.; Gunjal, A.B.; Nawani, N.N.; Patil, N.N. Management of Floral Waste by Conversion to Value-Added Products and Their Other Applications. *Waste Biomass Valorization* **2016**, *9*, 33–43. [CrossRef]
46. Reed, S.D.; Armstrong, P.R.; Brusewitz, G.H.; Stone, M.L. Resistance od Marigold Flower to Airflow. *Am. Soc. Agric. Eng.* **2001**, *449*, 639–642.
47. Awan, M.O.U.; Khan, N.; Ali, M.; Junaid, M.; Sohail; Nawaz, F.; Khan, R.; Shah, J.A.; Ali, A.; Mehmood, N.; et al. Response of Cut Rose Cv. Cardinal to Sucrose and NaOCL Concentration. *Pure Appl. Biol.* **2017**, *6*, 171–179. [CrossRef]
48. Kaiser, C.S.; Rompp, H.; Schmidt, P.C. Supercritical Carbon Dioxide Extraction of Chamomile Flowers: Extraction Efficiency, Stability, and In-line Inclusion of Chamomile-Carbon Dioxide Extract in β-cyclodextrin. *Phytochem. Anal.* **2004**, *15*, 249–256. [CrossRef]
49. Chen, Y.; Jiang, X.Y.; Tong, X.; Chen, X.Q. Comparison on Volatile Components from Marchantia Convoluta Obtained by Microwave Extraction and Phytosol Extraction. *J. Chil. Chem. Soc.* **2008**, *53*, 1518–1522.
50. Pollien, P.; Chaintreau, A. Simultaneous Distillation−Extraction: Theoretical Model and Development of a Preparative Unit. *Anal. Chem.* **1997**, *69*, 3285–3292. [CrossRef]
51. Mohanty, S.K.; Behera, S.; Swain, M.R.; Ray, R.C. Bioethanol Production from Mahula (*Madhuca latifolia* L.) Flowers by Solid-State Fermentation. *Appl. Energy* **2009**, *86*, 640–644. [CrossRef]
52. Kanamarlapudi, S.L.R.K.; Chintalpudi, V.K.; Muddada, S. Application of Biosorption for Removal of Heavy Metals from Wastewater. In *Biosorption*; IntechOpen: London, UK, 2018.
53. Ribeiro, B.S.; Ferreira, M.d.F.; Moreira, J.L.; Santos, L. Simultaneous Distillation–Extraction of Essential Oils from *Rosmarinus officinalis* L. *Cosmetics* **2021**, *8*, 117. [CrossRef]
54. Geitner, F.K.; Bloch, H.P. Chapter 12—Life-cycle cost analysis, Practical Machinery Management for Process Plants. *Gulf Prof. Publ.* **2006**, *5*, 201–228.
55. Junckers. What Is "Press Drying", and Why Did Junkers Invent it? Available online: https://www.junckers.com/about/blogposts/technical-bulletins/what-is-press-drying-and-why-did-junckers-invent-it (accessed on 31 March 2022).
56. FirstDayofHome. How to Dry Flowers with Silica Gel: An Amazing Technique. 2022. Available online: https://www.firstdayofhome.com/drying-flowers-with-silica-gel/ (accessed on 31 March 2022).
57. Flower Forever & Bellabeads. Freeze Dried Flowers: Freeze Dried Floral Preservation. Available online: https://myflowersforeverjewelry.com/pages/freeze-dried-flowers#:~{}:text=FREEZE%20DRIED%20FLORAL%20PRESERVATION,of%2010%20to%2015%20days (accessed on 31 March 2022).
58. Deyin. Price for Edible Flowers. 2022. Available online: https://www.alibaba.com/product-detail/Flower-Edible-Bulk-Wholesale-Natural-Quality_1600589019831.html?spm=a2700.galleryofferlist.normal_offer.d_title.22a44a30CHNIfx&s=p (accessed on 28 March 2022).
59. Liaoning. Price for Flower Garnishing. 2022. Available online: https://www.alibaba.com/product-detail/Factory-Supply-Amazon-Custom-Package-Biodegradable_1600466871476.html?spm=a2700.galleryofferlist.normal_offer.d_title.1a795056AUHTif (accessed on 28 March 2022).
60. AMAZY. Price for Medical Flower. 2022. Available online: https://www.alibaba.com/product-detail/Bulk-Wholesale-Bulk-Natural-Tea-Golden_1600428372790.html?spm=a2700.galleryofferlist.topad_classic.d_title.5eb8a6553FhrnD (accessed on 28 March 2022).
61. GlobalPetraolPrices. Ethanol and Fuel Prices. 2022. Available online: https://www.globalpetrolprices.com/ethanol_prices/ (accessed on 28 March 2022).
62. Weifang. Price for Organic Compost. 2022. Available online: https://www.alibaba.com/product-detail/New-High-Quality-Compost-Accelerator-Bacteria_1600479322856.html?spm=a2700.galleryofferlist.normal_offer.d_title.5f717edfQTFqC3&s=p (accessed on 28 March 2022).
63. Model N0.1, Price for Biosorbent and Biogas. Available online: https://web.iitd.ac.in/~{}vkvijay/files/biogas%20entrepreneurship.pdf (accessed on 28 March 2022).
64. Lau, S. Price for Polyphenols. 2022. Available online: https://joysun-group.en.made-in-china.com/product/CdgTwzLEbtpv/China-Best-Price-Natural-Green-Tea-Extract-Green-Tea-Polyphenols-25-98-.html (accessed on 28 March 2022).
65. Cano, M.E.; García-Martin, A.; Comendador Morales, P.; Wojtusik, M.; Santos, V.E.; Kovensky, J.; Ladero, M. Production of Oligosaccharides from Agrofood Wastes. *Fermentation* **2020**, *6*, 31. [CrossRef]
66. VINEVIDA. Price for Essential Oil. 2022. Available online: https://www.vinevida.com/?gclid=CjwKCAjwqJSaBhBUEiwAg5W9p3fo0MLWq59vF6EHTGXniKcoBB4_7SHNVGQpdiV995Y0whg_V_-FohoCO8kQAvD_BwE (accessed on 28 March 2022).

67. VIGON. Price for Geranyl Acetate. 2022. Available online: https://www.vigon.com/product/geranyl-acetate-natural/ (accessed on 28 March 2022).
68. Mech Extrmeme Design with Innovation. Energy Required for Press Drying. 2017. Available online: https://mechextreme.blogspot.com/p/manufacturing.html (accessed on 20 March 2022).
69. Huang, L.; Zhang, M.; Mujumdar, A.S ; Sun, D.; Tan, G.; Tang, S. Studies on Decreasing 549 Energy Consumption for a Freeze-Drying Process of Apple Slices. *Dry. Technol.* **2009**, *27*, 938–946. [CrossRef]
70. Fraunhofer, IFAM. Energy Required for Simultaneous Distillation. 2022. Available online: https://www.ifam.fraunhofer.de/en/Aboutus/Locations/Dresden/HydrogenTechnology/hydrolysis.html (accessed on 28 March 2022).
71. Fiori, L. Supercritical Extraction of Grape Seed Oil at Industrial-scale: Plant and Process Design, Modeling, Economic Feasibility. *Chem. Eng. Process. Process Intensif.* **2010**, *49*, 866–872. [CrossRef]
72. Suarez Garcia, E.; Van Leeuwen, J.; Safi, C.; Sijtsma, L.; Eppink, M.H.M.; Wijffels, R.H.; van den Berg, C. Selective and Energy Efficient Extraction of Functional Proteins from Microalgae for Food Applications. *Bioresour. Technol.* **2018**, *268*, 197–203. [CrossRef]
73. Tzen, E.; Zaragoza, G.; Padilla, E.A. *Comprehensive Renewable Energy (Second Edition)—Solar Desalination*; Elsevier: Amsterdam, The Netherlands, 2022; Volume 3, pp. 590–637.
74. Ranieri, E.; Giuliano, S.; Ranieri, A.C. Energy Consumption in Anaerobic and Aerobic Based Wastewater Treatment in Plants in Itali. *Water Pract. Technol.* **2021**, *16*, 851–853. [CrossRef]
75. Fertilizer Institute. Energy Required for Composting and Biosorbent. Available online: https://www.tfi.org/our-industry/state-of-industry/environment-energy (accessed on 20 March 2022).
76. Ntihuga, J.; Senn, T.; Gschwind, P.; Kohlus, R. Estimating Energy- and Eco-Balances for Continuous Bio-Ethanol Production Using a Blenke Cascade System. *Energies* **2013**, *6*, 2065–2083. [CrossRef]
77. Minowa, T.; Hanaoka, T.; Yokoyama, S. Process Evaluation of Biomass to Liquid Fuel Production System with Gasification and Liquid Fuel Synthesis. *Stud. Surf. Sci. Catal.* **2004**, *153*, 79–84. [CrossRef]
78. Office of Energy Efficiency & Renewable Energy. Energy Required for Hydrolysis Process. Available online: https://www.energy.gov/eere/fuelcells/hydrogen-production-electrolysis (accessed on 20 March 2022).
79. Pradhan, A.; Shrestha, D.S.; Van Gerpen, J.; Duffield, J. The Energy Balance of Soybean Oil 578 Biodiesel Production: A Review of Past Studies. *Trans. ASABE* **2008**, *51*, 185–194. [CrossRef]
80. World Benchmarking Alliance. Greenhouse Gases Emission Based on per Unit of Electricity 581 Used. 2019. Available online: https://www.worldbenchmarkingalliance.org/publication/electric-utilities/companies/tenaga-nasional/ (accessed on 15 March 2022).

processes

Article

A Novel Graphical Targeting Technique for Optimal Allocation of Biomass Resources

Dominic C. Y. Foo

Centre of Excellence for Green Technologies, University of Nottingham Malaysia, Broga Road, Semenyih 43500, Selangor, Malaysia; dominic.foo@nottingham.edu.my

Abstract: Biomass has gained global attention as one of the most important renewable energy resources that reduces greenhouse gas emissions. Various research works have been dedicated to *biomass supply chain* in the past decade as to continuously support the deployment of biomass resources for regional applications. In this work, a novel graphical method based on *process integration* is proposed for targeting the amount of biomass resources needed for a power generation problem. Apart from having a good visualized interface, the graphical method provides good insights to stakeholders on the macro-level planning of biomass allocation. Two examples are solved to demonstrate the newly proposed methods.

Keywords: pinch analysis; process integration; power generation; bioenergy; composite curves

Citation: Dominic C. Y. Foo A Novel Graphical Targeting Technique for Optimal Allocation of Biomass Resources. *Processes* **2022**, *10*, 905. https://doi.org/10.3390/pr10050905

Academic Editor: Alok Kumar Patel

Received: 28 February 2022
Accepted: 27 April 2022
Published: 4 May 2022

1. Introduction

The Global Energy Review 2021 [1] projected that the global energy demand was expected to increase by 4.6% in year 2021. This rise in global energy demand is primarily fuelled by fossil-based sources. Nevertheless, the global awareness of reducing greenhouse gas emissions (particularly CO_2) has encouraged the development of low-CO_2 renewable energy resources. In year 2020, the renewable energy sector reported a contribution of 29% to the global electricity generation, i.e., a growth of 3% despite the global lock down due to the COVID-19 pandemic [1]. Along with solar and hydropower, biomass is among the most important renewable energy sources for sustainable electricity generation.

Biomass supply chain consists of various activities involving the supply of biomass, their transportation, storage, conversion, and delivery of their value-added products [2]. One of the most important value-added biomass products is arguably biofuel/bioenergy. As reported by Lim et al. [3], various challenges are accounted for in the biomass supply chain for biofuel production; these include the variation in biomass availability, distinct characteristics of each biomass species, uncertain technology performance, logistics and transportation issues. Hence, various *process system engineering* tools were developed in the past two decades to address the various challenges encountered in biofuel and biomass supply chain. For instance, some earlier works which are based on mathematical programming models were proposed to synthesise regional bioenergy supply chain [4,5]. In the work by Ling et al. [6], centralized and decentralized technologies were considered for bioelectricity supply chain. In a more recent work, a stochastic model was proposed for co-firing biomass supply chain networks [7]. In some recent works, optimization models were developed with the objective to reduce CO_2 footprint [8,9].

Apart from the above-mentioned techniques, a widely accepted group of systematic tools for optimum planning of resources is arguably *process integration*. The latter consists of some useful graphical techniques that were commonly utilised for the conservation of materials [10,11] and energy resources [12,13] in the chemical processing industries. In recent years, these graphical tools have also been extended for optimal synthesis of biomass supply chain. In the seminal work of Lam et al. [14], the *Regional Energy Surplus–Deficit*

Curves were proposed to synthesise a biomass supply chain with the aim to minimise its carbon footprint. In another later work by Tan et al. [15], a graphical pinch diagram was extended for the optimal planning of a biochar network. Graphical approaches are always regarded as handy tools welcomed by industrial sectors, as they provide good insights to the problem due to its intuitive nature. Furthermore they are often used to facilitate discussion among team members. To date, however, no graphical approach has been reported for the allocation of biomass resources for power generation. Note that the earlier developed graphical techniques (e.g., [14,15]) cannot be used directly for biomass allocation, as they do not consider the unique characteristics of biomass that are important for power generation, e.g., moisture content, calorific values, etc. Hence, a new graphical technique that incorporates biomass characteristics is to be developed. This is the main subject of this work.

In this work, a novel graphical pinch diagram is presented to identify the optimal allocation of biomass resources for power generation. In particular, the novel graphical tool helps to identify the exact amount of biomass resources needed to fulfil the targeted power output of some power plants. The paper is structured as follows. In the next section, a formal problem statement is given. This is then followed by the power generation model, and the procedure for plotting the graphical pinch diagram. Two examples on bioenergy generation are used for demonstrating the novel graphical diagram.

2. Problem Statement

The problem to be addressed is formally stated as follows:

- Given a set of *biomass sources* $i \in I$. Each biomass type has its specific calorific value CV_i, moisture content MC_i and maximum availability S_i.
- The biomass sources are to be allocated to a set of *biomass demands* $j \in J$, which are power plants that require biomass for power generation. Each plant has its power output P_j that has to be fulfilled and can only handle a maximum capacity D_j of biomass.

The biomass allocation problem can be described by a superstructure diagram in Figure 1. The objective of this work is to determine the optimum allocation of biomass source i to power plant j.

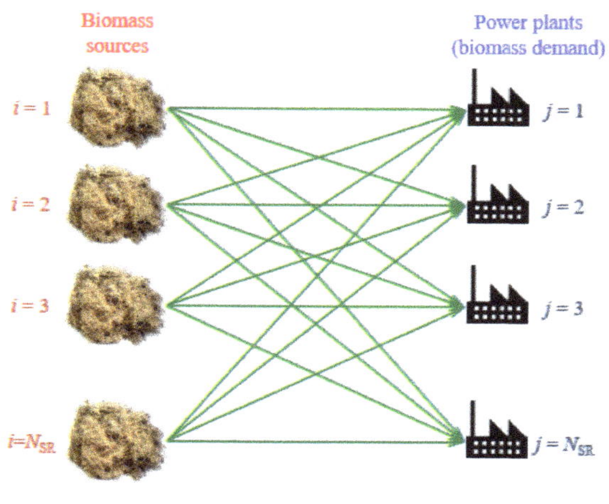

Figure 1. Superstructure representation of biomass source-demand model (adapted with permission from [16]).

3. Power Generation Model

To determine the biomass requirement for power generation, the model in Foo et al. [5] is adopted. For power plant j with output P_j, its steam requirement for the turbine (STM_j) may be calculated using Equation (1).

$$STM_j = \frac{P_j}{\eta_{Turb}\widehat{H}_{Turb}} \quad (1)$$

where η_{Turb} and $\widehat{H}_{Turb}$ are efficiency (%) and enthalpy (kJ/kg) for turbine calculation.

To generate the required amount of steam for the turbine, a boiler is to be used. The biomass requirement for power plant j (D_j) is hence calculated using Equation (2).

$$D_j = \frac{STM_j\widehat{H}_{\mathrm{Boil}}}{CV_{\mathrm{Biom}}SC_{\mathrm{Biom}}\eta_{\mathrm{Boil}}} \quad (2)$$

where η_{Boil} and $\widehat{H}_{\mathrm{Boil}}$ are efficiency (%) and enthalpy (kJ/kg) for boiler calculation, while CV_{Biom} and SC_{Biom} are average calorific value (kJ/kg) and solid content (wt%) of biomass, calculated based on average value of the various biomass types that are fed to the power plant. Note also that solid content can be calculated from moisture content (MC_{Biom}, wt%) that is more commonly used in the biomass industry.

Equations (1) and (2) may be combined and rearranged to the form in Equation (3).

$$P_j = C_jD_j \quad (3)$$

In Equation (3), C_j is characterised as the *power generation factor* for power plant j, given as in Equation (4).

$$C_j = \frac{\eta_{\mathrm{Turb}}\widehat{H}_{\mathrm{Turb}}CV_{\mathrm{Biom}}SC_{\mathrm{Biom}}\eta_{\mathrm{Boil}}}{\widehat{H}_{\mathrm{Boil}}} \quad (4)$$

Similar correlations may be expressed for power output (P_i, Equation (5)) and the generation factor (C_i, Equation (6)) for biomass i:

$$P_i = C_iS_i \quad (5)$$

$$C_i = \frac{\eta_{Turb}\widehat{H}_{Turb}CV_iSC_i\eta_{Boil}}{\widehat{H}_{Boil}} \quad (6)$$

where CV_i and SC_i are calorific value (kJ/kg) and solid content (wt%) of biomass i. Note also that solid content of biomass can be calculated from its moisture content (MC_i, wt%) easily.

Graphical Targeting Method

A novel graphical tool is presented here, known as the *bioenergy pinch diagram* (BEPD). Steps for plotting the BEPD are given as follows.

1. A *demand composite curve* is first plotted on a power versus biomass capacity diagram (Figure 2a). The demand composite curve consists of the individual power plants that require biomass feed. Its horizontal distance represents the maximum total capacity of biomass that can be handled by these plants (Σ_jD_j), while its vertical distance represents their total power output (Σ_jP_j). Note that the individual segments in the demand composite curve correspond to power plant j (PP1 and PP2 in Figure 2a) which have been arranged according to the descending order of their power generation factor C_j (slope of the segment), the latter may be calculated using Equation (4).
2. A *source composite curve* is next plotted on the same diagram as the demand composite curve, but is interpreted as power versus biomass handling capacity. The source composite curve may consist of one or more biomass sources, plotted according to the

descending order of their power generation factor C_i (determined using Equation (6)). The BEPD is considered feasible when the source composite curve is located to the left of the demand composite curve and has, at least, the same vertical distance as the latter, such as that shown in Figure 2a. For this case, the source composite curve will generate a total power of $\Sigma_i P_i$, which matches the total output of the power plants ($\Sigma_j P_j$), and yet is lower than their maximum total handling capacity (i.e., $\Sigma_i S_i \leq \Sigma_j D_j$).

3. In cases where the source composite curve is found on the right and/or below the demand composite curve (such as that in Figure 2b), the BEPD is considered infeasible. Additional biomass with a higher power generation factor is to be supplied in order to restore its feasibility. As shown in Figure 3a, additional biomass with higher power generation factor is added; the latter is characterised by its locus of steeper slope. The source composite curve is then slid along this locus until it stays completely above and to the left of the demand composite curve and touches the former at the *pinch*. The opening on the left of the BEPD represents the minimum amount of additional biomass to be added (F_{BIOM}). Its amount is to be minimised, as it is usually more expensive due to its higher power generation factor. Conversely, the opening on the right of the BEPD represents excess biomass (F_{EXC}) that is beyond the handling capacity of the power plant. This excess biomass can be utilised for other commercial purposes. Note that there are cases where the source composite curve is comprised of several source segments. Besides, there are also cases where the pinch occurs in the middle section of the composite curves. Both of these cases are shown in Figure 3b. The same principles are applied here. The source composite curve is slid along the locus of biomass with higher power generation factor, until it stays completely above and to the left of the demand composite curve. The composite curves touch each other at the pinch. For this case, excess biomass (F_{EXC}) is determined from the horizontal distance of the segment extended beyond the demand composite curve (represented by the rectangular box in Figure 3b).

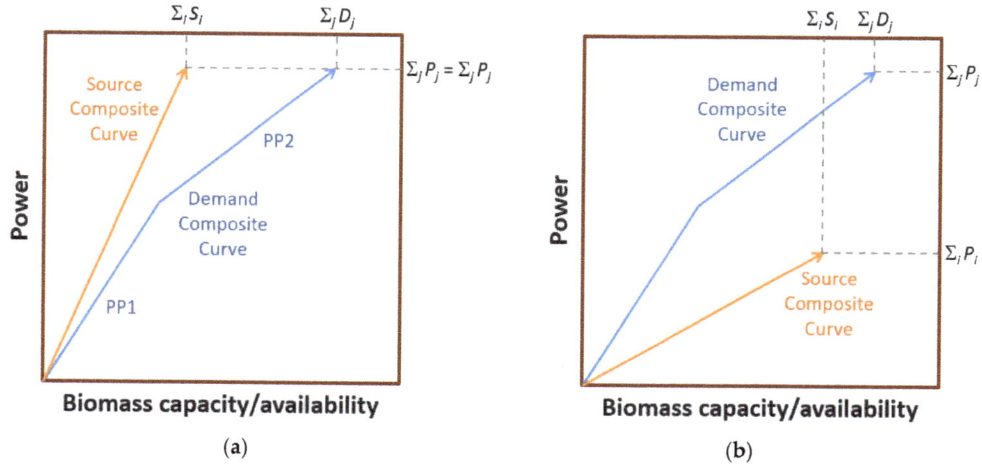

Figure 2. (**a**) Feasible BEPD; (**b**) infeasible BEPD.

The BEPD is next demonstrated with two examples on biomass allocation planning.

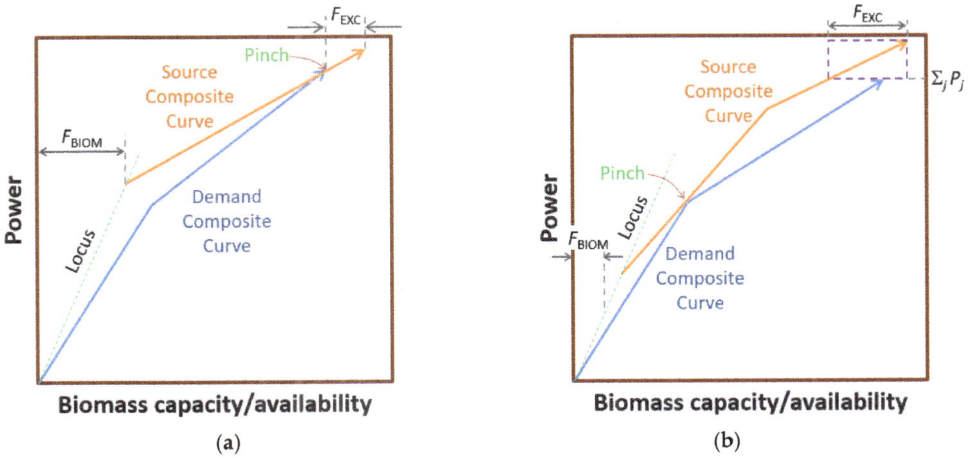

Figure 3. Feasible BEPD with (**a**) pinch occurs at the end of the demand composite curve; (**b**) pinch occurs at the middle section of the demand composite curve.

4. Illustrative Examples

Two examples are used here to elucidate the newly proposed graphical method. For both examples, the important parameters for power generation are given in Table 1, while the moisture content tolerated by the power plants is given in the respective examples.

Table 1. Important parameters for power generation.

Parameters	Values
Turbine	
Turbine efficiency, η_{Turb}	19.8%
Enthalpy of steam, $\hat{H}_{Turb}$	3140 kg/kg steam
Boiler	
Boiler efficiency, η_{Boil}	85%
Enthalpy of steam, $\hat{H}_{Boil}$	2669 kJ/kg steam
Average calorific value of biomass, CV_{Biom}	19,000 kJ/kg biomass

4.1. Example 1—Single Biomass Source

In Example 1, two biomass power plants (PP1 and PP2) are analysed, with their data given in Table 2. As shown, the individual plants were designed based on their respective average moisture content (MC_{Biom}), which may be converted as solid content (SC_{Bio} = 100% − MC_{Bio}) to be used in Equation (2). Their demand of biomass (D_j) can be calculated using Equation (1), while their power generation factors are calculated using Equation (4), listed in the last two columns of Table 2.

Table 2. Data for power plants in Example 1.

Power Plants	P_j (MW)	MC_{Biom} (%)	D_j (t/h)	C_j (MWh/t)
PP1	22	47.4	40	0.55
PP2	8	69.4	25	0.32
Total	30		65	

Three types of biomass are available for use, i.e., palm kernel shell (PKS), empty fruit bunch (EBB) and palm mesocarp fibre (PMF), with their data shown in Table 3. As shown, each biomass type has its CV and MC values. With given flowrate, their power output and generation factors can be determined using Equation (6), as shown in the last two columns of Table 3. Note however for this case, only two biomass types should be considered due to logistic concern. Among them, the EFB with the lowest CV value (lowest cost) is prioritised. The task is to determine the minimum amount of biomass with higher power generation factor, i.e., PKS or PMF.

Table 3. Data for biomass in Example 1.

Biomass Types	S_i (t/h)	CV_i (kJ/kg)	MC_i (%)	P_i (MW)	C_i (MWh/t)
PKS	To be determined	19,700	23	To be determined	1
PMF		19,000	36		0.67
EFB	60	18,700	61	24	0.4

Next, the BEPD was plotted following steps 1 and 2 of the procedure, with only EFB being used to construct the source composite curve. Figure 4 shows that the resulting BEPD is infeasible, as the source composite curve stays at the right of the demand composite curve. As shown, the EFB can only generate a power output of 24 MW (a total of 30 MW is required), but its supply is beyond the capacity limit of the power plant.

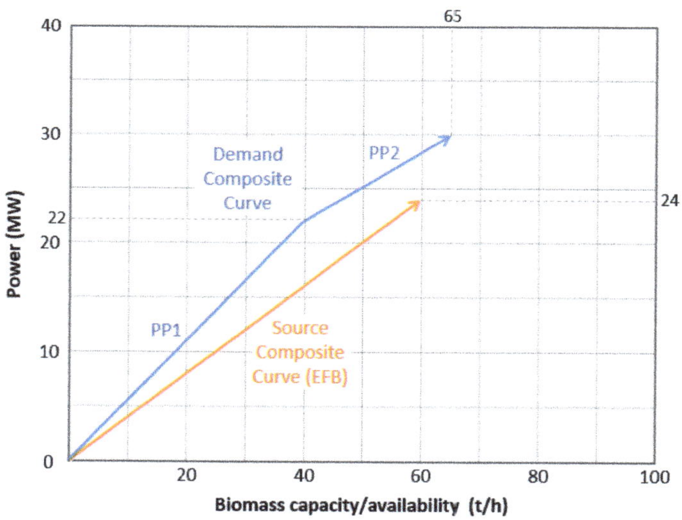

Figure 4. Infeasible BEPD due to insufficient EFB.

To restore the feasibility, the PMF is used. Its locus is first added to the BEPD, with a slope corresponding to its power generation factor (0.67 MW/t/h). Step 3 of the BEPD procedure is then followed. The source composite curve is slid along the locus until it stays entirely above and to the left of the demand composite curve, and touches the latter at the pinch. This results in a feasible BEPD (Figure 5), with minimum use of PMF (F_{PMF}), i.e., 22.5 t/h. The excess biomass (F_{EXC}), i.e., EFB, is determined from the horizontal distance of the rectangular box beyond the demand composite curve, i.e., 22.5 t/h (= 82.5 − 60 t/h); this excess biomass can be used for other commercial purposes. The overlapping region of the composite curves determines the amount of EFB to be utilised for power generation, i.e., 37.5 t/h (= 60 − 22.5 t/h).

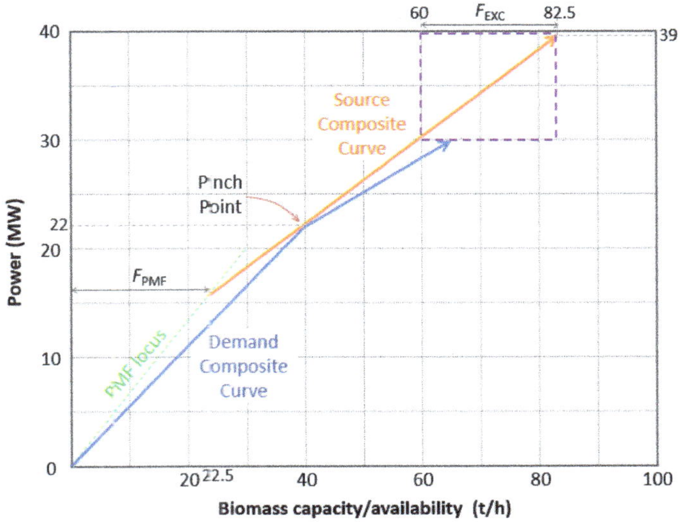

Figure 5. BEPD with minimum PMF.

One may also explore the use of PKS that has a higher CV value (and is more expensive). Replotting the BEPD by sliding the source composite curve on the steeper locus of PKS (due to its higher power generation factor of 1 MW/t/h), results in a feasible BEPD (Figure 6). As shown, both the minimum use of PKS (F_{PKS}) and excess EFB (F_{EXC} = 70 − 60 t/h) are determined as 10 t/h, which are both lower than the case in Figure 5. Furthermore, a higher amount of EFB (60 − 10 = 50 t/h) is utilised for power generation in this case. Detailed evaluation may be carried out to determine which allocation scheme is to be adopted based on their economic performance.

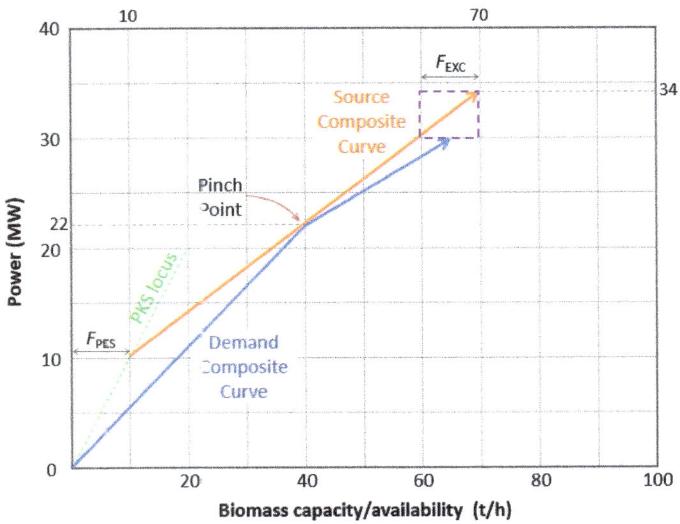

Figure 6. BEPD with minimum PKS.

4.2. Example 2—Multiple Biomass Sources

In this example, four power plants are analysed, with their data shown in Table 4. Three scenarios are analysed here, each with a different amount of palm biomass used, with data shown in Table 5.

Table 4. Data for power plants in Example 2.

Power Plants	P_j (MW)	MC_j (%)	D_j (t/h)	C_j (MWh/t)
PP1	8	49.0	15	0.53
PP2	12	54.0	25	0.48
PP3	10	60.0	24	0.42
PP4	10	63.2	26	0.38
Total	40		90	

Table 5. Data for biomass in Example 2.

Scenario	Biomass Types	S_i (t/h)	CV_i (kJ/kg)	MC_i (%)	P_i (MW)	C_i (MWh/t)
1	PKS	40	19,700	23	40	1
	EFB	100	18,700	61	40	0.4
2	PMF	To be determined	19,000	36	To be determined	0.67
	EFB	72.5	18,700	61	29	0.4
3	PKS	To be determined	19,700	23	To be determined	1
	PMF	18	19,000	36	12	0.67
	EFB	20	18,700	61	8	0.4

In Scenario 1, only single biomass is to be used for power generation. Figure 7 shows that the BEPD is infeasible, if EFB is used. Although sufficient EFB (100 t/h) can be used to generate the required power of 40 MW, the amount of biomass is beyond the handling capacity of the power plants (90 t/h). Conversely, if 40 t/h PKS is used, the BEPD shows that the power plants can fulfil the required power output of 40 MW, while the biomass supply rate (40 t/h) is lower than their handling capacity.

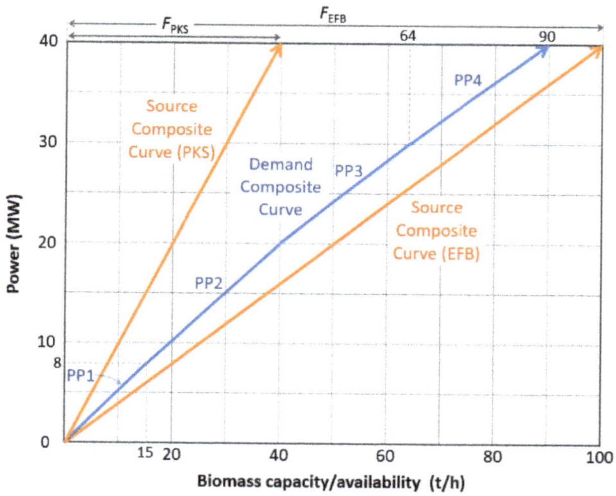

Figure 7. BEPD when single biomass is used (Scenario 1).

In Scenario 2, two types of biomass may be used, with EFB being prioritised. The BEPD in Figure 8 shows that 16.5 t/h of PMF is to be used, as the EFB alone is insufficient to cater the desired power output (40 MW) from the power plants. A similar situation also occurs in Scenario 3 where three biomass types are used. As shown in the BEPD in Figure 9, EFB and PMF are completely consumed, while 20 t/h of PKS is added to produce a total power output of 40 MW. For Scenario 3, no excess biomass sources are reported.

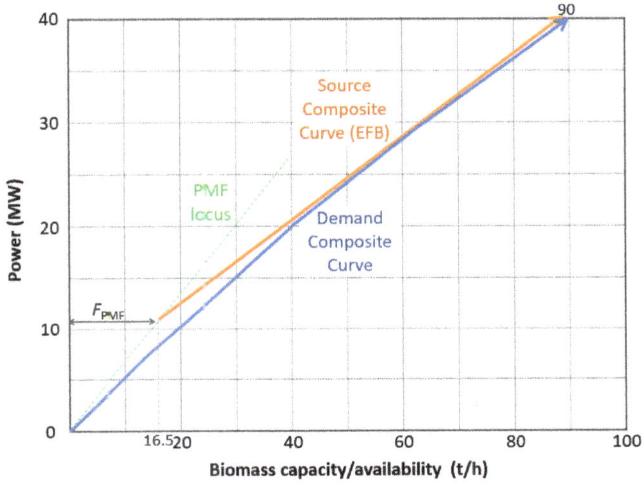

Figure 8. BEPD when two biomass types are used (Scenario 2).

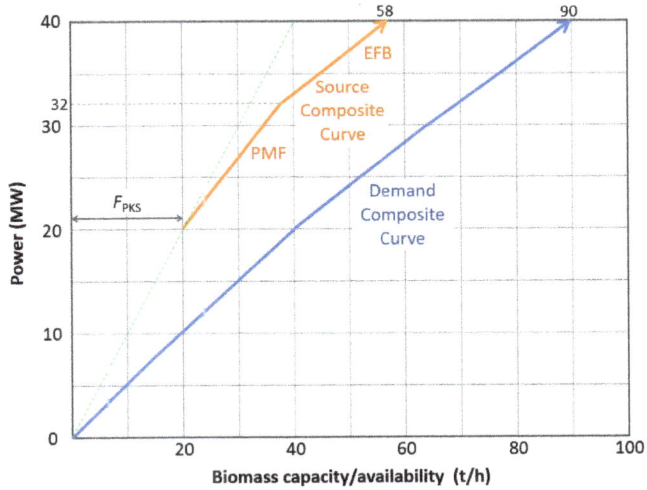

Figure 9. BEPD when three biomass types are used (Scenario 3).

5. Practical Implications

The new graphical targeting technique in this work serves as a handy planning tool for biomass industrial practitioners in their day-to-day operation. Although both examples make use of palm biomass resources, other types of biomass resources (e.g., wood, rice rusk) may also be used, as long as their biomass characteristics (calorific value, moisture content, etc.) are given. Finally, note that the above problems may be solved using the superstructural model as presented in Appendix A.

6. Conclusions

A novel graphical targeting method is proposed in this work for the optimal allocation of biomass resources, based on process integration principles. Two examples based on palm biomass were used to elucidate the newly proposed method. In both examples, biomass with lower power generation factor were prioritised, while those of higher power generation factor were minimised. Although the examples are based on palm biomass, the same principles are applied to other types of biomass resources that may be used for power generation. Future works should look at other environmental aspects of biomass resources, such as water and land footprints.

Funding: This research received no external funding.

Institutional Review Board Statement: Not applicable.

Informed Consent Statement: Not applicable.

Data Availability Statement: Not applicable.

Acknowledgments: Technical advice provided by Albert Y. H. Law, which is gratefully acknowledged.

Conflicts of Interest: The authors declare no conflict of interest.

Appendix A—Superstructural Model

The biomass allocation problem in this work may be solved using the following superstructural model, which was extended from Foo [16]. Equation (A1) described that the net output to be generated by power plant j is to be contributed by total of biomass source i ($f_{i,j}$), with has power generation potential C_i. Each power plant j can only handle a maximum capacity of biomass (D_j), as described by Equation (A2). In Equation (A3), the unutilised biomass i (u_i) is given by the difference between its availability (S_i) and its total allocation to the power plants. All variables in this model must take non-negative values, as indicated by Equation (A4).

$$\sum_i f_{i,j}C_i \geq P_j \;\forall j \tag{A1}$$

$$\sum_i f_{i,j} \leq D_j \;\forall j \tag{A2}$$

$$u_i = S_i - \sum_j f_{i,j} \;\forall i \tag{A3}$$

$$f_{i,j} \geq 0;\; u_i \geq 0 \;\forall i\, \forall j \tag{A4}$$

The objective of the model can be set to minimise a specific type of biomass resource, due to its scarcity; this is given in Equation (A5). Furthermore, one may also make use of the superstructural model to minimise the overall cost of the biomass allocation problem, which is beyond the scope of this work.

$$\min = \sum_i f_{i,j} \tag{A5}$$

References

1. IEA. *Global Energy Review 2021. Assessing the Effects of Economic Recoveries on Global Energy Demand and CO_2 Emissions in 2021*; IEA: Paris, France, 2021.
2. Ng, W.P.Q.; How, B.S.; Lim, C.H.; Ngan, S.L.; Lam, H.L. Biomass supply chain synthesis and optimization. In *Value-Chain of Biofuels*; Yusup, S., Rashidi, N.A., Eds.; Elsevier: Amsterdam, The Netherlands, 2021; pp. 445–479. [CrossRef]
3. Lim, C.H.; Ngan, S.L.; Ng, W.P.Q.; How, B.S.; Lam, H.L. Biomass supply chain management and challenges. In *Value-Chain of Biofuels*; Yusup, S., Rashidi, N.A., Eds.; Elsevier: Amsterdam, The Netherlands, 2021; pp. 429–444. [CrossRef]
4. Čuček, L.; Lam, H.L.; Klemeš, J.J.; Varbanov, P.S.; Kravanja, Z. Synthesis of regional networks for the supply of energy and bioproducts. *Clean Technol Environ. Policy* **2010**, 12, 635–645. [CrossRef]
5. Foo, D.C.Y.; Tan, R.R.; Lam, H.L.; Abdul Aziz, M.K.; Klemeš, J.J. Robust models for the synthesis of flexible palm oil-based regional bioenergy supply chain. *Energy* **2013**, 55, 68–73. [CrossRef]

6. Ling, W.C.; Verasingham, A.B.; Andiappan, V.; Wan, Y.K.; Chew, I.M.L.; Ng, D.K.S. An integrated mathematical optimisation approach to synthesise and analyse a bioelectricity supply chain network. *Energy* **2019**, *178*, 554–571. [CrossRef]
7. Aranguren, M.; Castillo-Villar, K.K.; Aboytes-Ojeda, M. A two-stage stochastic model for co-firing biomass supply chain networks. *J. Clean Prod.* **2021**, *319*, 128582. [CrossRef]
8. Mohd Yahya, N.S.; Ng, L.Y.; Andiappan, V. Optimisation and planning of biomass supply chain for new and existing power plants based on carbon reduction targets. *Energy* **2021**, *237*, 121488. [CrossRef]
9. Duc, D.N.; Meejaroen, P.; Nananukul, N. Multi-objective models for biomass supply chain planning with economic and carbon footprint consideration. *Energy Rep.* **2021**, *7*, 6833–6843. [CrossRef]
10. Foo, D.C.Y. *Process Integration for Resource Conservation*; CRC Press: Boca Raton, FL, USA, 2012. [CrossRef]
11. El-Halwagi, M.M. *Sustainable Design through Process Integration*, 2nd ed.; Elsevier: Waltham, MA, USA, 2017.
12. Linnhoff, B.; Townsend, D.W.; Boland, D.; Hewitt, G.F.; Thomas, B.E.A.; Guy, A.R.; Marshall, R.H. *A User Guide on Process Integration for the Efficient Use of Energy*; Institution of Chemical Engineers: Rugby, UK, 1982.
13. Smith, R. *Chemical Process Design and Integration*, 2nd ed.; John Wiley & Sons, Inc.: Chichester, UK, 2016.
14. Lam, H.L.; Varbanov, P.; Klemeš, J. Minimising carbon footprint of regional biomass supply chains. *Resour. Conserv. Recycl.* **2010**, *54*, 303–309. [CrossRef]
15. Tan, R.R.; Bandyopadhyay, S.; Foo, D.C.Y. Graphical Pinch Analysis for Planning Biochar-Based Carbon Management Networks. *Process Integr. Optim. Sustain.* **2018**, *2*, 159–168. [CrossRef]
16. Foo, D.C.Y. A Simple Mathematical Model for Palm Biomass Supply Chain. In *Green Technologies for the Oil Palm Industry*; Foo, D.C.Y., Aziz, M., Eds.; Springer: Singapore, 2019; pp. 115–130. [CrossRef]

processes

MDPI

Article

Computer-Aided Framework for the Design of Optimal Bio-Oil/Solvent Blend with Economic Considerations

Jia Wen Chong [1], Lik Yin Ng [2], Omar Anas Aboagwa [1], Suchithra Thangalazhy-Gopakumar [1], Kasturi Muthoosamy [3] and Nishanth G. Chemmangattuvalappil [1,*]

[1] Department of Chemical and Environmental Engineering, University of Nottingham Malaysia, Jalan Broga, Semenyih 43500, Selangor, Malaysia; keby5cjw@nottingham.edu.my (J.W.C.); omaranas155@gmail.com (O.A.A.); suchithra.thangalazhy@nottingham.edu.my (S.T.-G.)
[2] School of Engineering and Physical Sciences, Heriot-Watt University Malaysia, No. 1, Jalan Venna P5/2, Precinct 5, Putrajaya 62200, Wilayah Persekutuan Putrajaya, Malaysia; l.ng@hw.ac.uk
[3] Nanotechnology Research Group, Centre of Nanotechnology and Advanced Materials, University of Nottingham Malaysia, Semenyih 43500, Selangor, Malaysia; Kasturi.Muthoosamy@nottingham.edu.my
* Correspondence: Nishanth.C@nottingham.edu.my

Abstract: A major obstacle in utilising pyrolysis bio-oil as biofuel is its relatively low heating value, high viscosity, and non-homogeneity. Solvent addition is a simple yet practical approach in upgrading pyrolysis bio-oil. However, most solvents are often manufactured as specialty chemicals, and thus, this leads to a high production cost of solvents. It is crucial for the designed solvent-oil blend to achieve both fuel functionality and economic targets to be competitive with the conventional diesel fuel. Hence, the objective of this work is to generate feasible solvent candidates by solving this multi-objective optimisation (MOO) problem via a computer-aided molecular design (CAMD) approach. Initially, an optimisation model was developed to identify potential solvents that satisfied the predefined targeted properties. Next, a MOO model was developed via a fuzzy optimisation approach to identify the trade-off between profitability and heating value of the solvent-oil blend. A pricing model was employed to estimate the profitability of the solvent-oil blend. The production of bio-oil in a pyrolysis plant was used to illustrate the applicability of the pricing model. Lastly, phase stability analysis was conducted to ensure the stability and miscibility of the solvent-oil blend. With the developed framework, a promising and cost-effective solvent-oil blend can be generated while displaying optimal biofuel properties.

Keywords: pyrolysis bio-oil; CAMD; solvent design; fuzzy multi-objective optimisation; cost optimisation

Citation: Jia Wen Chong, Lik Yin Ng, Omar Anas Aboagwa, Suchithra Thangalazhy-Gopakumar, Kasturi Muthoosamy and Nishanth G. Chemmangattuvalappil Computer-Aided Framework for the Design of Optimal Bio-Oil/Solvent Blend with Economic Considerations. *Processes* **2021**, *9*, 2159. https://doi.org/10.3390/pr9122159

Academic Editor: Gade Pandu Rangaiah

Received: 29 October 2021
Accepted: 26 November 2021
Published: 29 November 2021

1. Introduction

The conversion of lignocellulosic biomass material, also called second generation biomass, into usable intermediates has attracted the attention as a potential renewable energy source. While the extent of carbon neutrality of biomass is debatable, the need to study alternative fuels produced from biomass conversion is critical in response to fast depleting resources and the rising demand for sustainable development, as well as to counter volatilities in crude petroleum oil prices. Various methods including thermochemical and biochemical techniques can be employed to convert biomass into liquid fuels. Among the thermochemical methods, fast pyrolysis received great attention and has been successful in its commercialisation due to the high yield of the main product pyrolysis bio-oil. High-quality pyrolysis bio-oil can be applied in engines, turbines, boilers, and in some cases, qualified as hydrocarbon and transportation fuels, although further improvement is required of the properties and cost-efficiency of bio-oil [1].

A pyrolysis bio-oil mixture is vaguely defined as one with water and oxygenated organic compounds, including carboxylic acids, aromatics, acetals, alcohols, aldehydes, esters, hemiacetals, olefins, phenolics, proteins, and sulphur compounds. The actual composition of a bio-oil is an intricate function of feedstock, oil production and purification techniques, a condensation system, and storage conditions [2]. The direct use of pyrolysis bio-oil as a fossil fuel substitute suffers from several drawbacks such as low heating value, high water content, high viscosity, low miscibility, high acidity, as well as storage instability due to product aging. The properties of pyrolysis bio-oils differ for different feedstock sources and production parameters, but the products are generally of poor quality compared to conventional fuels. Upgrading technologies have been actively studied and implemented in the effort to upgrade bio-oil properties, such as catalytic cracking, emulsification, esterification, hydrodeoxygenation, hydrogenation, steam reforming, and the manipulation of supercritical fluids [3]. Pidtasang et al. [4] found that adding a small proportion of alcohol solvents to bio-oil could improve its heating value, stability, and viscosity, although the values still could not meet the quality of conventional gasoline and diesel. Organic solvents such as acetone, acetone-methanol, ethanol, isopropanol, and methanol were also used for bio-oil upgrading [5].

The identification and selection of solvents often involve tedious experimentation through trial-and-error, and it could be time consuming to test all potential solvents and ineffective in optimising blend performance to meet the property targets. In response to these challenges, computer-aided molecular design (CAMD) techniques were commonly employed to identify potential candidates that satisfy a set of property targets and constraints in the design and optimisation of solvents [6]. CAMD is a reverse engineering approach which predicts, estimates, and constructs molecules from a given set of molecular building blocks based on predefined target properties [7]. To develop a CAMD framework, it is essential to have reliable property prediction models. One of the most widely used methods for the molecular property predictions is the Group Contribution (GC) methods [8]. GC methods predict the property of a molecule by summing up the contributions from the molecular groups in the compound according to the frequency of their appearance in the structure [8]. The ability of CAMD techniques to generate all possible non-intuitive solutions can be exploited to identify promising molecular structures for various applications. With these computer-aided methods, the search was focused on analysing the alternatives and rejecting infeasible solutions at the initial stages of the design process. The experiments were only performed during the final selection stage for verification.

In the past, the CAMD approach has been widely employed in solvent design problems for different applications [9]. Detailed review articles from Austin et al. [10] and Ng et al. [11] discussed the solution techniques, applications, and future opportunities of the CAMD tools. In addition, Chemmangattuvalappil [12] reviewed the recent development of CAMD applications in the design of solvents. Conventionally, the selection of solvents had formerly been tackled using a combination of fundamental physio-chemical knowledge, as well as heuristic and empirical approaches based on expert judgements. Using CAMD techniques, studies were able to identify solvents ranging from simple hydrocarbons such as methanol and ethanol [13] to more complex mixtures such as the methanol, *N,N*-dimethylformamide (DMF), and acetone blend [14]. A comprehensive review on the process system engineering tools, such as CAMD and group contribution (GC) models, for the design of ionic liquids and integrated biorefineries was presented by Chemmangattuvalappil et al. [15]. In one of the recent developments, CAMD tools were employed in a multi-objective optimisation (MOO) problem targeted at designing a green chemical product and integrated biorefinery process that incorporated green manufacturing while fulfilling the customer requirements [16]. In addition, the CAMD approach was also integrated with the miscibility characteristics to predict possible structures that would form stable blends with bio-oil [17,18].

A major obstacle in the commercialisation of biofuels is their high production cost as compared to that of the conventional diesel fuel. Their low heating value and the high cost of raw materials make biofuels more costly for heat generation [19]. As the biofuel ratio increases in the fuel, the fuel's energy density decreases. Furthermore, the addition of solvents is often required to improve the biofuel's properties. However, the solvents are generally associated with high cost as most of them are manufactured as specialty chemicals. This further increases the production cost of biofuels. Further, the existing and progressing biofuel legislative framework sparks the urge to assess the cost associated with upgrading bio-oil.

Various recent contributions have included the economic aspects such as product pricing, profitability, market share, and operating cost in the product design. In the design of a traditional Chinese medicinal supplement, customers' preference on product quality and economic considerations was taken into account while developing the chemical product design framework [20]. Zhang et al. [21] provided an overview of chemical product design in the context of a multidisciplinary hierarchical framework including design issues such as project management, market study, economic analysis, product design, and process design. Moreover, an activated carbon production plant from industrial waste nutshells was proposed [22]. In the plant design, an economic analysis was developed by considering the cost of the main equipment, the price of the raw materials, basic services, and operations. On the other hand, application of neural network approach is often reported in cost estimation and investment evaluation. A neural network growth model was proposed to estimate potential of investment in renewables in Japan by Jehan, and Sultonov [23]. In addition to the existing growth model, the extended growth model considered both environmental risk diversion and risk mitigation. With the developed neural network-based approach, the establishment and operation of a renewable investment opportunity is possible. Further, an artificial neural network method was used to develop a cost estimation model for the tendering of engineering services [24]. However, previous research on the design of bio-oil solvent mainly focused on the functionality of the solvent itself. It is also important to incorporate the economic aspects into the development of bio-oil solvent for the designed solvent-oil blend to be competitive with the conventional diesel fuel. Generally, the heating value of pyrolysis bio-oil increases with the addition of solvent. However, as the amount of solvent increases, the cost of solvent-oil blend increases as well, resulting in lower profitability. Thus, MOO approach was adapted in this work to investigate the trade-off between high heating value and high profitability of solvent-oil blend.

The weighted sum method is a more common approach for handling MOO problem. In weighted sum method, each objective function was allocated with a weighting factor to convert different objectives into an aggregated scalar objective function [25]. However, in CAMD problem, the weighting factor of each objective is not always definable. In most cases, the relative importance of each objective is fuzzy or uncertain. In addition, these objectives might be contradictory to each other in nature [26]. Thus, a fuzzy optimisation approach was employed in this study to solve the MOO problem under the fuzzy environment. Zadeh [27] first established fuzzy set theory for decision making problems. Later, the fuzzy set theory was extended to address linear mathematical programming problems involving multiple objectives [28]. Within the fuzzy optimisation algorithms, the trade-off between the objective functions to be optimised can be identified by introducing the fuzzy membership function. As a result, an optimal compromised solution can be identified by achieving near optimality for all the objectives.

In the past, fuzzy optimisation approach has been applied in various CAMD applications. A systematic fuzzy optimisation-based method was developed to design molecules for chemical processes with both property superiority and robustness optimised [26]. Khor et al. [29] adapted the fuzzy optimisation via max-min aggregation in the CAMD of alternative solvents for oil extraction from palm pressed fibre. The developed approach was able to optimise the physical properties of the solvent simultaneously with the safety and health aspects. A similar approach was employed and extended with the introduction of disjunctive programming in the work by Ten et al. [30] to design solvent for gas sweetening process. Recently, a max-min aggregation fuzzy optimisation approach was employed along with CAMD method in the design of green solvents for pyrolysis bio-oil upgrading with consideration of environmental, health and safety aspect while ensuring minimal compromise on the fuel functionality [31]. The goal of this study is to design a solvent which is capable to enhance the properties of bio-oil with minimal addition of solvent via CAMD approach. Fuzzy optimisation approach was employed to investigate the trade-off between the functionality and profitability of the solvent-oil blend. With the developed methodology, the final solvent-oil blend should demonstrate promising property targets that fulfils the standards of biofuel while displaying desirable profit margin.

2. Methodology

In this work, a MOO framework was established to solve the CAMD problem using fuzzy optimisation approach. The developed framework can be divided into four main stages (nine steps): Problem definition, CAMD formulation, Multi-objective optimisation problem formulation and Phase stability analysis. The detailed methodology for each stage will be further discussed in this section. Figure 1 demonstrates the overview of methodology for this CAMD problem.

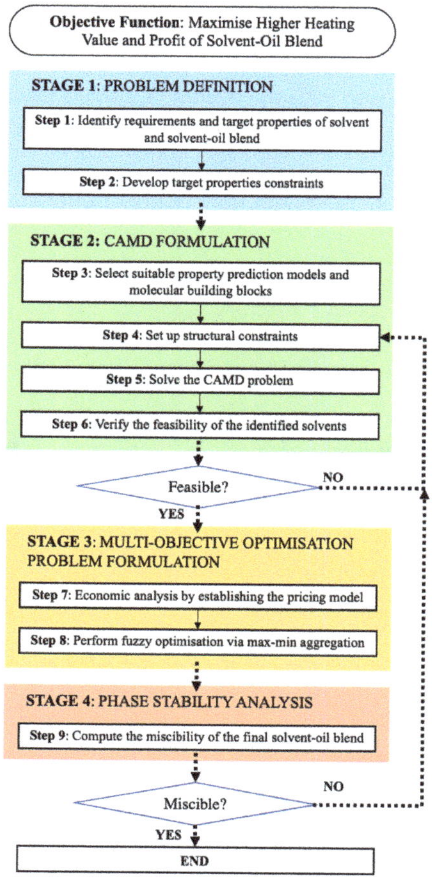

Figure 1. Graphical summary of the work carried out in this research.

2.1. *Stage 1: Problem Definition*

Identification of Product Requirements and Property Constraints

In the first stage, the product needs of solvent and the final solvent-oil blend were identified to infer the target properties and set their constraints based on the requirements from various regulations and specifications. The final requirement of the design problem is to have an alternative solvent-oil blend that can blend well with diesel, or to be applied directly to standard diesel engines. This is the principal product functionality and therefore, from these, the property targets can be developed based on commercial standards of biodiesel. In addition, the general safety and environmental regulatory requirements were also embedded in the target properties at this stage. Generally, the constraints in a product design problem took form as shown in Equation (1), bounded by upper and lower bounds [32].

$$v_p^L \leq V_p \leq v_p^U;\ p = 1,\ 2,\ \ldots P \tag{1}$$

Based on Equation (1), V_p is the value of the target property and p is the property index. The lower and upper bound value is represented by v_p^L and v_p^U, respectively. The ASTM D7544 biofuel standard was used to develop the constraints on the target properties [33]. The standard governed the pyrolysis liquid biofuel, which served as a guideline for the final solvent-oil blend's targeted properties. To maintain the solvent as liquid at room temperature, the boiling point was set to be above 60 °C and the melting point was restricted at an upper bound of 10 °C, while the flash point was set to be greater than 45 °C. On the other hand, the lower bound for the density of mixture was set below that specified by the standard. That is to give more solutions, since eventually the solvent-oil blend is to be blended with other fuels such as diesel, which can have lower densities than those in the ASTM standard. The critical high heating value (HHV) property was restricted at a lower bound of 30 MJ/kg. Table 1 summarises the requirements of the solvent and the solvent-oil blend, which are then translated into the target properties along with its respective property constraint.

Table 1. Target properties identified and their respective constraints.

Requirement	Target Property	Lower Bound	Upper Bound	ASTM D7544 Standard
Solvent				
Liquid at room conditions	Normal boiling point	60 °C	–	–
	Normal melting point	–	10 °C	–
Safety	Flash point	45 °C	–	–
Bio-oil blend				
Consistency of fuel flow	Density	800 kg/m^3	1200 kg/m^3	1100–1300 kg/m^3
	Kinematic viscosity	1.0 mm^2/s	5.0 mm^2/s	Maximum 5.0 mm^2/s
Fuel combustion quality	Higher heating value	30 MJ/kg	–	Minimum 15 MJ/kg
Homogeneous form	Gibbs energy of mixing		To be determined	

2.2. Stage 2: CAMD Formulation

2.2.1. Selection of Property Prediction Models

In the next stage, suitable property prediction models were selected to estimate the solvent's target properties via a GC approach. In the GC approach, the property of a compound was defined as a function of structurally dependent parameters, which can be estimated by summing contributions per structural group according to their recurrence in the solvent molecule. The method was mainly attributed to the research done by Joback and Reid [8]. The general property estimation model via GC methods can be seen in Equation (2).

$$f(X) = \sum_i N_i C_i + w \sum_j M_j D_j + z \sum_k O_k E_k \tag{2}$$

here, C_i is the contribution of the first order group of type i that occurs N_i times, while D_j is the contribution of the second order group of type j that occurs M_j times. E_k is the contribution of the third order group of type k that occurs O_k times. In addition, mixing rules were applied to determine the final values of the targeted solvent-oil blend. The selected property prediction models and mixing rules for targeted property estimation can be found in Appendix A: Table A1. In this study, crude bio-oil derived via the fast pyrolysis of palm kernel shells (PKS) was used as the basis [34]. However, only the organic phase of pyrolysis bio-oil was considered. The properties and components of the pyrolysis bio-oil applied in this study are summarised and listed in Appendix A: Table A2.

2.2.2. Structural Constraints

Other than the abovementioned property constraints, structural constraints were also included in the CAMD model to ensure the formation of a feasible molecule by limiting the molecule to that which favours the targeted properties. The molecular building blocks forming the solvent's structure also need to be carefully tailored. Molecular design can utilise many functional group families and, in this work, several groups were selected, and are shown in Appendix A: Table A3, whereby their group-contribution variables for property estimation were available from the work of Marrero and Gani [34].

In addition, for completeness of the designed molecule, the final molecular structure must not have any free bonds. In other word, the free bond number (FBN) of the final solvent molecule must equal to zero, to be identified as a feasible molecule [35]. FBN can be expressed mathematically as Equation (3), where N_r is the number of rings in the structure and N_i is the number of acyclic groups in the molecule. Constraints were set for the parameter N_r and N_i to ensure there is a feasible length of groups included to form an aliphatic chain or cyclic ring.

$$FBN = \Sigma_i N_i FBN_i - 2(\Sigma_i N_i - 1) - 2N_r = 0 \tag{3}$$

For acyclic structures: $N_i \geq 0$. For cyclic structures: $N_i \geq 3$. For ring molecules: $N_r \geq 1$.

2.2.3. Formulation of CAMD Model

The CAMD model was formulated with generalised mathematical expressions as shown from Equations (4)–(8) [10]:

$$F_{obj} = \max/\min F(x,p) \tag{4}$$

$$h_1(p,x) \leq 0 \tag{5}$$

$$s_1(x) \leq 0 \tag{6}$$

$$p_k^L \leq p_k \leq p_k^U \ \forall k \tag{7}$$

$$x_g^L \leq x_g \leq x_g^U \ \forall g \tag{8}$$

here, F_{obj} (Equation (4)) is the objective function, which is to minimise or maximise one or more parameters. Meanwhile, the $F(x,p)$ is the vector to the objective function, which evaluate the performance of the designed solvent based on its property p. The target properties constraints can be formulated as Equation (5), which is the general function that correspond to the solvent design specification. As the properties of each solvent molecule are highly dependent on the presence of GC building blocks, this constraint can limit the number of appearances of specific GC groups in the designed solvent molecule. On the other hand, Equation (6) can be referred to as the general function that relates the molecular structure generation, to ensure structure feasibility of the generated solvent molecule. In Equations (7) and (8), p_k indicates the property values for each property k and x_g indicates the number of occurrences of each GC group g. Both equations represented the boundaries set on p_k and x_d. Here, p_k^L and x_g^L are the lower bounds for parameter p_k and x_g, respectively. On the other hand, p_k^U and x_g^U are the upper bounds for parameter p_k and x_g, respectively. With the developed CAMD framework, the list of solvent candidates that were feasible, available, and had a relatively established commercial or industrial scale presence were identified and applied in the next stage.

2.2.4. Database Verification

All solvent candidates generated from the CAMD model were verified by conducting a database search on online platforms such as PubChem, ChemSpider, etc. The main purpose of this step was to ensure that all the generated solvent candidates were feasible and practical enough to be applied in real life applications. For solvent candidates that could be found in the databases, property values estimated from the design problem were compared to validate the CAMD results. However, for solvent candidates that could be found in the database or proved to be infeasible, the previous step was revisited by revising the property attributes and constraints.

2.3. Stage 3: Multi-Objective Optimisation Problem Formulation

2.3.1. Formulation of Pricing Model

After the potential solvent candidates were identified in the previous stage, a detailed economic analysis was conducted to determine the selling price of the new solvent-oil blend based on the current market demand, availability and existing competitors identified via a comprehensive background study. The pricing model proposed by Bagajewicz [36] was employed to relate the product quality to demand and price of the product [37]. In the past, the pricing model has been incorporated in various product design such as wine [38], carpet deodorizers/disinfectants [39], skin moisturising lotion [40], die-attach adhesive [37] and dry-cleaning solvent [41]. The mathematical expression for the pricing has been shown in Equations (9) and (10).

$$A^P T^P = A^C \left(T^P\right)^{\delta} \left(\frac{\alpha}{\beta}\right)^{\delta} \left(\frac{Y - A^P T^P}{A^C}\right)^{1-\delta} \quad (9)$$

$$Y \geq A^P T^P + A^c T^c \quad (10)$$

here, A^P and T^P refer to the price and demand of the new solvent-oil blend while A^C and T^C refer to the price and demand of the competitor's product. In this study, A^P can be obtained by summing up the cost of bio-oil production, cost of solvent and the profit obtained by selling the solvent-oil blend (Equation (11)).

$$A^p = Cost_{Bio-oil} + Cost_{Solvent} + Profit \quad (11)$$

Y is the total market size for the solvent-oil blend and δ is the elasticity of substitution, which is an adjustable parameter that measures the change in the ratio of products demand in response to a change in the ratio of their prices. On the other hand, α is expressed as the consumer's awareness on the new product, which can be raised by allocating higher budget in the marketing of new product. The value of parameter α ranges between 0 and 1, where α with value 0 indicates that the consumers have no knowledge about the new product, and vice versa. Lastly, β is the consumer preference coefficient that relates the consumer's interest in the new product over the competing product, which can be determined using Equation (12), where the λ^C and λ^P are the consumer's preference function of competitor's product and new product, respectively. In this study, the consumer's preference was related to the HHV of the solvent-oil blend, which possess a significant influence on the functionality of the solvent-oil blend.

$$\beta = \frac{\lambda^C}{\lambda^P} \quad (12)$$

From Equation (12), the λ^C and λ^P refer to the consumer's preference function of competitor's product and new product, respectively. The new solvent-oil blend is said to be preferred by consumers if β is smaller than 1. However, the competitor's product is preferred by consumers when the value of β greater than 1.

Based on the market analysis conducted, the total market size, Y of solvent-oil blend was identified to be USD 500 million annually [42]. The price elasticity was defined between the range of 0.11 to 0.33, based on the previous studies for diesel fuel. The demand of bio-oil blend was said to be price inelastic when the parameter δ lies between 0.1 to 1. Thus, in this work, the parameter δ was assumed to be 0.1 [43]. On the other hand, the parameter α was estimated to be at the value 0.85. These values should be revised and updated based on the response received once the solvent-oil blend was introduced into the market. The benchmark for this work is the reported solvent-oil blend consisting of 50 wt.% pyrolysis bio-oil and 50 wt.% of iso-propanol, with a HHV of 27.55 MJ/kg [44]. The cost of iso-propanol and bio-oil was assumed to be USD 1336.57 per tonne of iso-propanol and USD 354 per tonne of pyrolysis bio-oil [45,46]. The cost of the competitor's solvent-oil blend can be calculated using Equation (13), where x_i and C_i are the ratio and costs of the solvent and bio-oil, respectively.

$$C_{blend} = \Sigma_i x_i C_i \tag{13}$$

Next, the selling price of competitor's blend A^C can be calculated by summing up the cost of solvent-oil blend and the profit obtained, which was assumed to be USD 50 per tonne of solvent-oil blend, in this case. Table 2 summarises the parameters and its respective values obtained from this market analysis.

Table 2. Parameters and values from market analysis.

Parameters	Values
Total market size, Y	USD 500,000,000 per year
Consumer's awareness coefficient, α	0.85
Elasticity of substitution, δ	0.10
Price of competitor's product, A^C	USD 895.29 per tonne solvent-oil blend
HHV of competitor's product, λ^C	27.55 MJ/kg

2.3.2. Formulation of Fuzzy Multi-Objective Optimisation via Max-Min Aggregation

The HHV of pyrolysis bio-oil can be increased with the addition of solvent. The higher the mass fraction of solvent in the solvent-oil blend, the higher is the energy content. However, a higher amount of solvent was often associated with higher cost, and thus, lower profitability obtained from the solvent-oil blend. In this study, a MOO problem was developed to investigate the trade-off between high HHV and high profitability. Most of the current CAMD techniques focus on optimising a single objective or property of the chemical product [29], but having a multi-objective problem necessitates the use of more complex optimisation methods.

Thus, fuzzy mathematical programming was applied to solve the MOO design problem. The satisfaction degree of fuzzy, λ is introduced to both property functions that were to be optimised. The degree λ is a continues variable that lies between the value 0 and 1, where 0 indicates unsatisfactory and 1 is completely satisfactory (Equation (14)).

$$0 \leq \lambda \leq 1 \tag{14}$$

The objective function of the fuzzy optimisation model was to maximise the overall satisfaction degree of fuzzy constraint λ as shown in Equation (15). The max-min aggregation was applied to the fuzzy programming, where every fuzzy constraint should be satisfied partially at least to the degree λ.

$$f_{obj} = \max\lambda \tag{15}$$

Fuzzy goals for the HHV of solvent-oil blend and the profitability were expressed using a linear membership function, as shown in Equations (16) and (17).

$$\lambda_{p(\max)} = \begin{cases} 0, & \left| V_p \leq v_p^L \right. \\ \frac{V_p - v_p^L}{v_p^L - v_p^L}, & \left| v_p^L \leq V_p \leq v_p^U \right. \\ 1, & \left| V_p \geq v_p^U \right. \end{cases} \quad \forall p \in P \tag{16}$$

$$\lambda_p \geq \lambda \tag{17}$$

where V_p is the target property values bounded by the lower and upper bounds, v_p^L and v_p^U, respectively. The values for the lower and upper bound can be obtained by performing single objective optimisation for both objective functions.

2.4. Stage 4: Phase Stability Analysis

With the identification of optimal solvent-oil blend ratio from the previous step, phase stability analysis was carried out to ensure the miscibility of the designed solvent-oil blend at the targeted mixing ratio. In this work, the phase stability analysis was conducted via computation of the tangent plane distance. With fixed temperature and pressure, Gibbs tangent plane distance function was employed for the phase stability analysis of N-component mixture (Equation (18)) [47]:

$$d(x) = \sum_{i=1}^{n} x_i[\ln x_i\gamma_i(x) - \ln z_i\gamma_i(z)] \tag{18}$$

From Equation (18), z refers to the compositions of component i in mole fractions of the tested phase, x is the composition component i of a trial phase and γ is the activity coefficient of component i in respective phases. For a solvent-oil blend that is stable and demonstrates homogenous single-phase, Equation (19) can be followed [47]:

$$d(x) \geq 0 \tag{19}$$

Additional information on the computation of tangent plane distance can be found in the Appendix A: Equations (A1)–(A11). The solvent-oil blend can be concluded as stable if the tangent plane distance is non-negative. If otherwise, the previous steps are repeated by revising the property attributes and constraints.

3. Results and Discussion

A case study on solvent design for bio-oil applications was conducted to illustrate the application of this proposed methodology. The fast pyrolysis process considered in this work is related to an application in Malaysia. All pricing in this study was converted to U.S. Dollar at the exchange rate of RM 1 = USD 0.24 and adjusted to 2021 values using appropriate indices.

3.1. Identification of Feasible Solvent Candidates

In the initial part of CAMD optimisation, 32 feasible solvent candidates, which are mostly petroleum and natural gas-based solvents that are commonly used as lubricants, lubricant additives, and food additives, were identified according to the pre-defined target properties constraints. The candidates list included mostly higher alkanes and alkenes, with few esters and aromatic compounds. The list was also comprised of several alcohols and nitriles, which are known to be miscible in water. However, there are only a few that can be identified as common chemicals. Most of the molecules are complex and may be challenging to even validate in a lab-scale process, and may not be available at chemical suppliers. Among the favourable candidate was benzyl acetate which is a readily available ester often employed in the food industry as a flavouring agent. 1-Pentanol is also a well-known alcohol employed in the food industry and used as a solvent for lubricants.

Among the candidates for alkenes were 1-octadecene, 1-tetradecene, 1-hexadecene, and 1-dodecene. Most of these alkenes were produced through the oligomerisation of ethylene using triethyl-aluminium catalyst, followed by fractional distillation of the resulting alpha-olefin mixture. In other words, these chemicals were produced from downstream processing of petroleum- or natural gas-based raw materials. Therefore, their availability can be guaranteed, and the price variation could be related to that of the hydrocarbons. Comparatively, more alkanes were chosen as they are commonly used as industrial solvents and additives. Among those were octadecane, N-tridecane, dodecane and undecane. Only one nitrile molecule was included, which is decanenitrile. Among the chemicals used in the additives and food flavouring industry were three ketones: 1-octanone, 2-undecanone and 2-nonanone. Two aldehydes were included too: 1-nonanal and octanal. Hexyl acetate and nonyl acetate were some of the esters included in the list as well. Esters are particularly important for any further research into the reaction system in the blend as they may react with some of the components of pyrolysis bio-oil. Table 3 summarises the 32 solvent candidates identified along with their respective target properties estimated from the GC prediction models. Based on the potential solvent candidates identified in the previous stage, a thorough search was conducted on the cost of solvent as listed in Appendix A: Table A4. The solvents and chemicals were of analytical grade and the costs were obtained from chemical vendors.

Table 3. Feasible solvent candidates generated from Stage 1 Optimisation.

No.	Compound Name	Chemical Formula	Melting Point (K)	Boiling Point (K)	Flash Point (K)	Density ρ (kg/m^3)	Viscosity v (mm^2/s)	HHV (MJ/kg)
S1	1-Pentanol	$C_5H_{12}O$	221.44	408.63	320.35	805.74	3.83	37.97
S2	1-Hexanol	$C_6H_{14}O$	229.47	432.62	334.83	812.14	4.70	39.31
S3	1-Octanone	$C_8H_{16}O$	234.67	449.51	325.07	817.42	1.05	40.24
S4	Octanal	$C_8H_{16}O$	243.73	453.46	322.20	818.04	1.45	40.90
S5	Hexyl acetate	$C_8H_{16}O_2$	208.70	449.66	329.02	872.42	1.14	34.75
S6	Pentyl propionate	$C_8H_{16}O_2$	205.48	449.57	329.02	870.32	0.96	34.75
S7	Phenylacetaldehyde	C_8H_8O	265.52	468.63	336.74	1025.68	1.74	35.36
S8	Benzyl Acetate	$C_9H_{10}O$	243.14	483.92	358.03	1056.67	1.75	31.21
S9	2-Nonanone	$C_9H_{18}O$	242.03	469.65	339.54	820.95	1.29	40.98
S10	1-Nonanal	$C_9H_{18}O$	250.66	473.27	336.68	821.52	1.79	41.58
S11	Hexyl propionate	$C_9H_{18}O_2$	214.41	469.71	343.49	868.91	1.19	35.91
S12	Benzyl acetone	$C_{10}H_{12}O$	270.12	501.17	368.55	987.48	1.99	37.14
S13	Decanenitrile	$C_{10}H_{19}N$	238.23	521.90	387.44	822.90	2.45	43.60
S14	Octyl acetate	$C_{10}H_{20}O_2$	225.69	488.25	357.97	869.50	1.76	36.87
S15	Hexyl butyrate	$C_{10}H_{20}O_2$	222.82	488.18	357.97	867.74	1.48	36.87
S16	4-tert-Butyltoluene	$C_{11}H_{16}$	246.79	462.25	335.65	858.44	0.26	43.19
S17	2-Undecanone	$C_{11}H_{22}O$	255.73	505.18	368.49	826.34	1.96	42.10
S18	Undecanal	$C_{11}H_{22}O$	263.62	508.26	365.63	826.82	2.72	42.60
S19	Nonyl acetate	$C_{11}H_{22}O_2$	233.50	505.29	372.44	868.37	2.18	37.69
S20	Undecane	$C_{11}H_{24}$	191.06	466.22	324.06	737.02	1.18	48.49
S21	1-Dodecene	$C_{12}H_{24}$	209.17	484.87	334.48	754.08	1.19	48.46
S22	Dodecanal	$C_{12}H_{24}O$	269.70	523.89	380.11	828.88	3.35	42.99
S23	Dodecane	$C_{12}H_{26}$	200.87	484.96	338.53	745.48	1.44	48.43
S24	N-Tridecane	$C_{13}H_{28}$	210.08	502.25	353.01	752.80	1.76	48.38
S25	1-Tetradecene	$C_{14}H_{28}$	226.11	518.22	363.43	766.99	1.78	48.36
S26	Tetradecane	$C_{14}H_{30}$	218.74	518.30	367.48	759.21	2.16	48.34
S27	N-Pentadecane	$C_{15}H_{32}$	226.92	533.26	381.96	764.85	2.65	48.30
S28	1-Hexadecene	$C_{16}H_{32}$	241.30	547.21	392.38	776.96	2.69	48.28
S29	N-Hexadecane	$C_{16}H_{34}$	234.67	547.28	396.44	769.87	3.26	48.27
S30	N-Heptadecane	$C_{17}H_{36}$	242.03	560.47	410.91	774.35	4.01	48.24
S31	1-Octadecene	$C_{18}H_{36}$	255.07	572.86	421.34	784.90	4.08	48.23
S32	Octadecane	$C_{18}H_{38}$	249.04	572.92	425.39	778.39	4.93	48.21

3.2. Multi-Objective Optimisation Model

Here, a multi-objective optimisation model was developed via fuzzy max-min aggregation approach to optimise the higher heating value (HHV) and the profitability of the solvent-oil blend, simultaneously. Two case studies were presented to investigate the effect of different constraints on the outcome while optimising both objective functions.

3.2.1. Estimation of Pyrolysis Bio-Oil Production Cost

In this study, a pyrolysis plant was proposed to aid the estimation of pyrolysis bio-oil production cost. The pyrolysis plant was designed to produce 120 tonne of pyrolysis bio-oil from 200 dry tonne of PKS biomass daily via fast pyrolysis. The overall pyrolysis bio-oil yield was assumed to be 60%. It is expected for the pyrolysis plant to operate on a continuous operation daily for 24 h and 300 days, with a plant lifetime of 30 years. The production costs of the pyrolysis plant include the biomass, capital, labour, electrical and other operational costs. Assumption was made that the PKS biomass used in the pyrolysis plant were supplied by a palm oil mill at no cost.

On the other hand, the capital cost of the pyrolysis plant was estimated based on the sizing curve developed in Rogers et al. [48] which relates both the total plant cost and the plant capacity. In this case, the total plant cost of the designed pyrolysis plant was estimated to be USD 16 million. In addition, the capital cost for the biomass pre-processing plant was included, with an estimated cost of USD 2.98 million. As for the labour cost estimation, the following rough scenario was assumed where the designed pyrolysis plant operates on a shift-work basis, with 5 operators and 1 supervisor per shift. Three 8-h shifts pattern was implemented with 4 teams to provide 24/7 coverage. An average annual salary of USD 13 K was allocated for each employee, which cover the employers' insurance cost, pension contribution, anti-social hours payments, training and administration charges [49].

A total electrical consumption of 240 kWh per dry tonne of biomass was estimated for both the biomass pre-processing plant and the pyrolysis plant [50]. The electric tariff of E1 for general industry with medium voltage as defined by Malaysia's energy provider (Tenaga Nasional Berhad) was considered in this study. The price of tariff E1 is USD 0.08/kWh [51] was used to calculate the total cost of electricity. Lastly, an allowance of 4% of the total plant cost (USD 771.48 K per year) has been made to cover other miscellaneous cost such as repair, maintenance, insurance and business costs [50]. Thus, the total cost to produce 1 tonne of pyrolysis bio-oil was calculated to be USD 80.37, as shown in Table 4.

Table 4. Summarised pyrolysis bio-oil production cost.

Production Cost	Cost (USD/Tonne of Bio-Oil)
Biomass Cost	N/A
Capital Cost	17.86
Labour Cost	8.67
Electrical Cost	32.35
Other Operating Cost	21.43
Total Production Cost	80.37

3.2.2. Fuzzy Optimisation

In case study 1, the constraint on solvent fraction added to the blend was relaxed to allow higher HHV value of the generated solvent-oil blend. The parameter β was set to be lesser than 0.75 in this case to achieve HHV of at least 35 MJ/kg. In case study 2, the constraint on consumer preference coefficient was relaxed, thus lowering the HHV requirement of solvent-oil blend to allow higher profitability. As mentioned above, the competitor's product consisted of 50 wt.% solvent. Hence, the solvent fraction was set to be less than 0.5. Table 5 summarises the constraints defined in case study 1 and 2, respectively.

Table 5. Comparison of constraints for case study 1 and 2.

Case	Consumer Preference Coefficient, β	Solvent Fraction
Case 1	<0.75	<0.99
Case 2	<0.9	<0.5

Firstly, single objective optimisation was conducted to generate the upper and lower bounds for both the objective functions, as shown in Table 6. The values obtained was then substituted into Equation (16) to solve the multi-objective fuzzy optimisation problem. The max-min aggregation approach was employed to study the trade-off between high HHV and high profitability.

Table 6. Results from single objective optimisation of HHV and profitability of solvent-oil blend.

Case	Objective Function	Max HHV_{blend}	Max $Profitability_{blend}$
Case 1	HHV_{blend} (MJ/kg)	47.93	36.73
	$Profitability_{blend}$ (USD/tonne of blend)	0.00	3627.62
Case 2	HHV_{blend} (MJ/kg)	33.61	30.61
	$Profitability_{blend}$ (USD/tonne of blend)	3086.00	4438.22

Among the 32 solvents identified in the previous stage, only 4 solvents, including the octadecane, 1-octadecene, 1-tetradecene and 2-octanone, demonstrated promising performance in terms of functionality and economics. Table 7 shows the results obtained from case study 1 and 2. As the constraints on the HHV of the solvent-oil blend were relaxed in case study 1, higher HHV can be observed, ranging from 37.11 to 44.65 MJ/kg. However, a large amount of solvent was required to be blended with pyrolysis bio-oil, thus leading to the increased cost and low profitability. From Table 7, higher profit was obtained from case 2, which is a 1.6-fold increase as compared to the profit in case 1. Nonetheless, this was compensated with the lower HHV of solvent-oil blend ranging from 31.53 to 32.93 MJ/kg. The lowest profit was that of 2-octanone at USD 122.77 per tonne of solvent-oil blend. 2-octanone is a flavouring ingredient naturally present in apple, apricot, banana, papaya, wheat bread and alcoholic beverages. The ketone solvent was available on an industrial scale and should be delivered at a slightly higher cost than 1-tetradecene or octadecane, thus the lower profitability observed.

Table 7. Results for solvent blend candidates.

Case Study	Solvent	HHV_{blend} (MJ/kg)	Profit (USD/Tonne of Blend)	Solvent Ratio	Miscibility
Case 1	Octadecane	44.65	2564.74	0.88	No
	1-Octadecene	41.45	1527.30	0.77	No
	1-Tetradecene	38.16	2174.28	0.99	No
	2-Octanone	37.11	122.77	0.85	Yes
Case 2	Octadecane	32.93	4132.13	0.48	No
	1-Octadecene	31.53	3498.17	0.43	No

Apart from improving the HHV of solvent-oil blend, the miscibility of the final blend can also be improved with the addition of solvent candidates. Instead of dispersing in aqueous and organic phase, the strong intermolecular forces between the molecules in the crude pyrolysis bio-oil will attract each other [52]. However, the dispersion of bio-oil can be improved with addition of solvent candidates due to its amphiphilic properties, and thus improving the phase separation of bio-oil. In this work, the phase stability analysis was carried out by computing the tangent plane distance against the identified solvent candidates. Except for 2-octanone, the remaining solvent candidates identified in both case 1 and 2 (octadecane, 1-octadecene and 1-tetradecene) were immiscible with pyrolysis bio-oil at their respective solvent ratio. This could be explained by the existence of non-polar hydrocarbon groups in the solvent molecule. Figure 2 illustrates the Gibbs energy and tangent plot for 2-octanone-oil blend. The X-axis of the graph represents the mass fraction of 2-octanone solvent in solvent-oil blend, while the Y-axis indicates the calculated Gibbs Energy. As shown in Figure 2, the blend of 2-octanone and pyrolysis bio-oil is stable and demonstrated homogenous single-phase as the tangent line was completely plotted below the Gibbs energy curve. This may be due to the polar carbonyl (C=O) functional group found in the 2-octanone which helps in promoting the miscibility of the solvent-oil blend. As the final solvent-oil blend was expected to be homogenous while demonstrating promising properties, it can be concluded that solvent-oil blend with 35 wt.% of 2-octanone is the most promising blend with HHV of 37.11 MJ/kg and profit of USD 122.77/tonne of blend.

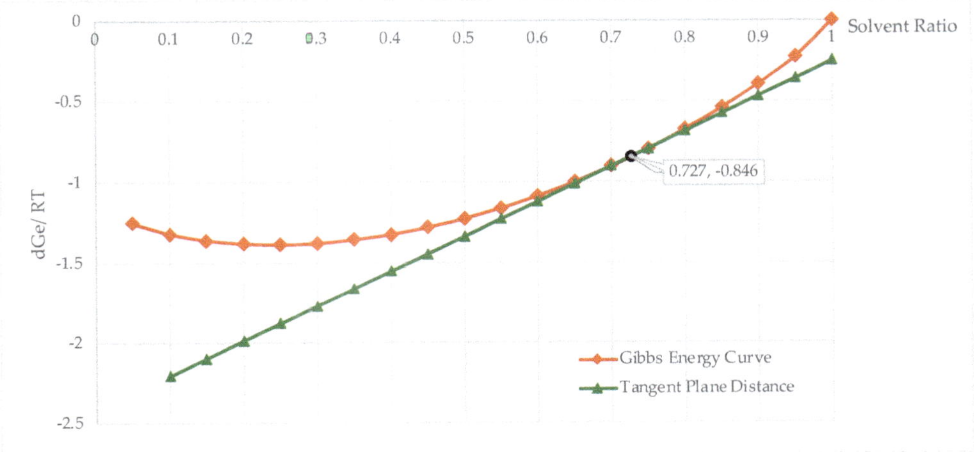

Figure 2. Gibbs energy and tangent plot for 2—Octanone in case study 1.

3.3. Economic Study on the Bio-Oil-Diesel Blend

Based on the optimal solvent-oil blend identified in the previous section, an economic analysis was carried out to investigate the relationship between the ratio of bio-oil-diesel blend, the price and HHV of the bio-oil-diesel. The blend of bio-oil and petroleum diesel was commonly referred to as BX, where X refers to the volume percent of bio-oil in the blend. For example, B5, B10 and B100 consist of 5%, 10% and 100% bio-oil, respectively. Generally, bio-oil-diesel blend has a lower energy content and higher fuel consumption as compared to that of the conventional diesel. Although the bio-oil-diesel blend provides sufficient environmental advantages, the price of this blend is costlier than the conventional diesel fuel. As of September 2021, the average price of diesel around the world is USD 1.07 per litre [53]. Meanwhile, the energy content of the conventional diesel fuel generally ranged between 44 to 48 MJ/kg [54]. From the results obtained, the price of 2-octanone-oil blend was observed to cost USD 6249.38 per tonne of solvent-oil blend, with a HHV of 37.11 MJ/kg. Table 8 summarises the price and HHV of the solvent-oil blend and diesel fuel used in this study.

Table 8. Price and HHV for both solvent-oil blend and diesel fuel.

Properties	Price (USD/Tonne)	HHV (MJ/kg)
Solvent-oil blend	6249.38	37.11
Diesel fuel	1258.82	46.00

Figure 3 illustrates the effects of solvent-oil blend ratio on the price and HHV of the bio-oil-diesel blend. It was observed that as the ratio of solvent-oil blend increases, the price of bio-oil-diesel blend increases proportionately. However, the HHV of the bio-oil-diesel blend decreases as the amount of solvent-oil blend increases. In this study, biodiesel with HHV of 40 MJ/kg was used as benchmark to determine the desired ratio of solvent-oil-diesel blend. As shown in Figure 3, blending with at least 40 wt.% of diesel fuels, or 60 wt.% of solvent-oil blend was required to generate bio-oil-diesel with HHV of at least 40 MJ/kg. However, blending with 60 wt.% solvent-oil blend will cost approximately USD 4.2 K per tonne bio-oil-diesel, which is equivalent to a 3.4-fold increase as compared to pure diesel fuel. To be competitive with conventional diesel fuel, substantial subsidies and tax incentives from government are crucial. In addition, the demand for bio-oil-diesel could be stimulated with the introduction of legislation mandating the blending of biofuel in conventional diesel fuel, thus making the bio-oil-diesel demand independent of the diesel fuel price [55].

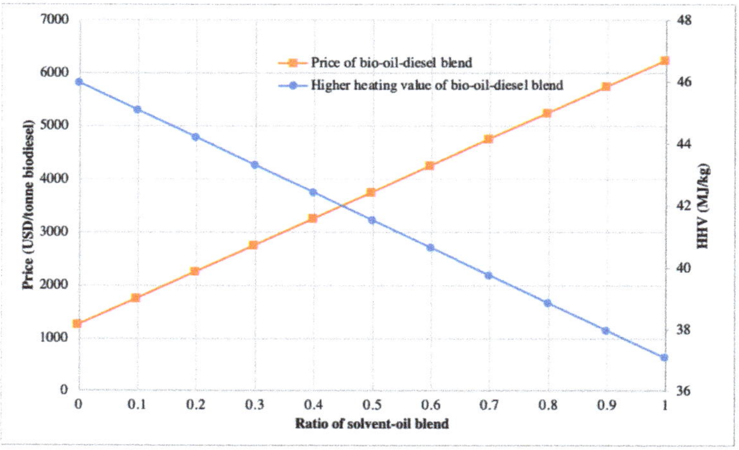

Figure 3. The relationship between the price, HHV and ratio of the biodiesel blend.

4. Conclusions

A CAMD framework was developed to design solvent molecules that can upgrade the properties of bio-oil upon blending, while achieving a low mixing ratio with pyrolysis bio-oil and maintaining profitability. At the initial stage, the requirements of the solvent and solvent-oil blend were identified and translated into target properties. Suitable property prediction models were used to build the structures of 32 promising solvent molecules. In the second stage, a MOO model was developed to investigate the trade-off between high HHV and high profitability of the solvent-oil blend. The HHV and profitability of the solvent-oil blend were optimised simultaneously via the fuzzy max-min aggregation approach. Meanwhile, a pricing model was introduced to evaluate the profitability of the solvent-oil blend. In addition, a pyrolysis plant was proposed to aid the estimation of pyrolysis bio-oil production cost. Solvent-oil blend with octadecane, 1-octadecene, 1-tetradecene and 2-octanone demonstrated positive performance in terms of functionality and economical. Among the four solvent-oil blends, the blend with 85 wt.% of 2-octanone was selected as the most promising solvent-oil blend with a HHV of 37.11 MJ/kg and profit of USD 122.77/tonne of blend, while displaying other desirable attributes. As a conclusion, the developed framework in this work can be applied in the design of bio-oil solvents with different bio-oil types and compositions. However, financial and legislative supports from government are also critical in the commercialisation of bio-oil-diesel blend. In addition, further upgrading of bio-oil via other approaches than solvent addition needs to be considered to add value on this developed framework. It is recommended that the stability of said blends be experimentally verified and the results to be validated. Further investigation on the life cycle sustainability assessment is recommended to compare the sustainability to conventional diesel fuel.

Author Contributions: Conceptualization, J.W.C., N.G.C. and L.Y.N.; methodology, J.W.C., N.G.C. and S.T.-G.; software, O.A.A.; validation, S.T.-G. and K.M.; formal analysis, J.W.C.; investigation, J.W.C., S.T.-G. and K.M.; resources, N.G.C.; data curation, O.A.A. and S.T.-G.; writing—original draft preparation, J.W.C. and L.Y.N.; writing—review and editing, S.T.-G. and K.M.; visualization, L.Y.N.; supervision, N.G.C., S.T.-G. and K.M.; project administration, N.G.C., S.T.-G. and K.M.; funding acquisition, N.G.C. All authors have read and agreed to the published version of the manuscript.

Funding: This research was funded by the Ministry of Higher Education, Malaysia, Grant number FRGS/1/2019/TK02/UNIM/02/1.

Institutional Review Board Statement: Not applicable.

Informed Consent Statement: Not applicable.

Data Availability Statement: Not applicable.

Conflicts of Interest: The authors declare no conflict of interest.

Appendix A

Table A1. Property prediction models and mixing rules.

Property	Property Prediction Model
Normal boiling point, T_b (K) [34]	$\exp\left(\frac{T_b}{T_{b0}}\right) = \Sigma_i N_i T_{b1,i}$ $T_{b0} = 222.543\ K$
Normal melting point, T_m (K) [34]	$\exp\left(\frac{T_m}{T_{m0}}\right) = \Sigma_i N_i T_{m1,i}$ $T_{m0} = 147.450\ K$
Dynamic viscosity, μ (mPa.s) [56]	$\ln(\mu) = \Sigma_i N_i \mu_{(1,i)}$
Kinematic viscosity, ν (mm^2/s)	$\nu = 1000 \times \left(\frac{\mu}{\rho}\right)$
Molar volume, V_1 ($m^3/kmol$) [57]	$V_1 - d = \Sigma_i N_i v_{1i}$ $d = 0.01211\ m^3/kmol$
Density, ρ (kg/m^3)	$\rho = \frac{N_i M_i}{V_1}$
Flash Point, FPT (K) [58]	$\frac{FPT-180.594}{23.3514} = \Sigma_i N_i FPT_i$
Higher Heating Value, HHV (MJ/kg) [59]	$HHV = \frac{\Sigma_i N_i H_i}{\Sigma_i N_i M_i}$
Mixing Rules	
Kinematic viscosity	$\ln(\nu_{mix}) = \Sigma_i x_i \ln(\nu_i)$
Density	$\frac{1}{\rho_{mix}} = \Sigma_i \left(\frac{x_i}{\rho_i}\right)$
Higher heating value	$HHV_{mix} = \Sigma_i x_i HHV_i$

Table A2. Pyrolysis bio-oil properties.

Property	Values
Moisture content	16 wt.%
Kinematic viscosity	17.4 mm^2/s
Higher Heating Value	19.0 MJ/kg
Density	1150 kg/m^3
Components	**Mole Fraction**
Phenol	0.62
2,6-dimethoxyphenol	0.11
2-methoxyphenol	0.11
Furfural	0.08
1,2-benzenediol	0.08
Acetic acid	0.07

Table A3. The functional groups considered for the solvent molecular design.

Family	Functional Group
Alkanes	CH_3, CH_2, CH, C
Alkenes	$CH_2 = CH,\ CH = CH, CH_2 = C, CH = C$
Aromatics	$aCH, aC - CH_3, aC - CH_2, aC - CH, aC - C, aC - CG = CH_2, aC - CH = CH,$ $aC - C = CH_2, aC - OH$
Alcohols	OH
Ketones	CH_3CO, CH_2CO
Aldehydes	CHO
Esters	$CH_3COO, CH_2COO, CHCOO, CCOO$
Ethers	CH_3O, CH_2O, CHO
Carboxyl	$COOH, aC - COOH$
Nitriles	$CH_3CN, CH_2CN, CHCN, CCN, aC - CN$
Amides	$CONHCH_3, CONHCH_2, CON(CH_3)_2$

Table A4. Cost of solvent candidates.

No.	Compound Name	Cost (USD/g Solvent)
S1	1-Pentanol	0.09
S2	1-Hexanol	0.40
S3	1-Octanone	0.07
S4	Octanal	0.45
S5	Hexyl acetate	0.26
S6	Pentyl propionate	0.04
S7	Phenylacetaldehyde	0.22
S8	Benzyl Acetate	0.07
S9	2-Nonanone	0.24
S10	1-Nonanal	0.48
S11	Hexyl propionate	0.12
S12	Benzyl acetone	0.12
S13	Decanenitrile	0.70
S14	Octyl acetate	0.09
S15	Hexyl butyrate	0.49
S16	4-tert-Butyltoluene	0.13
S17	2-Undecanone	0.13
S18	Undecanal	0.24
S19	Nonyl acetate	0.22
S20	Undecane	0.26
S21	1-Dodecene	0.28
S22	Dodecanal	0.15
S23	Dodecane	0.22
S24	*N*-Tridecane	0.57
S25	1-Tetradecene	0.04
S26	Tetradecane	0.33
S27	*N*-Pentadecane	0.58
S28	1-Hexadecene	0.19
S29	*N*-Hexadecane	0.21
S30	*N*-Heptadecane	0.41
S31	1-Octadecene	0.06
S32	Octadecane	0.04

Phase Stability Analysis

To estimate the activity coefficients in non-ideal liquid mixture, group contribution estimation approach developed by [60] was applied. In this work, the GC prediction model combines the solution-of-functional-groups concept with a model for activity coefficient based on UNIQUAC. In a multi-component mixture, the UNIQUAC equation for the activity coefficient of component i is given by:

$$\ln \gamma_i = \ln \gamma_i^C + \ln \gamma_i^R \quad \text{(A1)}$$

In Equation (A1), C represent the combinatorial part while the residual part is denoted as R. Here, Equations (A2) and (A3) calculates the value $\ln \gamma_i^C$ and $\ln \gamma_i^R$:

$$\ln \gamma_i^C = \ln \frac{\phi_i}{x_i} + 5q_i \ln \frac{\theta_i}{\phi_i} + l_i - \frac{\phi_i}{x_i} \sum_j x_j l_j \tag{A2}$$

$$\ln \gamma_i^R = \sum_k v_k^{(i)} (\ln \Gamma_k - \ln \Gamma_k^{(i)}) \tag{A3}$$

Equations (A4)–(A11) represents the calculation for terms in Equations (A2) and (A3):

$$l_j = 5(r_i - q_i) - (r_i - 1) \tag{A4}$$

$$\phi_i = \frac{r_i x_i}{\sum_j r_j x_j} \tag{A5}$$

$$\theta_i = \frac{q_i x_i}{\sum_j q_j x_j} \tag{A6}$$

$$r_i = \sum_k v_k^{(i)} R_k \tag{A7}$$

$$q_i = \sum_k v_k^{(i)} Q_k \tag{A8}$$

$$\ln \Gamma_k = Q_k \left[1 - \ln \sum_m \vartheta_m \psi_{m,k} - \sum_m \frac{\vartheta_m \psi_{m,k}}{\sum_n \vartheta_n \psi_{n,m}} \right] \tag{A9}$$

$$\vartheta_m = \frac{Q_m X_m}{\sum_n Q_n X_n} \tag{A10}$$

$$\psi_{m,n} = -\exp\left(\frac{a_{mn}}{T}\right) l_j = 5(r_i - q_i) - (r_i - 1) \tag{A11}$$

where γ_i = activity coefficient of component i,
ϕ_i = segment fraction (volume fraction) of component i,
θ_i = area fraction of component i,
x_i = mole fraction of component i,
r_i = pure component molecular van der Waals volume parameter,
q_i = pure component molecular surface areas parameter,
${v_k}^{(i)}$ = number of groups of type k in molecule i,
R_k = group volume parameters,
Q_k = group area parameters,
Γ_k = group residual activity coefficient,
${\Gamma_k}^{(i)}$ = residual activity coefficient of group k in pure component i,
ϑ_m = area fraction of group m,
$\psi_{m,k}$ = group interaction parameter,
X_m = mole fraction of group m in the mixture,
$a_{m,n}$ = group interaction parameters obtained from experimental phase equilibrium data.

References

1. Czernik, S.; Bridgwater, A.V. Overview of applications of biomass fast pyrolysis oil. *Energy Fuels* **2004**, *18*, 590–598. [CrossRef]
2. Diebold, J.P.; Czernik, S. Additives to lower and stabilize the viscosity of pyrolysis oils during storage. *Energy Fuels* **1997**, *11*, 1081–1091. [CrossRef]
3. Zhang, X.; Wang, T.; Ma, L.; Zhang, Q.; Jiang, T. Hydrotreatment of bio-oil over ni-based catalyst. *Bioresour. Technol.* **2013**, *127*, 306–311. [CrossRef] [PubMed]
4. Pidtasang, B.; Udomsap, P.; Sukkasi, S.; Chollacoop, N.; Pattiya, A. Influence of alcohol addition on properties of bio-oil produced from fast pyrolysis of eucalyptus bark in a free-fall reactor. *J. Ind. Eng. Chem.* **2013**, *19*, 1851–1857. [CrossRef]
5. Zhang, L.; Liu, R.; Yin, R.; Mei, Y.; Cai, J. Optimization of a mixed additive and its effect on physicochemical properties of bio-oil. *Chem. Eng. Technol.* **2014**, *37*, 1181–1190. [CrossRef]

6. Achenie, L.E.K.; Gani, R.; Venkatasubramanian, V. *Computer Aided Molecular Design: Theory and Practice*, 1st ed.; Elsevier: Amsterdam, The Netherlands, 2003.
7. Harper, P.M.; Gani, R. A multi-step and multi-level approach for computer aided molecular design. *Comput. Chem. Eng.* **2000**, *24*, 677–683. [CrossRef]
8. Joback, K.G.; Reid, R.C. Estimation of pure-component properties from group-contributions. *Chem. Eng. Commun.* **1987**, *57*, 233–243. [CrossRef]
9. Papadopoulos, A.I.; Tsivintzelis, I.; Linke, P.; Seferlis, P. Computer-aided molecular design: Fundamentals, methods, and applications. In *Reference Module in Chemistry, Molecular Sciences and Chemical Engineering*; Elsevier: Amsterdam, The Netherlands, 2018. [CrossRef]
10. Austin, N.D.; Sahinidis, N.V.; Trahan, D.W. Computer-aided molecular design: An introduction and review of tools, applications, and solution techniques. *Chem. Eng. Res. Des.* **2016**, *116*, 2–26. [CrossRef]
11. Ng, L.Y.; Chong, F.K.; Chemmangattuvalappil, N.G. Challenges and opportunities in computer-aided molecular design. *Comput. Chem. Eng.* **2015**, *81*, 115–129. [CrossRef]
12. Chemmangattuvalappil, N.G. Development of solvent design methodologies using computer-aided molecular design tools. *Curr. Opin. Chem. Eng.* **2020**, *27*, 51–59. [CrossRef]
13. Xiu, S.; Shahbazi, A. Bio-Oil production and upgrading research: A Review. *Renew. Sustain. Energy Rev.* **2012**, *16*, 4406–4414. [CrossRef]
14. Zhu, L.; Li, K.; Zhang, Y.; Zhu, X. Upgrading the storage properties of Bio-Oil by adding a compound additive. *Energy Fuels* **2017**, *31*, 6221–6227. [CrossRef]
15. Chemmangattuvalappil, N.G.; Ng, D.K.S.; Ng, L.Y.; Ooi, J.; Chong, J.W.; Eden, M.R. A review of process systems engineering (PSE) tools for the design of ionic liquids and integrated biorefineries. *Processes* **2020**, *8*, 1678. [CrossRef]
16. Lee, J.W.Y.; Ng, L.Y.; Andiappan, V.; Chemmangattuvalappil, N.G.; Ng, D.K.S. Inverse molecular design techniques for green chemical design in integrated biorefineries. *Processes* **2021**, *9*, 1569. [CrossRef]
17. Mah, A.X.Y.; Chin, H.H.; Neoh, J.Q.; Aboagwa, O.A.; Thangalazhy-Gopakumar, S.; Chemmangattuvalappil, N.G. Design of Bio-Oil additives via computer-aided molecular design tools and phase stability analysis on final blends. *Comput. Chem. Eng.* **2019**, *123*, 257–271. [CrossRef]
18. Chong, J.W.; Thangalazhy-Gopakumar, S.; Muthoosamy, K.; Chemmangattuvalappil, N.G. Design of Bio-Oil additives via molecular signature descriptors using a multi-stage computer-aided molecular design framework. *Front. Chem. Sci. Eng.* **2021**, *1*, 1–15. [CrossRef]
19. Clemente, J. *Why Biofuels Can't Replace Oil. Forbes*; Forbes: Jersey City, NJ, USA, 2015; p. 1.
20. Cheng, Y.S.; Fung, K.Y.; Ng, K.M.; Wibowo, C. Economic analysis in product design—A case study of a tcm dietary supplement. *Chin. J. Chem. Eng.* **2016**, *24*, 202–214. [CrossRef]
21. Zhang, L.; Fung, K.Y.; Wibowo, C.; Gani, R. Advances in chemical product design. *Rev. Chem. Eng.* **2018**, *34*, 319–340. [CrossRef]
22. León, M.; Silva, J.; Carrasco, S.; Barrientos, N. Design, cost estimation and sensitivity analysis for a production process of activated carbon from waste nutshells by physical activation. *Processes* **2020**, *8*, 945. [CrossRef]
23. Jehan, S.N.; Sultonov, M. Green investment policy initiatives in Japan. In Proceedings of the 15th International Scientific and Practical Conference of the Russian Society for Ecological Economics Strategies and Tools, Stavropol, Russia, 2–5 July 2019; pp. 21–32.
24. Matel, E.; Vahdatikhaki, F.; Hosseinyalamdary, S.; Evers, T.; Voordijk, H. An artificial neural network approach for cost estimation of engineering services. *Int. J. Constr. Manag.* **2019**, 1–14. [CrossRef]
25. Fishburn, P.C. Method of estimating additives utilities. *Manag. Sci.* **1967**, *13*, 435–453. [CrossRef]
26. Ng, L.Y.; Chemmangattuvalappil, N.G.; Ng, D.K.S. Robust chemical product design via fuzzy optimisation approach. *Comput. Chem. Eng.* **2015**, *83*, 186–202. [CrossRef]
27. Zadeh, L.A. Fuzzy sets. *Inf. Control* **1965**, *8*, 338–353. [CrossRef]
28. Zimmermann, H.-J. Fuzzy programming and linear programming with several objective functions. *Fuzzy Sets Syst.* **1978**, *1*, 45–55. [CrossRef]
29. Khor, S.Y.; Liam, K.Y.; Loh, W.X.; Tan, C.Y.; Ng, L.Y.; Hassim, M.H.; Ng, D.K.S.; Chemmangattuvalappil, N.G. Computer aided molecular design for alternative sustainable solvent to extract oil from palm pressed fibre. *Process. Saf. Environ. Prot.* **2017**, *106*, 211–223. [CrossRef]
30. Ten, J.Y.; Hassim, M.H.; Ng, D.K.S.; Chemmangattuvalappil, N.G. A molecular design methodology by the simultaneous optimisation of performance, safety and health aspects. *Chem. Eng. Sci.* **2017**, *159*, 140–153. [CrossRef]
31. Neoh, J.Q.; Chin, H.H.; Mah, A.X.Y.; Aboagwa, O.A.; Thangalazhy-Gopakumar, S.; Chemmangattuvalappil, N.G. Design of Bio-Oil additives using mathematical optimisation tools considering blend functionality and sustainability aspects. *Sustain. Prod. Consum.* **2019**, *19*, 53–63. [CrossRef]
32. Qin, X.; Gabriel, F.; Harell, D.; El-Halwagi, M.M. Algebraic techniques for property integration via componentless design. *Ind. Eng. Chem. Res.* **2004**, *43*, 3792–3798. [CrossRef]
33. ASTM D7544-12. *Standard Specification for Pyrolysis Liquid Biofuel*; ASTM International: West Conshohocken, PA, USA, 2017. [CrossRef]

34. Asadullah, M.; Ab Rasid, N.S.; Kadir, A.A.S.A.; Azdarpour, A. Production and detailed characterization of bio-oil from fast pyrolysis of palm kernel shell. *Biomass Bioenergy* **2013**, *59*, 316–324. [CrossRef]
35. Marrero, J.; Gani, R. Group-contribution based estimation of pure component properties. *Fluid Phase Equilibria* **2001**, *183-184*, 183–208. [CrossRef]
36. Eljack, F.T.; Eden, M.R.; Kazantzi, V.; Qin, X.; El-Halwagi, M.M. Simultaneous process and molecular design—A property based approach. *AIChE J.* **2007**, *53*, 1232–1239. [CrossRef]
37. Bagajewicz, M.J. On the role of microeconomics, planning, and finances in product design. *AIChE J.* **2007**, *53*, 3155–3170. [CrossRef]
38. Fung, K.Y.; Ng, K.M.; Zhang, L.; Gani, R. A grand model for chemical product design. *Comput. Chem. Eng.* **2016**, *91*, 15–27. [CrossRef]
39. Whitnack, C.; Heller, A.; Frow, M.T.; Kerr, S.; Bagajewicz, M.J. Financial risk management in the design of products under uncertainty. *Comput. Chem. Eng.* **2009**, *33*, 1056–1066. [CrossRef]
40. Street, C.; Woody, J.; Ardila, J.; Bagajewicz, M. Product design: A case study of slow-release carpet deodorizers/disinfectants. *Ind. Eng. Chem. Res.* **2008**, *47*, 1192–1200. [CrossRef]
41. Bagajewicz, M.; Hill, S.; Robben, A.; Lopez, H.; Sanders, M.; Sposato, E.; Baade, C.; Manora, S.; Hey Coradin, J. Product design in price-competitive markets: A case study of a skin moisturizing lotion. *AIChE J.* **2011**, *57*, 160–177. [CrossRef]
42. Lai, Y.Y.; Yik, K.C.H.; Hau, H.P.; Chow, C.P.; Chemmangattuvalappil, N.G.; Ng, L.Y. Enterprise decision-making framework for chemical product design in integrated biorefineries. *Process. Integr. Optim. Sustain.* **2019**, *3*, 25–42. [CrossRef]
43. Energy Industry Review. *Waste-Derived Pyrolysis Oil Market to Close in on USD 500 Mn Valuation by 2029*; Energy Industry Review: Ploiesti, Romania, 2020.
44. Dahl, C.A. Measuring global gasoline and diesel price and income elasticities. *Energy Policy* **2012**, *41*, 2–13. [CrossRef]
45. Omar, S.; Alsamaq, S.; Yang, Y.; Wang, J. Production of renewable fuels by blending bio-oil with alcohols and upgrading under supercritical conditions. *Front. Chem. Sci. Eng.* **2019**, *13*, 702–717. [CrossRef]
46. China Market Price: Monthly Avg: Organic Chemical Material: Propan 2 ol, Isopropyl Alcohol 99%. Available online: https://www.ceicdata.com/en/china/china-petroleum--chemical-industry-association-petrochemical-price-organic-chemical-material/cn-market-price-monthly-avg-organic-chemical-material-propan-2-ol-isopropyl-alcohol-99 (accessed on 27 September 2021).
47. EUBIA. Pyrolysis. Available online: https://www.eubia.org/cms/wiki-biomass/pyrolysis-and-gasification/pyrolysis/ (accessed on 27 September 2021).
48. Prausnitz, J.M. *Molecular Thermodynamics of Fluid-Phase Equilibria*; Prentice-Hall: Michigan, MI, USA, 1969.
49. Rogers, J.G.; Brammer, J.G. Estimation of the production cost of fast pyrolysis Bio-Oil. *Biomass Bioenergy* **2012**, *36*, 208–217. [CrossRef]
50. Factory and Manufacturing Average Salaries in Malaysia 2021. Available online: https://www.salaryexplorer.com/salary-survey.php?loc=130&loctype=1&job=33&jobtype=1 (accessed on 27 September 2021).
51. Pricing & Tariffs—Industrial Tariffs. Available online: https://www.tnb.com.my/commercial-industrial/pricing-tariffs1 (accessed on 27 September 2021).
52. Manara, P.; Bezergianni, S.; Pfisterer, U. Study on phase behavior and properties of binary blends of Bio-Oil/Fossil-based refinery intermediates: A step toward Bio-Oil refinery integration. *Energy Convers. Manag.* **2018**, *165*, 304–315. [CrossRef]
53. Diesel Prices, Liter, 27 September 2021. Available online: https://www.globalpetrolprices.com/diesel_prices/ (accessed on 27 September 2021).
54. Heat Values of Various Fuels. Available online: https://world-nuclear.org/information-library/facts-and-figures/heat-values-of-various-fuels.aspx (accessed on 27 September 2021).
55. Chin, M. *Biofuels in Malaysia: An Analysis of the Legal and Institutional Framework*; CIFOR: Bogor, Indonesia, 2011. [CrossRef]
56. Conte, E.; Martinho, A.; Matos, H.A.; Gani, R. Combined group-contribution and atom connectivity index-based methods for estimation of surface tension and viscosity. *Ind. Eng. Chem. Res.* **2008**, *47*, 7940–7954. [CrossRef]
57. Constantinou, L.; Gani, R.; O'Connell, J.P. Estimation of the acentric factor and the liquid molar volume at 298 k using a new group contribution method. *Fluid Phase Equilibria* **1995**, *103*, 11–22. [CrossRef]
58. Albahri, T.A. MNLR and ANN structural group contribution methods for predicting the flash point temperature of pure compounds in the transportation fuels range. *Process. Saf. Environ. Prot.* **2015**, *93*, 182–191. [CrossRef]
59. Walters, R.N. Molar group contributions to the heat of combustion. *Fire Mater.* **2002**, *26*, 131–145. [CrossRef]
60. Fredenslund, A.; Jones, R.L.; Prausnitz, J.M. Group-contribution estimation of activity coefficients in nonideal liquid mixtures. *AIChE J.* **1975**, *21*, 1086–1099. [CrossRef]

processes

Article

Inverse Molecular Design Techniques for Green Chemical Design in Integrated Biorefineries

Jamie W. Y. Lee [1], Lik Yin Ng [1,*], Viknesh Andiappan [1], Nishanth G. Chemmangattuvalappil [2] and Denny K. S. Ng [1]

1 School of Engineering and Physical Sciences, Heriot-Watt University Malaysia, Putrajaya 62200, Malaysia; jamieyinn@gmail.com (J.W.Y.L.); V.Murugappan@hw.ac.uk (V.A.); Denny.Ng@hw.ac.uk (D.K.S.N.)
2 Department of Chemical and Environmental Engineering, University of Nottingham Malaysia, Jalan Broga, Semenyih 43500, Malaysia; Nishanth.C@nottingham.edu.my
* Correspondence: L.Ng@hw.ac.uk

Abstract: Over the past decades, awareness of the increase in environmental impact due to industrial development and technological advancement has gradually increased. Green manufacturing is one of the key approaches that begin to address environmental issues. With the current global attention, methodologies to incorporate green manufacturing into the design of green products through the green process route are much needed. However, it is challenging for industries to achieve this, as there is no definite definition of green. This work presents a systematic approach that provides a clear and consistent green manufacturing definition with a measurement method in terms of both product and process. With the consistent green definitions, the developed approach designs a product that satisfies green property and other product performance properties. In addition, the developed approach synthesises the production process that fulfils green manufacturing definitions and financial considerations for the designed product. A case study on the design and production of green biofuel is solved to illustrate the efficacy of the approach. Green product design is obtained by identifying molecular building blocks that fulfil the targeted properties using an inverse molecular design technique. The goal is to design a chemical product that is environmentally friendly while fulfilling customer requirements. Moreover, a superstructural mathematical optimisation approach is used to determine optimal conversion pathways that have minimal environmental impact on the production of the identified green product. The utilisation of multi-objective optimisation allows the design of product and process to strike a good balance between operational and environmental performances.

Keywords: green manufacturing; integrated biorefineries; multi-objective optimisation; chemical product design; inverse design techniques

Citation: Jamie W. Y. Lee, Lik Yin Ng, Viknesh Andiappan, Nishanth G. Chemmangattuvalappil and Denny K. S. Ng Inverse Molecular Design Techniques for Green Chemical Design in Integrated Biorefineries. *Processes* **2021**, *9*, 1569. https://doi.org/10.3390/pr9091569

Academic Editor: Francesco Parrino

Received: 31 July 2021
Accepted: 30 August 2021
Published: 1 September 2021

1. Introduction

In recent decades, issues of increased environmental impact and excessive consumption of natural resources due to industrial development have become major concerns for the industrial and academic sectors [1]. These issues include chemical pollution caused by abuse or unintentional spread of hazardous pollutants to the atmosphere [2]. This can lead to a direct toxicological impact on mankind and the environment. Prolonged exposure to such pollutants can cause serious effects on humans such as shortness of breath, nausea, vomiting, and even death. Other than that, the issue of depletion of non-renewable natural resources will be devastating to mankind in the future [3]. According to EIA [4], the amount of oil remaining would last until 2050 due to the high oil exploitation rate. The adverse impacts of oil depletion include a fall in business, an increase in living costs, and uncertainties in the transportation sector. Therefore, awareness of the requirement for environmental protection has gradually increased. Actions from researchers and governments have been taken to address these issues by proposing a managing model as well as new law and regulation formulation [5]. For example, in Malaysia, Environmental Quality (Clean Air) Regulations have been implemented since 2014 with an emission standard of 150 mg/m^3

in total discharged gas [6]. In order to encourage compliance with emissions limits, a heavy penalty of up to RM 1,000,000 or two years of imprisonment is imposed on parties who fail to comply. In European countries, the European Parliament has adopted the "Industrial Emission Directive" since 2010 to regulate pollutant emissions from industrial installations by reviewing the whole environmental performance of the plant. This is aimed to achieve a high level of protection of the environment and human health by restricting harmful industrial emissions [7].

Aside from the laws and regulations, green manufacturing is one of the key approaches that initiates the production of materials and processes that minimise the effect on the environment together with human health. Green manufacturing plays an important role as issues related to excessive waste generation and energy consumption need to be resolved in the near future due to growing living standards [8]. Green manufacturing is widely known as a sustainable manufacturing approach that aims for profitability through environmentally friendly operating processes [9]. Generally, green manufacturing refers to processes of manufactured products that utilise natural resources to reduce impacts on the environment and conserve energy utilisation [10]. Traditionally, green manufacturing is defined as activities that attempt to decrease waste [11]. Mohanty and Deshmukh [11] classified waste as the generation of non-value-adding products and activities in the work process. Thereupon, the assessment of wastivity is introduced as an approach to monitor the amount of waste produced in an organisation. The concept of wastivity is defined as the ratio of waste produced to the total resource inputs. Sources of waste are identified from wastage in raw materials, the labour force, maintenance works, and the operational process to the final packaged product. Consequently, action plans suggested to reduce waste include pipeline leakage sealing, strict monitoring of operation parameters, and the utilisation of empty bulkers during the return trip. However, these suggestions only stressed waste detection and correction without taking waste prevention into consideration.

A decrease in waste generation alone is insufficient to prove or provide a safer and green process as a decrease in waste does not represent the safe disposal of waste. In addition, the presence of hazardous waste with a high level of toxicity without going through treatment would cause a negative effect on public health and the environment. Aside from reducing waste, the manufacturing process should involve hazardous waste management to lessen the number of hazardous substances produced and treat hazardous waste to reduce the toxicity to an acceptable level [12]. The disposal and emissions of harmful waste play a major role in causing environmental pollution and eventually lead to catastrophic consequences for mankind in the future. Therefore, green manufacturing is further defined as actions that prevent pollution. The risk assessment method is one of the actions used in pollution prevention, which helps to quantify the degree of environmental impact for different chemicals. This enables the estimation of environmental impacts caused by specific chemicals and the minimisation of the associated risk in all directions [13]. By considering the environmental impacts of both product and process, sources of pollution and contamination can be identified from raw materials to end-of-life product disposal [14]. The source of pollution is then used as an indexing parameter in the Green Manufacturing Index (GMI). Nukman et al. [15] proposed the use of GMI that utilises information such as the source of pollution and impacts of the pollution as its' indexing parameters to determine the effectiveness of pollution. Ultimately, the practice of pollution prevention practice is geared toward decreasing environmental damage arising from product- and process-related design.

In addition to pollution prevention, green manufacturing should be practised to reduce the damage caused by chemical products and processes on renewable and non-renewable natural resources. As reserves of certain resources will be exhausted at the current rate of consumption, activities that destroy natural resources such as uncontrolled consumption should be avoided to prevent overexploitation of non-renewable resources [16]. Likewise, a higher rate of harvesting than replenishment of renewable resources and pollution of biodiversity results in the deterioration of the environment. To reduce the environmental burden, the product life cycle is used as an important tool to promote design that leaves a smaller environmental footprint [17]. The entire product life cycle should involve environmental practises such as clean production, green design, and reuse with the aim to minimise the disposal of products. In this case, non-toxic and renewable feedstock should be used as raw material for manufacturing processes to promote healthy consumption patterns of resources.

From the abovementioned works, it can be seen that past research works on green manufacturing tend to focus more on reducing waste and environmental impact. Other than waste and environmental impact reduction, energy consumption plays an important role in the manufacturing process. According to Deif [18], green manufacturing is an approach to designing the product and process that requires less material and energy input. Maruthi and Rashmi [19] state that the goal of green manufacturing is to maximise resources and energy efficiency by conserving energy sources. Energy conservation is the application of devices that are able to reduce the consumption of energy by having low energy requirement functionalities. This results in a decrease in natural resources consumption used to provide energy generation. Li and Lin [20] adopted a range-adjusted measure-based data envelopment analysis (DEA) model developed by Wang et al. [21] as an approach to calculating energy efficiency and environmental performance using the directional distance function (DDF) model. However, most studies adopt the DDF model to assess only environmental performance. This is due to the model complexity and contradiction to the actual process, which may lead to the recommendation of the wrong strategies [20]. In view of the fourth industrial revolution, the internet of things (IoT) has become significant in industrial development by improving the machine's energy efficiency and cutting down energy waste [22]. Although rich research results are available for IoT, works that integrate IoT and green manufacturing are limited. Hence, manufacturing processes that consume less energy and are able to recycle heat waste in the first place are prioritised during the product and process design stage. As the demand for manufacturing is expected to grow in the near future, the improvement of the energy-efficient manufacturing process has become increasingly crucial for any industry [15].

In addition, green manufacturing is also defined as processes to produce a reusable sustainable product. The verb reuse is recognised as an action to divert material from the waste stream. One of the approaches to reuse material is material exchange by means of passing the materials that are no longer useful to other organisations to be used for other applications [23]. The material exchange concept is similar to industrial symbiosis, which was discovered by van Berkel [24]. Industrial symbiosis involves the collection of waste material from symbiosis partners, which act as alternative feedstock or energy sources for the organisation. Furthermore, remanufacturing and refurbishing are other ways of reusing material by breaking down waste material to recover or rebuild them into useful components [25]. These methods turn useless material into value-added material, which directly reduces unnecessary waste disposal and emission. Additionally, approaches to reuse offer a better and environmentally preferred alternative compared to the waste management method by limiting the need for new natural resources. However, it is noted that factors such as ease of recyclable material employment and potential hazards of the material should be considered while selecting the materials to be reused. To avoid unnecessary impacts after the end of a product or process's usable life, it is preferable for products to have shorter life cycles [26]. In this context, the life cycle refers to the usable life of a material. Material with longer usable life required more time and energy to allow the decomposition of the material. Additionally, the product life cycle relates to the biodegradability of the organic material. The material should be able to break down or be mineralised by microbial or fungal activity to avoid pollution and contamination to the environment [27].

From the abovementioned discussion on green manufacturing, it can be seen that specific action or measurement methods for green manufacturing cover a wide range of definitions and categories. In addition, the systematic approach to simultaneously measuring and assessing green products and processes is limited. Hence, an organised approach that is capable of identifying, quantifying, and assessing the performance of green is needed for long-term sustainability in the future.

2. Incorporation of Green Manufacturing into Product and Process Design

2.1. Integration of Process and Product Design

Green manufacturing encompasses a variety of aspects to be considered, from the design of the product to the synthesis of the production process. It is thus important to integrate process and product design aspects as an overall design problem. A hybrid design method that integrates mathematical modeling with heuristic approaches was proposed by Hostrup et al. [28] to simultaneously synthesise a separation process and select the suitable solvent. Eden et al. [29] presented a systematic approach that formulates the process and product design problem as two reverse formulations. The approach first identifies the process design targets by solving the product design problem, then synthesises the production process by solving the identified design targets. Papadopoulus and Linke 2006 [30] presented a systematic framework to solve an integrated solvent design and process synthesis problem. The molecular clustering approach was utilised in the framework to identify a Pareto optimal solvent that fulfils both the product and process performance targets. An approach that integrates computer-aided molecular design (CAMD) techniques and reaction network flux analysis was presented by Hechinger et al. [31]. The presented approach was utilised in identifying biofuels and the production pathways for the design of biofuels. Ng et al. [32] proposed a systematic methodology for the synthesis of biochemicals in integrated biorefineries. Biochemicals that satisfy customer requirements were designed using CAMD techniques, while biomass conversion pathways that produce the designed biochemicals were determined using a superstructural mathematical optimisation approach. This methodology was later extended by Ng et al. [33] for the design of a mixture in integrated biorefineries to address the product design challenge where the desired product properties cannot be fulfilled by a single-component chemical product. Meramo-Hurtado and González-Delgado [34] presented a hierarchical approach for the synthesis and design of a multi-product biorefinery. Through decision-making, multi-objective optimisation, and numerical methods, an optimal biorefinery was synthesised by considering different feedstock and final products. The concept of decision-making was also incorporated into the work presented by Lai et al. [35] to consider the cooperation among the corporate unit, business unit, research and development unit, and production unit in a chemical design and production enterprise. Restrepo-Flórez and Maravelias [36] developed a superstructural framework for the design of biorefineries that utilises ethanol to produce gasoline, jet fuel, and diesel. The framework considers aspects such as catalysis, process synthesis, and fuel property modeling in designing biorefineries that produce fuels with specified product properties. Recently, Tey et al. [37] proposed a comprehensive framework for the design of value-added pharmaceutical products from biomass. A chemical reaction pathway map (CRPM) was utilised to connect raw material, potential conversion pathways, and final products, while a mathematical optimisation approach was used to identify the compromised solution that considers the gross product and sustainability index. While these works discussed in detail the integration of process and product design, it is realised that the consideration of green manufacturing into the overall process and product design is limited.

2.2. Methodology

In order to incorporate green manufacturing into the overall product and process design, a systematic approach is developed to integrate product and process design with the measurement of green manufacturing. Definitions of green manufacturing in product and process design are first defined and categorised. This is followed by the design of an optimal green chemical product that fulfils the defined green and customer requirements by using the group-contribution method. Based on the designed product, an environmentally friendly conversion pathway is then determined by employing a superstructural mathematical optimisation approach. The step-by-step procedure to design a green chemical product and conversion pathway is presented as shown in Figure 1. The developed approach shown in Figure 1 can be separated into two stages, stage one that focuses on product design and stage two to identify the conversion pathway.

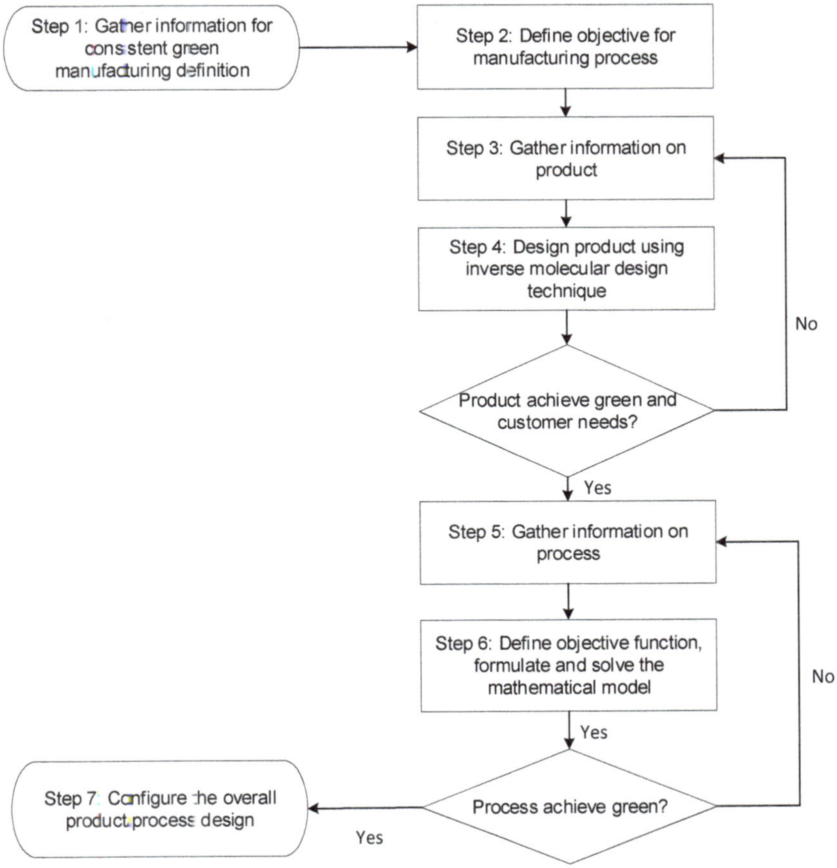

Figure 1. Procedure to incorporate green manufacturing into product and process design.

Step 1: Gather information for consistent green manufacturing definition.

The available green manufacturing definitions are reviewed and generalised into different categories. The purpose of generalisation is to group similar definitions that serve the same purpose and require the same green outcome on mankind and the environment. The generalised green definitions are further categorised in terms of product and process as shown in Tables 1 and 2. This is imperative to allow easier identification of distinct differences between measurement methods required for the product and process. Additionally, categorisation provides comprehensive and clearer judgements on approaches to achieve green manufacturing needs. For instance, one of the green manufacturing product definitions is generalised as avoidance of unnecessary impact or use of energy after the end of usable life. This definition is needed and essential to be defined because the definition indicates the design of products that are able to break down into innocuous degradation products. Added to that, this definition restricts the design of high-complexity products that persist in the environment. By incorporating this definition into the initial product design strategy, end-of-life burdens of the designed chemical product can be significantly reduced. Table 1 below shows a summary of the generalised green definitions for products.

Table 1. Green definitions of products with measurement/assessment method.

Generalised Definition	Measurement/Assessment	Purpose	Reference
Avoidance of unnecessary impact or use of energy after end of usable life	Biodegradability	Product can be broken down at the end of life without excessive energy.	De Wilde et al. [27]
	Octanol/water partition coefficient	Product readily adsorbs in soil to avoid contamination of water source.	Sangster [38]
Manufacturing of non-hazardous product	Flammability test	Avoid product that can spontaneously ignite and burn readily in air.	Albahri [39]
	Auto-ignition temperature	Avoid product with self-ignite properties at atmospheric condition.	Albahri [39]
	Latent heat of vaporisation	Ensure stability of product.	Yunus [40]
Harmless to human health and environment	Bioassays	Measure toxicity of the product by organisms' responses.	Schulze [41]
	Oral Rat LD_{50}	Measure the level of acute toxicity of product.	Hall et al. [42]
	Ozone depletion potential (ODP)	Avoid producing ozone-depleting product.	Singh et al. [43]
	Global warming potential (GWP)	Minimised greenhouse effect caused by product.	Singh et al. [43]
Utilisation of natural resources	Use of non-fossil-fuel-based natural resources as the feedstock for the manufacturing of product to conserve non-renewable resources.		Anastas and Warner [44]

Table 2. Green definitions of process with the measurement/assessment method.

Generalised Definition	Measurement/Assessment	Purpose	Reference
Minimisation of waste and material consumption	Percent atom economy	Limit produces of undesired product.	Wang et al. [45]
	E-factor	Minimise resource intensity and waste generated.	Tieves et al. [46]
	Percent yield	Maximise possible mass of a product produce.	Anastas and Warner [44]
Minimisation of energy requirement	Specific energy consumption	Minimise ratio of energy used for producing a product.	Lawrence et al. [47]
	Equipment efficiency	Avoid unnecessary use of energy.	Siegel et al. [48]
Pollution prevention	Emission rate	Minimise pollutant produce.	MPCA [49]
Minimisation of accident risk	Lost time injury frequency rate (LTIFR)	Minimise number of lost time injuries.	SWA [50]
Avoidance of unnecessary impact or use of energy after end of usable life	Equipment usable life value	Avoiding long usable life may cause negative impact on the environment.	CFI Education Inc. [51]

It should be noted that most of the green definitions of products are focusing on the chemical industry. For other manufacturing industries, the information in Table 1 should be updated to consider the important green performance indicators of the industry. Table 2 below shows the list of generalised green manufacturing definitions for processes. One of the green manufacturing definitions is minimisation of waste and material consumption. This definition emphasises the process pathway to restrict waste generation instead of treatment for created waste. Decisions about the process pathway should be able to reduce the need for raw materials and incorporation of all materials used into the final desired product. Hence, this definition is essential to employ in the process design as it is the key to initiating practice to minimise the impact of undesirable output from processes on the environment and human health.

Similar to green definitions of products, green definitions of process, as shown in Table 2, are developed for the chemical industry. While it is possible to utilise the definitions for other manufacturing industries, it is suggested to update Table 2 for a more correlated and straightforward assessment of green performance indicators.

Step 2: Define objective for manufacturing process.

With the generalised definitions, measurement methods are first identified according to the target properties of the product to solve the product design problem. This is performed by identifying the appropriate green definitions of green products. Green products cover the targeted chemical and physical properties to ensure the designed product fulfils green manufacturing needs with reference to Table 1. Apart from that, the properties of product requirement are identified to comply with customer needs. With these, the process pathway to produce the green product is determined with the definitions in Table 2 to ensure both product and process fulfil defined green manufacturing. Hence, the objective is to develop a measurement approach that can meet green manufacturing and customer needs.

Step 3: Gather information on product.

Once the objectives for green products and processes have been identified, the defined green manufacturing and customer needs are translated into measurable properties. Method of measurement is identified to quantify the green manufacturing definitions. For example, Oral Rat LD50 can be used to ensure the designed green product is safe to use and will not cause a negative impact on the environment, which fulfils the definition of being harmless to human health and the environment. On top of that, the method of measurement and targeted property constraints are determined to guarantee that the product design solution achieves green manufacturing and customer needs. The property constraints are written as a set of property ranges bounded by upper and lower limits. The limits can be extracted from current environmental regulations and industrial standard. For example, the designed fuel product viscosity should fall between 0.30 cp to 0.60 cp to allow consistent fuel flow and lower pumping power requirements that fulfil customer needs. The properties constraints for product design are generalised as shown in Equation (1).

$$v_p^L \leq V_p \leq v_p^U \qquad \forall\, p \in P \tag{1}$$

In Equation (1), p represents the index for the target property. V_p is the target value, while the constraints are represented as v_p^L, the lower limit, and v_p^U, the upper limit for the desired product property.

Step 4: Design product using inverse molecular design technique.

With the product information, the mathematical model using the group-contribution method is formulated to design a green chemical product. Molecular building blocks that are suitable for the product design problem are first determined in this step. The designed product with the combination of selected molecular building blocks should be able to replace the currently available product with similar functionality. For instance, CH- can be set as one of the molecular building blocks to represent the alkane functional group when the product design problem is to design a fuel product. Other than setting properties constraints in the model, as shown in step 3, structural constraints are also employed to allow the generation of a feasible chemical structure without the formation of free bonds. Assuming that only a single bond is considered, the structural constraint is illustrated as shown in Equation (2), based on Chemmangattuvalappil et al. [52]. n_1, n_2, n_3, n_4 refers to the number of degrees one, two, three, and four of a molecular building block that are available to bond with other molecular blocks, while N represents the total number of molecular blocks in a molecule.

$$\sum_{i=1}^{n_1} x_i + 2\sum_{n_1}^{n_2} x_i + 3\sum_{n_2}^{n_3} x_i + 4\sum_{n_3}^{n_4} x_i = 2\left[\left(\sum_{i=1}^{N} x_i\right) - 1\right] \tag{2}$$

Moreover, the objective function is applied in the model to obtain an optimal solution of the product for a targeted property. For example, maximisation of a higher heating value can be set as the objective function when designing biofuel in order to fulfil engine efficiency according to customer needs. Subsequently, the solution for product design can be obtained. However, targeting on single objective is insufficient to provide a green chemical product solution. Other properties of the product will also be the key to designing a green chemical product. In order to consider multiple objectives simultaneously on several targeted properties, the chemical product design problem is solved as a multi-objective optimisation problem. The traditional weighted-sum method in solving the multi-objective optimisation problem requires decision-maker(s) to assign a weighing factor to each objective using expert judgement. In addition, the weighted-sum method might be biased as the weighing factors assigned to each objective are heavily dependent on expert knowledge or the personal preferences of the decision-maker. To address this, the fuzzy optimisation approach is incorporated into this mathematical model. The degree of satisfaction λ_p of targeted property p is introduced as shown in Equations (3) and (4). λ_p ranges between the values of 0 to 1, which implies the level of satisfaction on targeted property value V_p within the property constraints. A higher value of λ_p indicates higher satisfaction of the targeted property. When minimisation of the property is required, the value of λ_p approaches 1 when the obtained property value approaches the lower limit. Equation (3) is used when property needs to be minimised.

$$\begin{array}{lll} \lambda_p = \frac{v_p^U - V_p}{v_p^U - v_p^L} & \text{where} \quad v_p^L \leq V_p \leq v_p^U & \forall\, p \in P \\ \lambda_p = 1 & \text{when} \quad V_p \rightarrow v_p^L & \\ \lambda_p = 0 & \text{when} \quad V_p \rightarrow v_p^U & \end{array} \tag{3}$$

Equation (4) is utilised when the property need to be maximised. The λ_p value approaches 1 when the obtained property value approaches the upper limit of the target property range.

$$\begin{array}{lll} \lambda_p = \frac{V_p - v_p^L}{v_p^U - v_p^L} & \text{where} \quad v_p^L \leq V_p \leq v_p^U & \forall\, p \in P \\ \lambda_p = 0 & \text{when} \quad V_p \rightarrow v_p^L & \\ \lambda_p = 1 & \text{when} \quad V_p \rightarrow v_p^U & \end{array} \tag{4}$$

The degree of satisfaction is split into three regions of above satisfactory, satisfactory, and below satisfactory, as shown in Figure 2.

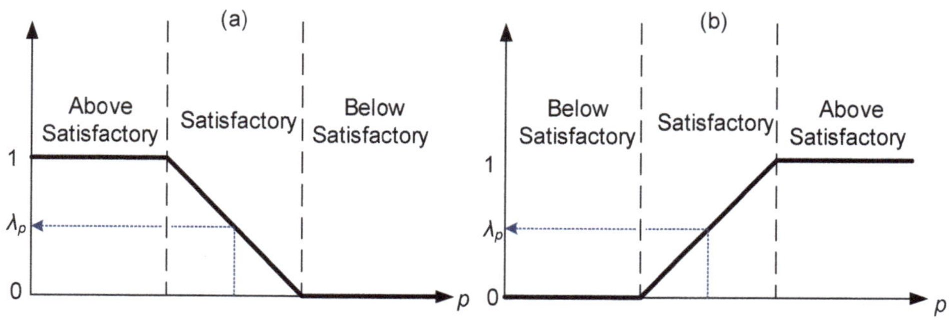

Figure 2. Degree of satisfaction for (**a**) property to be minimised and (**b**) property to be maximised.

Next, the max–min aggregation approach is used to maximise the least-satisfied degree of satisfaction λ. This is to ensure every targeted property is optimised simultaneously without any bias. To achieve this, the objective of the proposed method is to optimise the weakest/worst property among all targeted properties to be optimised. Thus, the least-satisfied λ is maximised where the overall objective function is formulated as shown in Equations (5) and (6).

$$\text{Maximise } \lambda \tag{5}$$

$$\lambda \leq \lambda_p \qquad \forall\, p \in P \tag{6}$$

It can be seen from Equations (5) and (6) that the fuzzy optimisation approach identifies the relative importance of each targeted property to be optimised without the presence of a decision-maker, hence minimising the influence of bias on the chemical product design problem. To generate additional alternative feasible solutions, integer cuts are utilised. Integer cuts are applied as an additional constraint on the proposed mathematical model to avoid the same solution being generated again. The application of integer cuts can be continued until no feasible solution is obtained. Once the designed product achieved green and customer needs, Step 5 can be proceeded to determine the process that produces the designed product. If not, Step 3 needs to be repeated for further research to amend or improve product information.

Step 5: Gather information on process.

With the identified green definitions of process, the measurement method is determined to quantify green manufacturing needs on the process pathway. Information on all possible conversion pathways that convert specific feedstock to intermediates, then from intermediates into the final desired product is identified. With all the identified conversion pathways, the superstructure can be constructed. Next, parameters to measure the process performance can be identified. For example, the yield of the desired product is aimed to maximise the achievement of the green definition on minimisation of waste generation. To calculate the yield of the process, the conversion rate on each conversion pathway need to be obtained, which can be found through the literature. Other than that, the measurement of specific energy consumption can be formulated to acquire a process pathway that fulfils the minimisation of energy requirements and the green manufacturing needs. Specific energy consumption can be calculated from the heat of reaction through changes in heat of formation of each substance where the heat of formation of substances can be easily found. On top of that, all parameters are determined to guarantee that the process design solution achieves green manufacturing without neglecting customer needs. Customer needs can refer to the economic performance of the process, which is calculated as gross profit. The total capital and operating costs required for gross profit calculation can be obtained through references.

Step 6: Define objective function and formulate and solve the mathematical model.

Once the information on the process is gathered, a mathematical model can be created by formulating the design problem using a superstructural mathematical optimisation approach. In order to obtain the optimal solution to the conversion pathway, the objective function is applied to targeted parameters, which measure the performance of the process. For example, minimising environmental burden is set as the objective function when selecting conversion pathways. This is to fulfil green manufacturing needs where the prevention of pollution is considered when designing the process. Subsequently, the solution to conversion pathway selection to produce the desired product as found in the product design problem can be obtained. As multiple objectives are needed to be optimised, a fuzzy optimisation approach is utilised. With the developed model, an optimal process pathway that is green and economically feasible can be obtained based on the desired green product design.

Step 7: Configure the overall product/process design.

When green manufacturing needs on product and process are fulfilled, the developed mathematical model can successfully propose an optimal green product design with a green process pathway. If the solution to the product design does not achieve green manufacturing and customer needs, the steps cannot proceed to determine the optimal process pathway. This is because the desired product design is not identified. Hence, Step 5 has to be repeated to refine process information until the desired green product is achieved. After that, the refinement of process information is needed when the green process did not improve the feasibility of determining the optimal process pathway. The developed approach solves the process and product design problem as two design problems. Although an iterative feedback loop might be needed, this developed approach lowers the computational complexity of the problem. This reduces the computational efforts required to solve the problem compared with solving the integrated process and product design problem simultaneously. If it is required to solve an integrated process and product design problem simultaneously to obtain one optimum overall result, an algebraic approach presented by Bommareddy et al. [53] can be utilised. However, simultaneous solution is not considered in this approach.

This work employs CAMD techniques and the fuzzy optimisation approach. It is noted that suitable property prediction models for product property estimation are not always available. Though data-driven techniques can be used in these situations, the accuracy of the identified correlation might be a concern to the user. In addition, it can be seen from Equations (3) and (4) that upper and lower limits for an objective are required in utilising the fuzzy optimisation approach, which might be challenging to obtain. Hence, this presented approach needs to be utilised judiciously by replacing the property prediction models and solution strategies depending on the nature of the design problem.

A case study is presented to show the efficacy of the developed two-stage optimisation approach.

3. Case Study

The application of the developed two-stage optimisation approach is demonstrated by solving the biobased fuel design and production problem taken from Ng et al. [32]. This case study is suitable to illustrate the developed approach as it involves designing a chemical product from biomass and identifying the biomass conversion pathway to produce the designed chemical product. In the first stage, the green biofuel is designed via a CAMD approach using the group contribution method. In the second stage, the optimum green manufacturing conversion pathway that produced the designed biofuel product from biomass in terms of different objectives is determined by the superstructural mathematical optimisation approach.

3.1. Design of Optimal Green Product

3.1.1. Identification of Target Properties, Target Property Range, and Property Prediction Model

The biobased fuel product is designed to fulfil the defined green manufacturing needs while complying with customer needs. Based on the developed green manufacturing definitions as explained in methodology in Table 1, properties of octanol/water partition coefficient ($LogK_{ow}$), flash point (T_f), lethal dose (LD_{50}), and latent heat of vaporisation (H_v) are considered in this case study. The properties are used to quantify green manufacturing definitions. Other than properties that define green, the bio-based fuel product should also comply with customer needs, which represent the product performance by satisfying the performance constraints. In this case study, properties that cover customer needs include viscosity (η), surface tension (σ), density (ρ), Hildebrand solubility parameter (δ), and higher heating value (HHV). The green biofuel product should not cause an unnecessary impact on the environment. This can be measured as $LogK_{ow}$, which is the tendency of a chemical product to bind with a living organism or soil. The value should remain low to prevent bioaccumulation. Furthermore, the flammability characteristic and toxicity of the biofuel are measured as T_f and LD50, respectively. This is to ensure that the product is safe to use as it is non-inflammable, non-hazardous, and non-toxic to human health and the environment. Additionally, the stability of biofuel is taken into consideration by H_v. As mentioned, the designed product should also comply with customer needs and the product design stage. Therefore, η is measured and preferred to be low to minimise the pumping power requirement as well as ensuring consistency of the fuel flow. Meanwhile, the energy requirement for combustion is measured as surface tension (σ) to assure the combustion quality of the biofuel. Other important properties are ρ and δ. A high density is expected to increase the fuel mass flowrate and hence reduce the production equipment size. As biofuel is also considered as biosolvent due to its properties; biofuel can be an effective diluent for liquid–liquid extraction other than an environmentally friendly fuel. Thus, δ of biofuel is part of fulfilling the customer needs to determine the behaviour of the fuel. In order to obtain high engine efficiency, HHV of the biofuel is measured and set to be high, which represents the high energy content of the fuel. Through using the assessment report by API [54] as the main reference, the target property range to be fulfilled for each of the abovementioned properties is shown in Table 3.

Table 3. Upper and lower limits for biofuel product design.

Category	Property	Target Property Range		Reference (GC Model)
		v_p^L	v_p^U	
Green	$LogK_{ow}$	-	4.50	Hukkerikar et al. [55]
	T_f (K)	294.00	-	Hukkerikar et al. [55]
	Oral Rat LD_{50} (mg/kg)	500.00	15,000.00	Hukkerikar et al. [56]
	H_v (kJ/mol)	25.00	70.00	Hukkerikar et al. [55]
Performance	η (cP)	0.10	1.50	Conte et al. [57]
	σ (mN/m)	15.00	40.00	Conte et al. [57]
	ρ (g/cm^3)	0.55	0.90	Qiao et al. [58]
	δ (MPa$^{1/2}$)	10.00	30.00	Hukkerikar et al. [55]
	HHV (kJ/kg)	4000.00	10,000.00	Yunus et al. [59]

After identifying the target properties for the biofuel product, property prediction models based on group contribution (GC) methods are utilised to estimate the target properties in the design problem. The estimation of properties by employing the GC method is illustrated using Equation (7). In Equation (7), $f(X)$ is the function of the targeted property; X. N_i, M_j, and O_k are the number of occurrences of the first-, second-, and third-order groups; C_i, D_j, and E_k represent the contributions of the first-, second-, and third-order groups; and w and z are binary coefficients.

$$f(X) = \sum N_i C_i + w \sum M_j D_j + z \sum O_k E_k \quad (7)$$

It is clear that property prediction models are developed with certain accuracy and uncertainties, which might affect the effectiveness in estimating the product property. From the property prediction models identified in Table 3, the coefficient of determination (R^2) of the models ranges from the reported value of 0.73 for LD^{50} to 0.9999 for HHV, with R^2 for most of the models above 0.95; the average relative error (ARE) of the models ranges from the published value of 16.40 for LD^{50} to 0.4855 for HHV, with ARE for most of the models below 5.00. Ng et al. [60] presented a systematic methodology for chemical product design by considering product property and accuracy of property prediction models. From the presented work, it was found that although the model accuracy affects the identification of the optimum chemical product, the optimum chemical product identified without considering the model accuracy is still among the best few products identified by considering the model accuracy. Hence, the model accuracy will not invalidate the identified results. In this work, the selected property prediction models as shown in Table 3 are the available models for each property category with the highest accuracy. It is noted that the property prediction model with higher accuracy can always be utilised in this developed approach. The best five solutions will be generated, as discussed in the following paragraphs.

3.1.2. Selection of Molecular Building Blocks

With the identification of property prediction models, appropriate molecular building blocks are selected for the design problem. As the goal of the design problem is to design biobased fuel, alkanes and alcohol are set as the target homologous series. Thus, only carbon (C), hydrogen (H), and oxygen (O) atoms are considered for the product design problem. In this case study, only molecular groups with single bond are considered. The considered molecular groups can be classified into the alkane first-order group with one carbon atom and zero to three hydrogen atoms (C-) (CH-) (CH_2-) (CH_3-) and the alcohol first-order group (-OH).

3.1.3. Formulation of Mathematical Model for Molecular Design Problem

The molecular design problem is then written as a mixed-integer linear programming (MILP) model. Although the objective of this design problem is to design a green biofuel, a single-objective function that targets only one green property is unlikely to guarantee that the designed product is the optimal green product. Additionally, other properties are equally important and should be considered in achieving the overall design goal that fulfils both green and customer needs. Hence, this green chemical product design problem is solved as a multi-objective optimisation problem. This is performed by addressing multiple target properties simultaneously during the product design stage. In this case study, H_v, η, ρ, and HHV are optimised simultaneously together with structural constraints to obtain solutions in terms of molecular groups. The properties considered for simultaneous optimisation cover the categories of both green and performance properties. Under the category of green product properties, H_v is considered; under the category of performance product properties, η and ρ are considered for their importance in reducing production operating costs, while HHV is considered for its importance in ensuring the quality of the biofuel product in terms of energy content.

To optimise multiple properties without any preference, the focus on optimisation will be the weakest property among all properties to be optimised. Hence, a degree of satisfaction λ is introduced to trade off the multiple properties. The overall objective of the proposed model is determined by maximising the least-satisfied property. This is to optimise the least-satisfied property among targeted properties and hence minimise the difference between each individual property. This approach fits multi-objective chemical product design problems where expert knowledge in assigning the weighing factor is unavailable, and each targeted property to be optimised is given equal importance. It makes sure that no targeted property is over-improved while neglecting the importance of other targeted properties. The objective function for this proposed model is shown in Equation (8). Additionally, the aforementioned properties to be optimised are written as linear membership functions as shown by Equations (9)–(12). H_v, ρ, and HHV are to be maximised while η is to be minimised in this case study.

$$\text{Maximise } \lambda \tag{8}$$

subject to,

$$\lambda \leq \frac{H_v - 25}{70 - 25} \tag{9}$$

$$\lambda \leq \frac{\rho - 0.55}{0.9 - 0.55} \tag{10}$$

$$\lambda \leq \frac{HHV - 4000}{10000 - 4000} \tag{11}$$

$$\lambda \leq \frac{1.5 - \eta}{1.5 - 0.1} \tag{12}$$

3.1.4. Result and Discussion on Molecular Design Problem

Optimum solutions to the product design problem are obtained in terms of combination of molecular groups. Additional feasible product solutions are obtained by utilising integer cuts. The best five solutions to the number of molecular group occurrences are shown in Table 4.

Table 4. List of solutions in terms of molecular groups.

Solution	Number of Occurrences of Molecular Groups				
	CH_3-	CH_2-	CH-	C-	-OH
A	5	3	3	0	0
B	6	2	2	1	0
C	4	5	0	1	0
D	3	6	1	0	0
E	2	7	0	0	0

From Table 4, it can be seen that the best five optimal biofuel products fall within the nonane, decane, and undecane alkanes groups. In the best five solutions, there is no alcohol (-OH) molecular group considered as part of the product. This indicates that the alcohol group product does not meet the requirement of defined green and customer needs, and hence products with the alcohol group are not suitable for this design problem. With the list of molecular groups obtained from solving the developed approach, the enumeration of the molecular structure of all five solutions as shown in Table 4 is performed. Molecular groups can be enumerated into feasible molecules that represent the designed product. The molecular structures and names of the biofuels identified from the solutions are shown in Table 5, while a list of product specifications for the best five solutions is summarised in Table 6.

From Table 6, all substituents' properties of the solutions fall within the target product range that represents green manufacturing and customer needs based on Table 3. This indicates that the design problem achieves green definitions while fulfilling other needs. Most importantly, the developed approach is able to design a bio-based fuel with optimised green manufacturing needs in conjunction with customer needs.

Table 5. Molecular structures of designed product.

Solution	Name	Molecular Structure
A	3,4,5-trimethyloctane	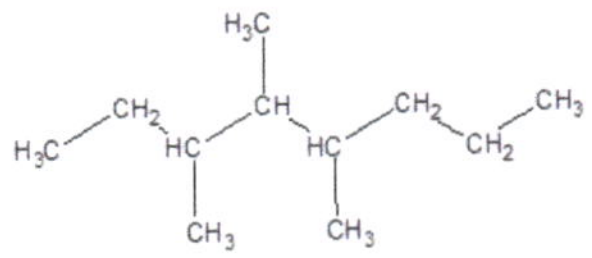
B	2,2,3,4-tetramethylheptane	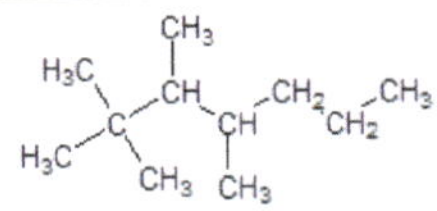
C	2,2-dimethyloctane	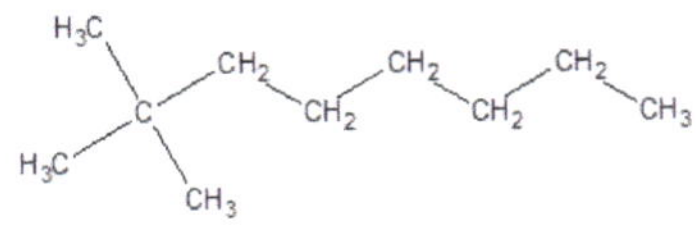
D	2-methylnonane	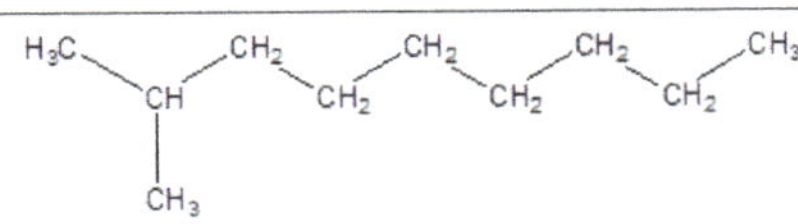
E	Nonane	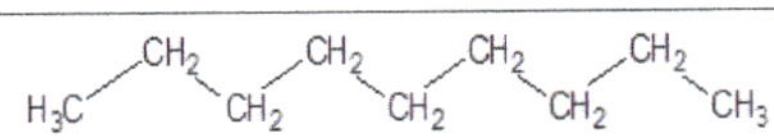

Table 6. Product specifications of designed product.

Sol.	Green Product Needs				Performance Needs				
	$LogK_{ow}$	T_f (K)	H_v (kJ/mol)	LD_{50} (mg/kg)	η (cP)	σ (mN/m)	ρ (g/cm³)	δ (MPa$^{1/2}$)	*HHV* (kJ/kg)
A	4.45	298.63	51.43	812.70	0.58	18.68	0.74	14.03	7401.02
B	4.46	294.23	49.51	751.32	0.75	17.48	0.74	12.92	7403.22
C	4.50	295.53	47.88	811.13	0.79	18.85	0.73	14.44	6778.69
D	4.38	299.93	49.81	877.38	0.61	20.04	0.73	15.54	6776.49
E	4.14	294.83	46.54	906.80	0.57	20.41	0.72	16.38	6137.80

As discussed previously, optimal solutions cannot be guaranteed with single property optimisation as there is a chance that other targeted properties can be further improved to produce a better solution. Therefore, the multi-optimisation approach is applied in this product design problem. The comparison of degrees of satisfaction between generated product solutions is shown in Table 7.

Table 7. Comparison of λ_p between different product solutions.

Sol.	Multi-Optimisation			
	λ_{Hv}	λ_η	λ_ρ	λ_{HHV}
A	0.5874	0.6583	0.5517	0.5668
B	0.5446	0.5375	0.5522	0.5672
C	0.5085	0.5062	0.5151	0.4631
D	0.5512	0.6340	0.5146	0.4627
E	0.4786	0.6667	0.4759	0.3563

In Table 7, the solutions obtained are ranked according to decreasing least-satisfied property. Therefore, the solution generated will be capable of replacing the currently available fuel as long as other important properties to be optimised are not overlooked. The optimal product to be produced in this design problem is identified as 3,4,5-trimethyloctane, which has the highest least-satisfied degree of satisfaction. 3,4,5-trimethyloctane is an alkane group product with a carbon number of 11, as shown in Table 5.

From Table 7, it is clear that solution A is the best biofuel product as the λ in solution A is the highest among other generated solutions. λ then decreases from solution B to E. As λ shows how much the product satisfies the criteria of property, it can be seen that λ_{HHV} appears to be the lowest in three solutions among the five generated solutions. In order to comply with green manufacturing and customer needs in biofuel products, HHV, which represents the energy content in the fuel, acts as a bottleneck in this product design problem. However, the least-satisfied property is not always the same property for all five generated solutions. For example, the least-satisfied property for solution A is ρ with λ_ρ of 0.5517, while the least-satisfied property for solution B is η with λ_η of 0.5375. This shows that the target properties were treated without any bias. The developed approach did not target only one property but identified the relative importance of each property to be optimised. Provided that the degree of satisfaction is maximised for the least-satisfied property, all five solutions can be replaced and produced as biofuel that fulfils green manufacturing and customer needs.

3.2. Selection of Optimal Conversion Pathway

After the identification of the optimal biofuel product in the first stage of the two-stage optimisation approach, the optimal green manufacturing conversion pathway to convert biomass into a designed biofuel product can be determined in the second stage of the methodology. In this case study, the empty fruit bunch (EFB), which is a palm-based biomass, is used as the feedstock in the integrated biorefinery process. The EFB composition is taken from Sukiran [61], as shown in Table 8. The use of EFB is part of the fulfilment of green manufacturing needs mentioned in Table 1 as non-fossil-fuel-based natural resources are utilised as feedstock and non-renewable resources are conserved.

Table 8. Composition of empty fruit bunch (EFP).

Component	Composition (Dry Matter %)
Cellulose	22.00
Hemicellulose	29.00
Lignin	39.00

3.2.1. Identification of Performance Parameters for Green Manufacturing

In the first stage, the optimal product is identified as 3,4,5-trimethyloctane, which represents alkane with carbon number of 11. Hence, undecane is set to be the end product of the integrated biorefinery process. As the goal of this case study is to design a green product with a green manufacturing process, the designed conversion pathway should fulfil green manufacturing definitions without neglecting the economic performance of the project. A process that fulfils green manufacturing definitions should have a minimal negative impact on the environment, waste production, and energy consumption. In this case study, the green manufacturing process is quantified in terms of total environmental burden (EB^{Total}), yield of designed product undecane (C_{11} Yield), and total energy consumption (H^{Total}) during the production. All the parameters considered are based on the developed green manufacturing definitions as explained in Table 2. EB^{Total} is considered for the green manufacturing definition of pollution prevention; C_{11} yield is calculated for the green manufacturing definition of minimisation of waste and material consumption while H^{Total} is assessed for the green manufacturing definition of minimisation of energy requirement. Additionally, the economic feasibility of the process is evaluated by assessing the gross profit (GP^{Total}) of the integrated biorefinery configuration.

The parameter considered in the green manufacturing process to minimise the negative impact on environment is by evaluation of the environmental burden. The change in environmental impact within a pathway can be determined using EB^{Total} as proposed by Andiappan et al. [62]. The change in environmental impact is measured between feedstock and product, assessing if the raw material is converted into a product that is less harmful or more harmful to the environment. EB^{Total} of an integrated biorefinery is determined using the environmental burden score E_{ijk}. E_{ijk} for a particular pathway j is calculated by the product of difference in potency factor (PF) and component to product ratio (α) within the reaction pathway, as shown in Equation (13). As every substance will contribute differently to the environmental burden, each substance has a different number of PF, which can be obtained through reference.

$$E_{ijk} = \sum \alpha^{product} PF^{product} - \sum \alpha^{reactant} PF^{reactant} \quad (13)$$

With the environmental burden score, EB^{Total} can be calculated via Equation (14). R^{I}_{ij} refers to the flowrate of feedstock i, i is then to be converted to intermediates k via pathway j at a given conversion rate V^{I}_{ijk}. Intermediates k with flowrate R^{II}_{kj} is then converted further to final product k' with a conversion rate of $V^{II}_{kj'k'}$.

$$EB^{Total} = \sum_{k}\sum_{j}\sum_{i}(R^{I}_{ij}V^{I}_{ijk}E_{ijk}) + \sum_{k'}\sum_{j'}\sum_{k}(R^{II}_{kj'}V^{II}_{kj'k'}E_{kj'k'}) \quad (14)$$

The next green manufacturing parameter considered for the conversion pathway selection is yield of undecane (C_{11} Yield). The conversion rate and selectivity of pathways to convert biomass into the designed biofuel product are obtained through references. With the conversion rate, the amount of C_{11} desired product to be produced can be measured. As the goal of the design problem is to fulfil the green manufacturing definition, the selected conversion pathway should give a higher amount of the desired product. With a fixed amount of input feedstock, this indirectly indicates a reduction in waste generated. Another two indicators are measured in this design problem to ensure that the selected pathways satisfy green manufacturing needs in reducing waste. The first indicator is the percent atom economy, which measures the ratio of the amount of raw material and the amount of useful products generated, as shown in Equation (15). A higher percentage of the atom economy implies a lower amount of undesired waste produced.

$$Atom\ Economy\ (\%) = \frac{Mass\ of\ desired\ product}{Mass\ of\ reactants} \times 100 \quad (15)$$

The second indicator is e-factor, which is used to estimate the resource intensity for the selected process and waste generated. E-factor can be calculated via Equation (16). In contrast to atom economy, a lower e-factor indicates a lower amount of waste produced. Both atom economy and e-factor are able to quantify a green manufacturing definition of minimisation of waste.

$$E - factor = \frac{Mass\ of\ waste\ (\mathrm{kg})}{Mass\ of\ product\ (\mathrm{kg})} \quad (16)$$

To satisfy a green manufacturing definition of the minimisation of the process energy requirement, heat of reaction (H_{ijk}) is used to evaluate the energy consumption for the conversion pathway. H_{ijk} for pathway j is determined by utilising the heat of formation H_f of each substance, as shown in Equation (17). n refers to the stoichiometry of the reaction pathway.

$$H_{ijk} = \sum n^{product} H_f^{product} - \sum n^{reactant} H_f^{reactnat} \quad (17)$$

With the calculated heat of reaction, H^{Total} for selected pathways can be determined using the summation of reaction heat in each pathway, as presented in Equation (18).

$$H^{Total} = \sum_k \sum_j \sum_i (R^I_{ij} V^I_{ijk} H_{ijk}) + \sum_{k'} \sum_{j'} \sum_k (R^{II}_{kj'} V^{II}_{kj'k'} H_{kj'k'}) \quad (18)$$

In addition, it is important to reduce energy intensity when designing a green manufacturing process. Hence, specific energy consumption (SEC), as shown in Equation (19), is measured to determine the energy required per unit mass of products produced. Lower SEC indicates a lower energy requirement to produce the product.

$$SEC = \frac{Energy\ used\ (\mathrm{kJ})}{Mass\ of\ product\ (\mathrm{kg})} \quad (19)$$

3.2.2. Construction of Superstructure Formulation of Mathematical Model

A list of available conversion pathways to produce alkane from biomass are compiled, as shown in Table S1 under Supplementary Material. Further, the information needed to measure and assess the green manufacturing parameters for all pathways is collected, as shown in Tables S2–S4. Based on the available conversion pathway, a superstructure is then developed, as presented in Figure 3. It should be noted that the superstructure can be modified according to different product requirements.

In this case study, a total of four scenarios with different production objectives and one scenario with multi-objective are considered in synthesising the integrated biorefinery. This is to demonstrate the ability of the proposed two-stage optimisation approach in determining the optimum pathway while accommodating different production objectives.

1. Design for maximum economic performance, Max GP^{Total}
2. Design for maximum product yield, Max C_{11} Yield
3. Design for minimum environmental burden, Min EB^{Total}
4. Design for minimum energy consumption, Min H^{Total}
5. Design for optimal conversion pathway with multi-objective optimisation approach

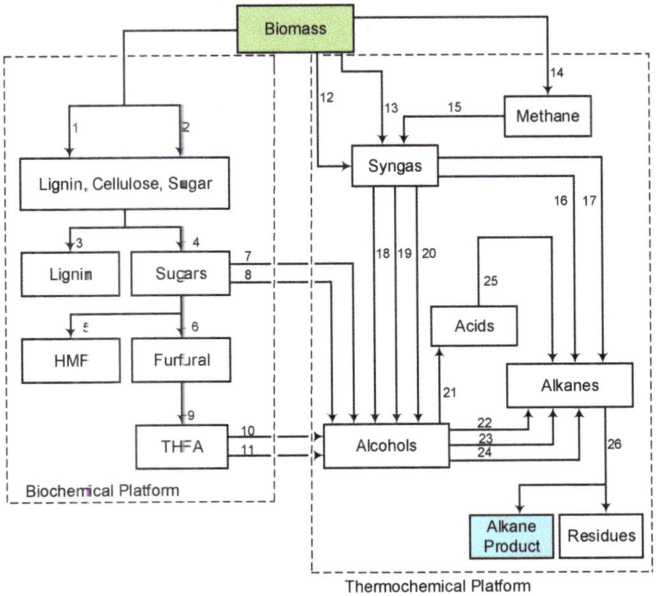

Figure 3. Conversion pathway of integrated biorefinery.

By utilising the superstructural mathematical optimisation approach, a different conversion pathway can be determined based on different design objectives. In this design problem, 50,000 tonnes per year of EFB are set as the basis of available feedstock.

3.2.3. Results and Discussion of Selection of Optimal Conversion Pathway

With the performance parameters for green manufacturing identified, each scenario is solved by considering different objective functions. The results for each scenario are discussed in the following subsections. This is followed by a detailed comparison, and discussion of all five scenarios is conducted.

Scenario 1: Design for Maximum Economic Performance

This scenario is solved by maximising the gross profit for an integrated biorefinery configuration. The objective function for this scenario is shown in Equation (20). Market prices of product, by-products, and biomass feedstock as well as capital and operating cost for each pathway are summarised in Tables S2 and S3, respectively.

$$\text{Maximise } GP^{Total} \tag{20}$$

Based on the optimisation result obtained, when economic performance is maximized, the gross profit is identified to be USD 23.40 million per annum with an annual production of 1181.82 tonnes/yr. As the main product for this design problem is alkane C_{11}, the production of other alkane and alcohol is sold as by-products. The optimal conversion pathway determined by this objective function is presented as synthesised integrated biorefinery, as shown in Figure S1 under Supplementary Material. In this scenario, the designed product is produced through ammonia explosion, organosoly separation, dehydration of sugars, and then hydrogenation of furfural and THFA 1. The main product is further produced by dehydration of alcohol and then fractional distillation of the desired separation.

Scenario 2: Design for Maximum Product Yield

The product to be produced is alkane C_{11}, which is represented by the optimum product determined in the first stage of the design problem, 3,4,5-trimethyloctane. The integrated biorefinery is synthesised by maximising the production of C_{11} in the optimisation model using Equation (21).

$$\text{Maximise } C_{11}\ \mathit{Yield} \tag{21}$$

By maximising the desired product yield, the maximum possible mass of the product can be produced. Hence, the green manufacturing need to minimise the waste generated from the process can be achieved. Based on the obtained result, the maximum yield of alkane C_{11} that can be produced in the integrated biorefinery is 1839.81 tonnes/yr with a gross profit of USD 13.20 million per annum. The optimal conversion pathway in this scenario is identified and shown in Figure S2. Most of the conversion pathways selected in Scenario 2 are similar to Scenario 1, which are mostly biochemical pathways. However, instead of dehydration of sugar, sugar is converted to alcohol via the yeast fermentation process. Alkane C_{11} is then refined and separated from other by-products through the fractional distillation process. The separation process performance is considered with identifying product yield.

Scenario 3: Design for Minimum Environmental Burden

In this scenario, the environmental burden in terms of global warming potential is minimised. With this objective, as shown in Equation (22), conversion pathways that help in reducing global warming potential can be determined. The unnecessary impact of product manufacturing on the environment can be reduced through minimisation of the environmental burden. This can be illustrated by reaction pathways that convert chemical reactants with high environmental impact into products with lower environmental impact.

$$\text{Minimise } EB^{Total} \tag{22}$$

Based on the optimisation result generated, minimum EB^{Total} achieved is 47.5 thousand tonnes of CO_2 equivalent per annum. This is insufficient to prove that the identified pathways are environmentally friendly. Ideally, it is preferred to have EB^{Total} of negative value. A positive value of EB^{Total} indicates that production pathways are contributing a negative impact on the environment while a negative value of EB^{Total} shows that the identified pathways are reducing the impact on the environment. In this case study, this situation cannot be avoided when the final product is fuel. However, the use of biofuel is more environmentally friendly than using fossil-based fuel. Although the identified pathways are not totally environmentally friendly, the synthesised integrated biorefinery from this scenario has the least impact caused to the environment. Aside from that, gross profit and product yield of C_{11} obtained are USD 1.26 million per annum and 805.58 tonnes/yr, respectively, when EB^{Total} is minimised. The optimal conversion pathway selected in this scenario is the same as Scenario 1, as shown in Figure S1.

Scenario 4: Design for Minimum Energy Consumption

The energy consumption of the pathways for integrated biorefinery is determined by using the summation of reaction heat, H^{Total}. In order to achieve green manufacturing needs, the energy requirement of the integrated biorefinery needs to be minimised to avoid the unnecessary use of energy. In this scenario, the integrated biorefinery is synthesised by minimising H^{Total} using the objective function, as shown in Equation (23).

$$\text{Minimise } H^{Total} \tag{23}$$

As the feedstock EFB biomass consists of different compositions of cellulose, hemicellulose, and lignin, the heat of formation for the feedstock is not available. Thus, the heat of formation of the biomass feedstock is estimated by using the heat of combustion of each composition. Based on the optimisation result, minimum H^{Total} obtained from the optimal pathway is -7.10×10^{11} million J/y. The negative value of reaction heat indicates that excess heat can be provided by the integrated biorefinery. Hence, the heating energy required when producing alkane C_{11} in this integrated biorefinery pathway can be reduced. For the minimisation of H^{Total}, gross profit and product yield of C_{11} acquired are

USD 23.40 million per annum and 1242.00 tonnes/yr, respectively. The selected conversion pathway is the same as Scenarios 1 and 3, as shown in Figure S1.

Scenario 5: Design for Optimal Conversion Pathway with the Multi-Objective Optimisation Approach

As mentioned previously, the designed conversion pathway should accommodate green manufacturing needs without neglecting the economic performance. Thus, the multi-objective approach is applied to synthesise the integrated biorefinery with maximum economic performance, maximum production yield, minimum environmental burden, and minimum energy consumption simultaneously. Similar to the product design problem, the fuzzy optimisation approach with max–min aggregation approach is applied in this process design problem. Thus, the degree of satisfaction λ is also introduced to trade-off between multiple objectives.

The multi-objective approach is adapted in process design by introducing λ for economic performance, production yield, environmental burden, and energy consumption objectives. However, the upper and lower limits for each objective are required in order to formulate the equation in maximising λ, as shown in Equation (8). Therefore, the upper and lower limits are obtained from Scenarios 1 to 4 by applying both maximisation and minimisation to the objective function of each scenario. For example, GP^{Total} in Scenario 1 is maximised to obtain the upper limit of this objective, while the lower limit of GP^{Total} is identified through minimising GP^{Total} in the same scenario. However, the negative value of USD -1.62×10^7 is found when GP^{Total} is minimised. This indicates that the process will not gain any profit even when the product is produced. To have a reasonable range for upper and lower limits, the lower limit of GP^{Total} is taken as USD 1.17×10^7, which is 50% of the maximum GP^{Total} of USD 2.34×10^7. In addition, maximum H^{Total} is found to be 1.16×10^9 J/yr, which indicates that heat is required to be provide to the system. The optimal case for energy consumption is that a portion of heat can be provided in the system itself to reduce energy intensity. Thus, the upper limit of H^{Total} is taken as -3.55×10^{11} J/yr, which is 50% of the minimum H^{Total} of -7.10×10^{11} J/yr. Upper and lower limits required for the multi-objective optimisation approach are summarised as shown in Table 9.

Table 9. Upper and lower limits.

Target Objective	Targeted Range	
	Lower Limit	Upper Limit
GP^{Total} (USD/yr)	1.17×10^7	2.34×10^7
Alkane C_{11} Product Yield (t/yr)	805.58	1839.81
EB^{Total} (tonne of CO_2 equivalent/yr)	4.03×10^4	1.96×10^6
H^{Total} (J/yr)	-7.10×10^{11}	-3.55×10^{11}

With the identified upper and lower limits of each target objective to be optimised, multi-objective optimisation can be performed. The objective function for this approach is to maximise the least-satisfied property λ, as shown in Equation (24). The targeted objectives to be optimised are written as Equations (25)–(28). GP^{Total} and C_{11} yield are to be maximised while EB^{Total} and H^{Total} are to be minimised.

$$\text{Maximise } \lambda \tag{24}$$

subject to,

$$\lambda \leq \frac{GP^{Total} - 1.17 \times 10^7}{2.34 \times 10^7 - 1.17 \times 10^7} \tag{25}$$

$$\lambda \leq \frac{C_{11}\ \text{Yield} - 805.58}{1839.81 - 805.58} \quad (26)$$

$$\lambda \leq \frac{1.96 \times 10^6 - EB^{Total}}{1.96 \times 10^6 - 4.03 \times 10^4} \quad (27)$$

$$\lambda \leq \frac{-3.55 \times 10^{11} - H^{Total}}{-3.55 \times 10^{11} - (-7.10 \times 10^{11})} \quad (28)$$

Based on the obtained results in this scenario, GP^{Total} is identified to be USD 2.25 × 10^7 million per annum with an annual production of 1242.00 tonnes/yr of undecane. EB^{Total} and H^{Total} obtained are 7.51 × 10^4 tonnes of CO_2 equivalent per annum and −5.05 × 10^{11} J/y. It can be seen that all parameters obtained in this scenario fall within the targeted range, as shown in Table 9. This indicates that the process design problem achieved an optimal conversion pathway in terms of green manufacturing definitions and economic performance. The optimal conversion pathway in this scenario is identified and shown in Figure S3. The conversion pathways selected in Scenario 5 are mostly similar to Scenarios 1, 3, and 4, with an additional pathway of yeast fermentation to convert sugar into alcohol. The degree of satisfaction λ of each parameter is shown in Table 10.

Table 10. Degree of satisfaction λ for Scenario 5.

Deg. of Satisfaction	λ_{Yield}	λ_{GP}	λ_H	λ_{EB}
Scenario 5	0.422	0.920	0.422	0.982

From Table 10, it can be noticed that there are two least-satisfied parameters λ_{Yield} and λ_H with an identical value of 0.422. This shows that both least-satisfied parameters are optimised. λ_{GP} and λ_{EB} show very high satisfaction with λ of 0.920 and 0.982, respectively. As λ_{GP} is the parameter to be maximised, 0.920 of λ indicates that GP_{Total} approaches its upper limit, while for λ_{EB}, 0.982 of λ implies that EB_{Total} approaches its lower limit as it is a parameter to be minimised. Table 11 below shows the summary of results obtained from Scenarios 1 to 5.

Table 11. Comparison between different scenarios.

Scenario	1	2	3	4	5
	Max GP^{Total}	**Max C_{11} Yield**	**Min EB^{Total}**	**Min H^{Total}**	**Multi-Objective**
GP^{Total} (USD/y)	2.34 × 10^7	1.32 × 10^7	1.26 × 10^6	2.34 × 10^7	2.25 × 10^7
EB^{Total} (tonne of CO_2 equivalent/y)	6.85 × 10^4	1.41 × 10^5	4.75 × 10^4	6.85 × 10^4	7.51 × 10^4
Alkane C_{11} Production Rate (t/y)	1181.82	1839.81	805.58	1181.82	1242.00
Alkane and Alcohol By-product Production Rate (t/y)	1.47 × 10^4	2.30 × 10^4	1.00 × 10^4	1.47 × 10^4	1.55 × 10^4
Waste Produced (t/y)	2.83 × 10^4	2.47 × 10^4	3.3 × 10^4	2.83 × 10^4	2.79 × 10^4
Atom Economy (%)	31.96	49.75	21.78	31.96	33.59
E-factor	1.78	0.99	2.80	1.78	1.67
H^{Total} (J/y)	−7.10 × 10^{11}	1.53 × 10^{12}	−2.42 × 10^{11}	−7.10 × 10^{11}	−5.05 × 10^{11}
SEC (kJ/t)	−4.47 × 10^7	6.15 × 10^7	−2.23 × 10^7	−4.47 × 10^7	−3.02 × 10^7

Comparison and Analysis of Different Scenarios

From Table 11, it can be observed that all scenarios have a positive value of EB^{Total}, which indicates the contribution of all selected conversion pathway toward generating a negative impact on the environment. In Scenario 1, the selected conversion pathways offer the most promising economic potential even though the process is not environmentally friendly. Moreover, it is observed that values from all parameters in Scenario 1 are the same as Scenario 4, which proves that when GP^{Total} is maximised, H^{Total} will be at its minimum. The energy required for conversion pathway for Scenarios 1 and 4 is the lowest where heat can be provided by the system itself. Hence, operating cost is lower, which indirectly brings a higher amount of GP^{Total}. Although the conversion pathway in Scenarios 1 and 4 is not the most environmentally friendly process, the selected pathway has achieved the green manufacturing definition of minimisation of the energy requirement.

The highest production of the desired product is determined in Scenario 2. When the yield of C_{11} is at a maximum, EB^{Total} is highest among all scenarios, which indicates that the conversion pathways bring the highest negative impact on the environment. When the production of the desired product is maximised, by-product production is increased as well. Although the desired product is at highest production, GP^{Total} is not the highest. This can be explained by the fact that, for this scenario, high operating costs and capital costs may be required for maximum C_{11} production. This can be further justified by the fact that, among all scenarios, Scenario 2 is the only scenario in which a positive value of H^{Total} is obtained. In addition, it can be observed that the atom economy for Scenario 2 is the highest among all scenarios. This signifies that the amount of material converted to the desired product is the highest for Scenario 2. Furthermore, the e-factor for Scenario 2 is the lowest among all scenarios, which shows that the mass of amount waste produced per unit mass of product is the lowest. Through the interpretation and analysis of these indicators, although the selected pathway is the least environmentally friendly among all scenarios, it is clear that the conversion pathway selected for Scenario 2 complies with the green manufacturing definition of minimisation of waste produced.

In Scenario 3, the chosen conversion pathway offers the lowest EB^{Total}. However, both C_{11} production and GP^{Total} are the lowest among all scenarios when the process is designed to have the lowest negative impact on the environment. This proves that to have a highly environmentally friendly process, the production of the designed product needs to be reduced, while additional expenditure is required. This explains the current dilemma faced by most of the manufacturing industries that the environmentally friendly process normally comes with a reduction in the gross profit gained. Although the designed product fulfils the green manufacturing needs of the minimisation of the negative impact on the environment, this might not be favoured by the stakeholders of the integrated biorefinery if their only interest is financial gain. Hence, a compromised solution is needed, which is shown in Scenario 5.

In Scenario 5, it is observed that the optimised solution, by considering all target objectives, has each individual objective fall within the upper and lower limits. This represents a trade-off between all individual objectives from Scenarios 1 to 4. While none of the individual objectives are the best if compared to other scenarios, the selected conversion pathway for Scenario 5 is the optimum if all objectives are seen as important. None of the objectives are over-improved while neglecting the other objectives. For integrated biorefinery stakeholders who are concerned with the negative impact that might be caused by the conversion pathways but at the same time not willing to sacrifice too much of their financial interest, Scenario 5 is the balanced solution that is suitable to them.

4. Conclusions

In this work, clear and consistent generalised definitions and quantification of green manufacturing are developed for product and process design. A two-stage optimisation approach is presented to design a green chemical product and integrated biorefinery process that meets both green manufacturing and customer needs. A case study in designing biobased fuel that incorporates green manufacturing with customer needs is solved to demonstrate the developed methodology. In the first stage, product properties are identified to quantify green products based on the developed green manufacturing definitions.

By incorporating the multi-objective optimisation approach into the methodology, the optimal green chemical product design can be obtained where all properties of interest are optimised simultaneously without any bias. In the second stage, conversion pathways are evaluated based on the targeted parameters considered from developed green manufacturing definitions of process performance. Five different scenarios are solved to represent different design considerations and philosophies of the integrated biorefinery stakeholders. While it has proven challenging to target the highest gross profit while aiming for a minimum negative impact on the environment as these two objectives are contradicting given the technological advancement today, a compromised solution can be obtained by solving the problem as a multi-objective optimisation problem. In addition, there might be situations where the integrated biorefinery is located in an area with a limited supply of resources and utilities. In this case, it is preferable to design conversion pathways that are able to provide excess heat that can be used as heat/power generation for an auxiliary system. In short, the proposed methodology is able to determine green conversion pathways that accommodate different production objectives on green manufacturing and customer needs.

As society progresses and advances, the definition of green manufacturing will be different. For future work, in order to enhance the comprehensiveness and accuracy of green definitions, more properties and parameters that quantify both green products and process can be considered. In addition, conversion pathways that convert biomass into a final product considered in the superstructure can be updated by including newly developed biomass conversion technologies when determining the optimal pathways. Furthermore, the completeness of the methodology of designing green processes can be further improved by considering the utilities required for the conversion pathway based on the desired operating conditions. For instance, the process will be highly energy-intensive when a vacuum or very high pressure is required for the process. Lastly, to further reduce the gap between outcomes from a developed approach and the actual process, any additional feedstock and side reactions required can be taken into consideration.

Supplementary Materials: The following are available online at https://www.mdpi.com/article/10.3390/pr9091569/s1, Table S1: List of Conversion Pathway to produce Alkane from Biomass, Table S2: Price List for Product and Feedstock, Table S3: List of Capital and Operating Cost for Conversion Pathways, Table S4: List of Component with Heat of Formation and Potency Factor, Figure S1: Conversion Pathway for Maximum Economic Performance, Minimum Environmental Burden and Minimum Energy Consumption, Figure S2: Conversion Pathway for Maximum Product C11 Yield, Figure S3: Conversion Pathway for Multi-objective Optimisation.

Author Contributions: Conceptualization, J.W.Y.L. and L.Y.N.; Methodology, J.W.Y.L. and L.Y.N.; Software, L.Y.N.; Validation, J.W.Y.L. and L.Y.N.; Formal analysis, J.W.Y.L.; Investigation, J.W.Y.L., L.Y.N. and V.A.; Resources, L.Y.N.; Data curation, J.W.Y.L.; Writing—original draft, J.W.Y.L.; Writing—review & editing, J.W.Y.L., L.Y.N., V.A., N.G.C. and D.K.S.N. Visualization, J.W.Y.L.; Supervision, L.Y.N., V.A., N.G.C. and D.K.S.N.; Project administration, L.Y.N. All authors have read and agreed to the published version of the manuscript.

Funding: This research received no external funding.

Institutional Review Board Statement: Not applicable.

Informed Consent Statement: Not applicable.

Conflicts of Interest: The authors declare no conflict of interest.

References

1. Paul, I.D.; Bhole, G.P.; Chaudhari, J.R. A Review on Green Manufacturing: It's Important, Methodology and its Application. *Procedia Mater. Sci.* **2014**, *6*, 1644–1649. [CrossRef]
2. Munthe, J.; Brorström-Lundén, E.; Rahmberg, M.; Posthuma, L.; Altenburger, R.; Brack, W.; Bunke, D.; Engelen, G.; Gawlik, B.M.; van Gils, J.; et al. An Expanded Conceptual Framework for Solution-Focused Management of Chemical Pollution in European Waters. *Environ. Sci. Eur.* **2017**, *29*, 1–16. [CrossRef] [PubMed]
3. Balaman, Ş.Y. *Decision-Making for Biomass-Based Production Chains*; Academic Press: London, UK, 2019; pp. 25–54.

4. United States Energy Information Administration (EIA). *Annual Energy Outlook 2019*; U.S. Energy Information Administration: Washington, DC, USA, 2019.
5. Ivanković, A. Review of 12 Principles of Green Chemistry in Practice. *Int. J. Sustain. Green Energy* **2017**, *6*, 39–48. [CrossRef]
6. Malaysia Department of Environment (DOE). *Environmental Quality (Clean Air) Regulations 2014*; Malaysia Department of Environment: Putrajaya, Malaysia, 2014.
7. European Commission. Industrial Emissions Directive. Available online: https://ec.europa.eu/environment/industry/stationary/ied/legislation.htm (accessed on 29 July 2021).
8. Morais, A.R.C.; Bogel-lukasik, R. Green Chemistry and the Biorefinery Concept. *Sustain. Chem. Process* **2013**, *1*, 1–3. [CrossRef]
9. Abdul-Rashid, S.H. The Impact of Sustainable Manufacturing Practices on Sustainability Performance Empirical Evidence from Malaysia. *Int. J. Oper. Prod. Man.* **2017**, *37*, 182–204. [CrossRef]
10. Moldavska, A.; Welo, T. The Concept of Sustainable Manufacturing and Its Definitions: A Content-analysis Based Literature Review. *J. Clean. Prod.* **2017**, *166*, 744–755. [CrossRef]
11. Mohanty, R.P.; Deshmukh, S.G. Managing green productivity: A Case Study. *Work. Study* **1999**, *48*, 165–169. [CrossRef]
12. Vallero, D.A. Hazardous Wastes. Hazardous Waste. In *Waste: A Handbook for Management*, 2nd ed.; Letcher, T.M., Vallero, D., Eds.; Academic Press: London, UK, 2011; pp. 585–630.
13. Ramachandran, P.A.; Shonnard, D.; Hesketh, R.; Fichana, D.; Stewart Slater, C.; Lindner, A.; Nguyen, N.; Engler, R. Green Engineering: Integration of Green Chemistry, Pollution Prevention, and Risk-Based Considerations. In *Handbook of Industrial Chemistry and Biotechnology*; Kent, J., Bommaraju, T., Barnicki, S., Eds.; Springer: Cham, Switzerland, 2012; pp. 1921–1994.
14. Snir, E.M. Liability as a Catalyst for Product Stewardship. *Prod. Oper. Manag.* **2009**, *10*, 190–206. [CrossRef]
15. Nukman, Y.; Farooqi, A.; Al-Sultan, O.; Alnasser, A.R.A.; Bhuiyan, M.S.H. A Strategic Development of Green Manufacturing Index (GMI) Topology Concerning the Environmental Impacts. *Procedia Eng.* **2017**, *184*, 370–380. [CrossRef]
16. Ferguson, M.E.; Toktay, L.B. The Effect of Competition on Recovery Strategies. *Prod. Oper. Manag.* **2006**, *15*, 3, 351–368. [CrossRef]
17. Wong, C.W.Y.; Lai, K.; Shang, K.; Lu, C.; Leung, T.K.P. Green Operations and the Moderating Role of Environmental Management Capability of Suppliers on Manufacturing Firm performance. *Int. J. Prod. Econ.* **2012**, *140*, 283–294. [CrossRef]
18. Deif, A.M. A System Model for Green Manufacturing. *J. Clean. Prod.* **2011**, *19*, 1553–1559. [CrossRef]
19. Maruthi, G.D.; Rashmi, R. Green Manufacturing: It's Tools and Techniques that can be Implemented in Manufacturing Sectors. *Mater. Today Proc.* **2015**, *2*, 3350–3355. [CrossRef]
20. Li, K.; Lin, B. Impact of Energy Conservation Policies on the Green Productivity in China's Manufacturing Sector: Evidence from a three-stage DEA model. *Appl. Energy* **2016**, *168*, 351–363. [CrossRef]
21. Wang, K.; Lu, B.; Wei, Y. China's regional energy and environmental efficiency: A Range-Adjusted Measure based analysis. *Appl. Energy* **2013**, *112*, 1403–1415. [CrossRef]
22. Meng, Y.; Yang, Y.; Chung, H.; Lee, P.; Shao, C. Enhancing Sustainability and Energy Efficiency in Smart Factories: A Review. *Sustainability* **2018**, *10*, 4779. [CrossRef]
23. Pakkanen, J.; Huhtala, P.; Juuti, T.; Lehtonen, T. Achieving Benefits with Design Reuse in Manufacturing Industry. *Procedia CIRP* **2016**, *50*, 8–13. [CrossRef]
24. van Berkel, R. Quantifying Sustainability Benefits of Industrial Symbioses. *J. Ind. Ecol.* **2010**, *14*, 371–373. [CrossRef]
25. Bauer, T.; Brissaud, D.; Zwolinski, P. Design for High Added-Value End-of-Life Strategies. In *Sustainable Manufacturing. Sustainable Production, Life Cycle Engineering and Management*; Stark, R., Seliger, G., Bonvoisin, J., Eds.; Springer: Cham, Switzerland, 2017; pp. 113–128.
26. Zhang, X.; Ming, X.; Liu, Z.; Qu, Y.; Yin, D. General Reference Model and Overall Frameworks for Green Manufacturing. *J. Clean. Prod.* **2019**, *237*, 117757. [CrossRef]
27. De Wilde, B.; Mortier, N.; Verstichel, S.; Briassoulis, D.; Babou, M.; Mistriotis, A.; Hiskakis, M. Report on Current Relevant Biodegradation and Ecotoxicity Standards. *Knowl. Based Bio-Based Prod. Pre-Stand.* **2013**, *1*, 14–20.
28. Hostrup, M.; Harper, P.M.; Gani, R. Design of environmentally benign processes: Integration of solvent design and separation process synthesis. *Comput. Chem. Eng.* **1999**, *23*, 1395–1414. [CrossRef]
29. Eden, M.R.; Jørgensen, S.B.; Gani, R.; El-Halwagi, M.M. A novel framework for simultaneous separation process and product design. *Chem. Eng. Process Process Intensif.* **2004**, *43*, 595–608. [CrossRef]
30. Papadopoulos, A.I.; Linke, P. Efficient integration of optimal solvent and process design using molecular clustering. *Chem. Eng. Sci.* **2006**, *61*, 6316–6336. [CrossRef]
31. Hechinger, M.; Voll, A.; Marquardt, W. Towards an integrated design of biofuels and their production pathways. *Comput. Chem. Eng.* **2010**, *34*, 12, 1909–1918. [CrossRef]
32. Ng, L.Y.; Andiappan, V.; Chemmangattuvalappil, N.G.; Ng, D.K.S. Novel Methodology for the Synthesis of Optimal Biochemicals in Integrated Biorefineries via Inverse Design Techniques. *Ind. Eng. Chem. Res.* **2015**, *54*, 5722–5735. [CrossRef]
33. Ng, L.Y.; Andiappan, V.; Chemmangattuvalappil, N.G.; Ng, D.K.S. A systematic methodology for optimal mixture design in an integrated biorefinery. *Comput. Chem. Eng.* **2015**, *81*, 288–309. [CrossRef]
34. Meramo-Hurtado, S.-I.; González-Delgado, Á.-D. Biorefinery synthesis and design using sustainability parameters and hierarchical/3D multi-objective optimization. *J. Clean Prod.* **2019**, *240*, 118134. [CrossRef]
35. Lai, Y.Y.; Yik, K.C.H.; Hau, H.P.; Chow, C.P.; Chemmangattuvalappil, N.G.; Ng, L.Y. Enterprise Decision-making Framework for Chemical Product Design in Integrated Biorefineries. *Process Integr. Optim. Sustain.* **2019**, *3*, 25–42. [CrossRef]

36. Restrepo-Flórez, J.M.; Maravelias, C.T. Advanced fuels from ethanol—A superstructure optimization approach. *Energy Environ. Sci.* **2021**, *14*, 493–506. [CrossRef]
37. Tey, T.O.; Chen, S.; Cheong, Z.X.; Choong, A.S.X.; Ng, L.Y.; Chemmangattuvalappil, N.G. Synthesis of a sustainable integrated biorefinery to product value-added chemicals from palm-based biomass via mathematical optimization. *Sustain. Prod. Consum.* **2021**, *26*, 288–315. [CrossRef]
38. Sangster, J. Octanol-Water Partition Coefficients of Simple Organic Compounds. *J. Phys. Chem. Ref. Data* **1989**, *18*, 1111–1227. [CrossRef]
39. Albahri, T.A. Flammability characteristics of pure hydrocarbons. *Chem. Eng. Sci.* **2003**, *58*, 3629–3641. [CrossRef]
40. Yunus, N.A.B. Systematic Methodology for Design of Tailor-Made Blended Products: Fuels and Other Blended Products. Ph.D. Thesis, Technical University of Denmark, Department of Chemical and Biochemical Engineering, Lyngby, Denmark, 2014.
41. Schulze, P. *Measure of Environmental Performance and Ecosystem Condition*; National Academy of Engineering: Washington, DC, USA, 1999.
42. Hall, L.H.; Kier, L.B.; Phipps, G. Structure-activity relationship studies on the toxicities of benzene derivatives: I. An additivity model. *Environ. Chem.* **1984**, *3*, 355–365. [CrossRef]
43. Singh, A.; Lou, H.H.; Yaws, C.L.; Hopper, J.R.; Pike, R.W. Environmental impact assessment of different design schemes of an industrial ecosystem. *Resour. Conserv. Recy.* **2007**, *51*, 294–313. [CrossRef]
44. Anastas, P.T.; Warner, J.C. *Green Chemistry: Theory and Practice*; Oxford University Press: New York, NY, USA, 1998.
45. Wang, W.; Lü, J.; Zhang, L.; Li, Z. Real atom economy and its application for evaluation the green degree of a process. *Front. Chem. Sci. Eng.* **2011**, *5*, 349–354. [CrossRef]
46. Tieves, F.; Tonin, F.; Fernández-Fueyo, E.; Robbins, J.M.; Bommarius, B.; Bommarius, A.S.; Alcalde, M.; Hollmann, F. Energysing the E-factor: The E^{+}-factor. *Tetrahedron* **2019**, *75*, 1311–1314. [CrossRef]
47. Lawrence, A.; Thollander, P.; Andrei, M.; Karlsson, M. Specific Energy Consumption/Use (SEC) in Energy Management for Improving Energy Efficiency in Industry: Meaning, Usage and Differences. *Energies* **2019**, *12*, 247. [CrossRef]
48. Siegel, R.; Antony, J.; Garza-Reyes, J.A.; Cherrafi, A.; Lameijer, B. Integrated green lean approach and sustainability for SMEs: From literature review to a conceptual framework. *J. Clean Prod.* **2019**, *240*, 118205. [CrossRef]
49. Minnesota Pollution Control Agency (MPCA). General Processes or Units Emission Calculations. Available online: https://www.pca.state.mn.us/air/general-processes-or-units-emission-calculations (accessed on 18 January 2020).
50. Safe Work Australia (SWA). Lost Time Injury Frequency Rates (LTIFR). Available online: https://www.safeworkaustralia.gov.au/statistics-and-research/lost-time-injury-frequency-rates-ltifr (accessed on 12 January 2020).
51. Corporate Finance Institute (CFI) Education Inc. Depreciation Methods. Available online: https://corporatefinanceinstitute.com/resources/knowledge/accounting/types-depreciation-methods/ (accessed on 26 January 2020).
52. Chemmangattuvalappil, N.G.; Solvason, C.C.; Bommareddy, S.; Eden, M.R. Reverse problem formulation approach to molecular design using property operators based on signature descriptors. *Comput. Chem. Eng.* **2010**, *34*, 2062–2071. [CrossRef]
53. Bommareddy, S.; Chemmangattuvalappil, N.G.; Solvason, C.C.; Eden, M.R. Simultaneous solution of process and molecular design problems using an algebraic approach. *Comput. Chem. Eng.* **2010**, *34*, 1481–1486. [CrossRef]
54. American Petroleum Institute (API). *Alcohols and Ethers: A Technical Assessment of Their Application as Fuels and Fuel Components*; American Petroleum Institute: Washington, DC, USA, 1988.
55. Hukkerikar, A.S.; Sarup, B.; Ten Kate, A.; Abildskov, J.; Sin, G.; Gani, R. Group-contribution + (GC +) based Estimation of Properties of Pure Components: Improved Property Estimation and Uncertainty Analysis. *Fluid Phase Equilib.* **2012**, *321*, 25–43. [CrossRef]
56. Hukkerikar, A.S.; Kalakul, S.; Sarup, B.; Young, D.M.; Sin, G.; Gani, R. Estimation of Environment-related Properties of Chemicals for Design of Sustainable Processes: Development of Group-contribution+ (GC +) Property Models and Uncertainty Analysis. *J. Chem. Inf. Model.* **2012**, *52*, 2823–2839. [CrossRef]
57. Conte, E.; Martinho, A.; Matos, H.A.; Gani, R. Combined Group-contribution and Atom Connectivity Index-based Methods for Estimation of Surface Tension and Viscosity. *Ind. Eng. Chem. Res.* **2008**, *47*, 7940–7954. [CrossRef]
58. Qiao, Y.; Ma, Y.; Huo, Y.; Ma, P.; Xia, S. A Group Contribution Method to Estimate the Densities of Ionic Liquids. *J. Chem. Thermodyn.* **2010**, *42*, 852–855. [CrossRef]
59. Yunus, N.A.; Gernaey, K.V.; Woodley, J.M.; Gani, R. A Systematic Methodology for Design of Tailor-made Blended Products. *Comput. Chem. Eng.* **2014**, *66*, 201–213. [CrossRef]
60. Ng, L.Y.; Chemmangattuvalappil, N.G.; Ng, D.K.S. Robust chemical product design via fuzzy optimization approach. *Comput. Aided Chem. Eng.* **2014**, *34*, 387–392.
61. Sukiran, M.A. Pyrolysis of Empty oil Palm Fruit Bunches Using the Quartz Fluidized-Fixed Bed Reactor. Master's Thesis, Universiti Malaya, Department of Chemistry, Kuala Lumpur, Malaysia, 2008.
62. Andiappan, V.; Ko, A.S.Y.; Lau, V.W.S.; Ng, L.Y.; Ng, R.T.L.; Chemmangattuvalappil, N.G.; Ng, D.K.S. Synthesis of Sustainable Integrated Biorefinery via Reaction Pathway Synthesis: Economic, Incremental Enviromental Burden and Energy Assessment with Multiobjective Optimization. *AIChE J.* **2015**, *61*, 132–146. [CrossRef]

processes MDPI

Article

Machine Learning for Ionic Liquid Toxicity Prediction

Zihao Wang [1], Zhen Song [2] and Teng Zhou [1,2,*]

1 Process Systems Engineering, Max Planck Institute for Dynamics of Complex Technical Systems, Sandtorstr. 1, D-39106 Magdeburg, Germany; zwang@mpi-magdeburg.mpg.de
2 Process Systems Engineering, Otto-von-Guericke University Magdeburg, Universitätsplatz 2, D-39106 Magdeburg, Germany; songz@mpi-magdeburg.mpg.de
* Correspondence: zhout@mpi-magdeburg.mpg.de

Abstract: In addition to proper physicochemical properties, low toxicity is also desirable when seeking suitable ionic liquids (ILs) for specific applications. In this context, machine learning (ML) models were developed to predict the IL toxicity in leukemia rat cell line (IPC-81) based on an extended experimental dataset. Following a systematic procedure including framework construction, hyper-parameter optimization, model training, and evaluation, the feedforward neural network (FNN) and support vector machine (SVM) algorithms were adopted to predict the toxicity of ILs directly from their molecular structures. Based on the ML structures optimized by the five-fold cross validation, two ML models were established and evaluated using IL structural descriptors as inputs. It was observed that both models exhibited high predictive accuracy, with the SVM model observed to be slightly better than the FNN model. For the SVM model, the determination coefficients were 0.9289 and 0.9202 for the training and test sets, respectively. The satisfactory predictive performance and generalization ability make our models useful for the computer-aided molecular design (CAMD) of environmentally friendly ILs.

Keywords: ionic liquid; toxicity; machine learning; neural network; support vector machine

Citation: Zihao Wang, Zhen Song and Teng Zhou Machine Learning for Ionic Liquid Toxicity Prediction. *Processes* **2020**, *9*, 65. https://doi.org/10.3390/pr9010065

Received: 10 December 2020
Accepted: 28 December 2020
Published: 30 December 2020

1. Introduction

Ionic liquids (ILs) are known as molten salts, consisting solely of organic cation and inorganic or organic anion. ILs present several unique and benign characteristics, such as low melting point, low volatility, and high thermal stability. For these reasons, ILs have attracted extensive attention in both academia and industry as alternatives to traditional organic solvents. In fact, they have been used in many chemical processes [1–4] to meet specific requirements while improving the process efficiency. In spite of their superior properties and large application prospects, many ILs present potential negative effects on the ecosystem, which could also impact human health. Therefore, toxicity is a significant factor that should be considered in IL selection, and therefore the toxicological assessment of ILs is essential for the development of a sustainable IL-based chemical process.

In order to find ILs showing desired low toxicity, experimental measurement is the most direct and effective way. However, the number of new IL structures is increasing rapidly due to numerous feasible cation–anion combinations. This makes the experimental measurement a time-consuming, resource-intensive, and even impractical method. In contrast, mathematical modeling methods based on existing IL structures and toxicity data become efficient substitutes. Researchers have built mathematical models for predicting various properties of ILs, including electrical conductivity [5], thermal decomposition temperature [6], critical properties [7], and water solubility [8]. These models are regressed from experimental property data and can provide satisfying predictions on the properties of new ILs that are not used in model training. Such well-established property models have been widely used in the computer-aided molecular design (CAMD) of IL solvents [9–14], where optimal ILs possessing desirable properties are identified based on the property models. Unfortunately, up to now, mathematical models for the toxicity properties of ILs have been explored more scarcely in comparison to other IL properties.

In literature, the mortality of leukemia rat cell line (IPC-81) is frequently used to quantitatively indicate the toxicity of ILs [15,16]. A few researchers have developed models to predict the toxicity of ILs against IPC-81 from IL structures. Based on a dataset of 173 experimental samples, Yan et al. [17] established a structure–activity relationship model for an IL to predict its toxicity against IPC-81 using multiple linear regression (MLR) and topological descriptors. The mean absolute error (MAE) was 0.226. Later, Sosnowska et al. [18] constructed a larger IPC-81 dataset containing 304 experimental samples, and on this basis, they used structure-related descriptors to develop another MLR model where the MAE was 0.3779. Due to the utilization of a larger database, this model, in principle, shows improved applicability. Based on this updated dataset, Wu et al. [19] recently reported another MLR model using IL fingerprint descriptors, which resulted in an improved predictive accuracy (MAE = 0.34). Despite the effectiveness of MLR, there is still a large potential for the further improvement of model accuracy.

As proven by many studies [20–25], machine learning (ML) is an efficient and promising approach for building quantitative structure–property relationship (QSPR) models to predict various properties for chemical compounds. Until now, ML techniques have obtained wide applications and great successes in predicting IL properties, including melting point [26], CO_2 solubility [27], viscosity [28], etc. Despite its popularity, there is still a lack of accurate ML models for predicting IL toxicity.

Considering the above aspects, we herein explore ML models to predict the toxicity of ILs against IPC-81 by applying a systematic procedure covering model framework construction, hyper-parameter optimization, training, and evaluation. Since ML is a data-driven method, an extended experimental dataset is always preferred for building a reliable ML model. In this work, we first collect experimental toxicity data for 355 ILs. Two different ML models are then developed, one using the feedforward neural network (FNN) algorithm and the other using the support vector machine (SVM) algorithm. The structures of ML models, characterized by their hyper-parameters, are optimized with the five-fold cross validation to improve the model robustness. The corresponding models are then established with the training dataset, followed by an evaluation with an external test set to measure their generalization abilities.

2. Experimental Data

A dataset of the toxicity of 355 ILs against the IPC-81 was collected from the literature [18,29]. The toxicity was evaluated in terms of the logarithm of the half maximal effective concentration ($logEC_{50}$), and the IL structures were depicted by SMILES (simplified molecular-input line-entry system) strings obtained from PubChem [30]. Therefore, the $logEC_{50}$ dataset for the 355 ILs was used for predictive modeling. The detailed data are tabulated in Table S1 (Supporting Information) accompanied with the corresponding IL full name and SMILES strings.

To translate the text-form IL structure (i.e., SMILES string) into a ML-acceptable type (i.e., numerical values), the IL descriptors were created by a feature extraction algorithm [31] to characterize the IL structure. Specifically, an IL was treated as a molecule in which the cation and anion are connected with each other, and its SMILES string was parsed by the RDKit cheminformatics tool [32] to obtain detailed structural and chemical information. A set of substructures (subgroups) were defined as descriptors, and their appearance frequencies in the IL molecule were used as model inputs to characterize the structure of ILs. After analyzing the structures of the 355 ILs, a total of 42 structural descriptors were resulted, including 9 cation descriptors, 9 anion descriptors, and 24 general descriptors. These structural descriptors and their occurrence frequencies in each IL are provided in Tables S2 and S3 (Supporting Information), respectively.

To develop the ML models for IL toxicity prediction, a systematic procedure covering framework construction (step A), hyper-parameter optimization (step B), model establishment (step C), and model evaluation (step D) was deployed (see Figure 1). The full dataset was randomly divided into two parts, a training set (80% of the dataset, 284 samples) to develop the ML model and a test set (20% of the dataset, 71 samples) to evaluate the predictive accuracy of the developed model on unseen data.

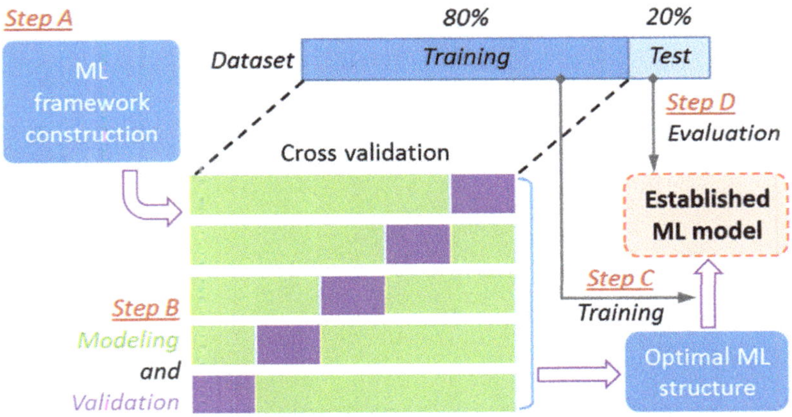

Figure 1. Systematic procedure for machine learning (ML)-assisted ionic liquid (IL) toxicity predictive modeling.

To obtain a ML model with high robustness, cross validation should be employed to optimize the structural parameters (i.e., hyper-parameters). Herein, the five-fold cross validation strategy was adopted to determine the optimal hyper-parameters (Step B in Figure 1). First, the training set was equally divided into five parts. Then, for each combination of different hyper-parameters, the modeling process was performed five times based on the selected four parts (marked green in Figure 1), and the validation process was also carried out five times on the remaining part (marked yellow). Optimal hyper-parameters could be determined from the variation of the averaged error of the validation results. Finally, for the fixed hyper-parameters, the ML models were developed using the entire training set (Step C) and evaluated with the test set (Step D in Figure 1). The ILs in the randomly divided training and test sets are provided in Table S1 (Supporting Information).

3. ML Modeling

3.1. FNN Modeling

Artificial neural networks are probably the most widespread ML technique in mathematical modeling, image recognition, process control, etc. due to their flexible and configurable structures and connections. Therein, a feedforward neural network (FNN) is a common type, which presents the regular layer structure and one-directional transmission of information. Due to its simpler structure and less adjustable parameters, a FNN is computationally efficient in learning and making predictions, compared to other neural networks, such as the convolutional neural network and recurrent neural network. Moreover, it is more suitable for handling one-dimensional time-independent input data, as is the case in this work.

In this section, a two-hidden-layer FNN (see Figure 2) is constructed to correlate the $\log EC_{50}$ values with the IL structures. The appearance frequencies of the 42 subgroups are loaded into the neurons in the input layer, and after being processed by two adjacent hidden layers, the predicted $\log EC_{50}$ is given by the output layer. The activation functions, namely "sigmoid" and "softplus", are assigned for the two hidden layers to achieve the non-linear data transformations, enabling the FNN to model complex mathematical relationships.

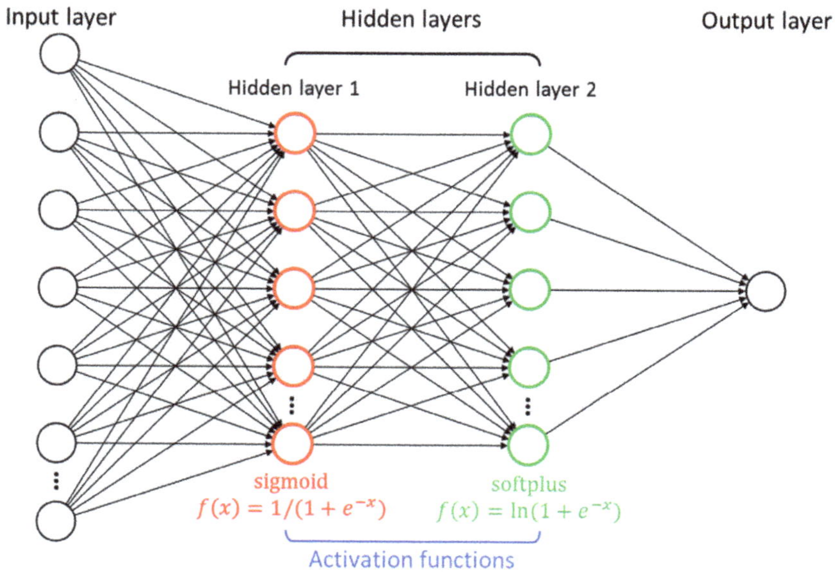

Figure 2. Structure of the two-hidden-layer feedforward neural network (FNN) with activation functions.

The FNN framework was implemented with the PyTorch [33] toolkit in Python. The root mean square error (RMSE) was employed as the loss function to monitor the model performance. Moreover, Adam optimizer [34] was used to adjust the learning rate and model parameters, coupling with the back-propagation algorithm [35] that computes the gradient of the loss function and guides the optimization process.

Since the number of hidden layers was pre-defined, the number of neurons in each layer is an important hyper-parameter in the FNN structure that determines the model performance. The number of neurons in the input layer depends on the dimension of the used descriptors (42 in this work), and the neuron number of the output layer equals one, corresponding to the scalar $\log EC_{50}$. In this case, the number of neurons in the two hidden layers needed to be optimized by the five-fold cross validation. As can be seen in Figure 3, the lowest average RMSE (0.5696) of the five-round independent validations was obtained when 10 and 6 neurons were used in hidden layer 1 and 2, respectively. This means that this FNN structure was the most robust and promising one in predicting the toxicity of ILs against IPC-81.

After the numbers of neurons in the two hidden layers were determined, the FNN model was established with all of the training samples and then evaluated with the test samples. Figure 4 plots the predicted $\log EC_{50}$ of samples in the training and test sets against the corresponding experimental $\log EC_{50}$. In addition to the RMSE, mean absolute error (MAE) and coefficient of determination (R^2) were also used to quantify the performance of the FNN model. The RMSE, MAE, and R^2 were 0.2906, 0.2111, and 0.9227 for the training set, and 0.3732, 0.3028, and 0.8917 for the test set, respectively. The satisfying and close metrics obtained for the training and test sets indicate a generally good predictive performance of the obtained model for the prediction of the toxicity of ILs against IPC-81. For the established FNN model, prediction results are provided in Table S1, and weight and bias values in each layer are summarized in Table S4 (Supporting Information).

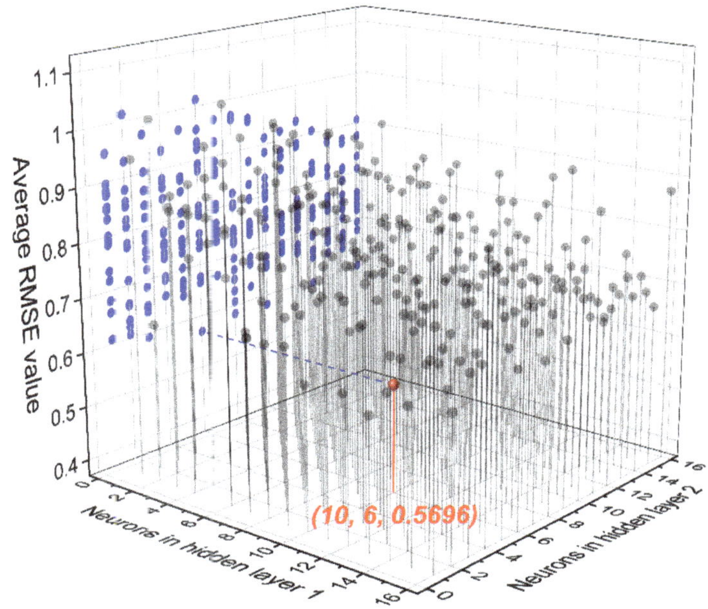

Figure 3. Average root mean square error (RMSE) values in the five-fold cross validation for the FNN model.

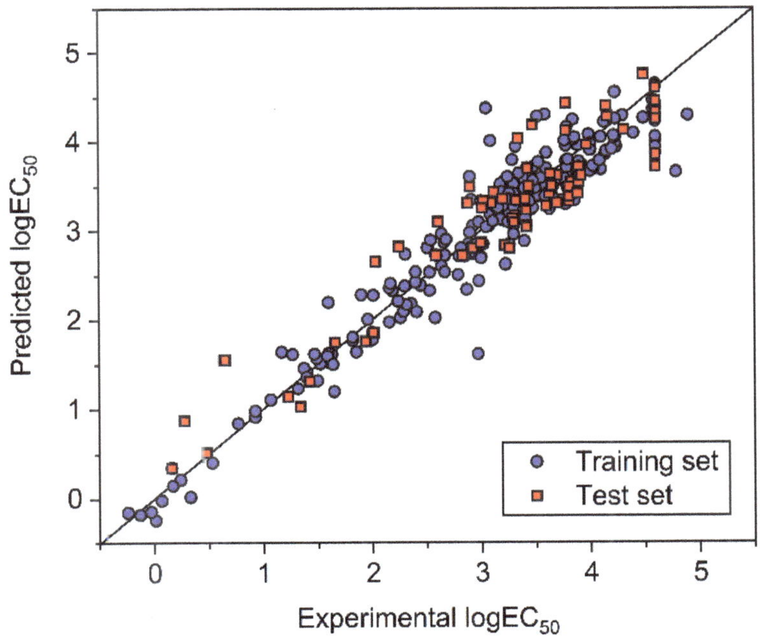

Figure 4. Comparison between experimental and FNN-predicted $logEC_{50}$ for ILs in the training and test datasets.

3.2. SVM Modeling

A support vector machine (SVM) is another popular ML algorithm in data analysis and predictive modeling. In terms of the regression task, the SVM algorithm is able to fit the training data by creating hyperplanes in the high-dimensional descriptor space. An accurate SVM model makes samples stay as close as possible to these hyperplanes.

The SVM framework was implemented with the scikit-learn [36] toolkit in Python. Gaussian radial basis function was adopted as the kernel function to map input descriptors into a high-dimensional space for complex nonlinear modeling. The RMSE was used again as the loss function to measure the model's predictive capability. Two important hyper-parameters in the SVM algorithm, regularization parameter C and tolerance ε, were optimized by the five-fold cross validation. As shown in Figure 5, the lowest average RMSE value (0.4591) of the five independent validations was achieved by the SVM algorithm when the regularization parameter C and tolerance ε were 30 and 0.11, respectively.

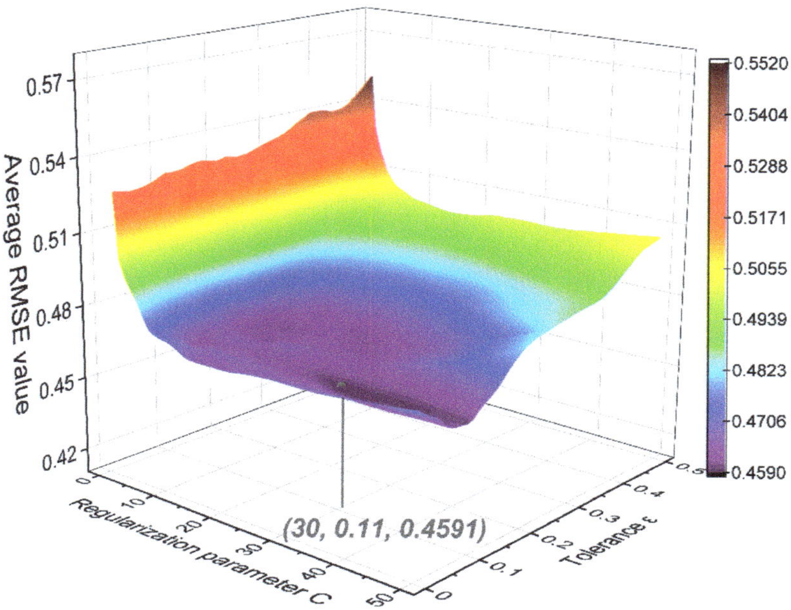

Figure 5. Average RMSE values in the five-fold cross validation for the support vector machine (SVM) model.

Based on the optimized hyper-parameters highlighted in Figure 5, the final SVM model was developed with all the training samples and then evaluated with the test samples. The comparison between experimental and SVM-predicted $logEC_{50}$ is shown in Figure 6. Except for a few outliers, most of the samples exhibited low deviations. The determined RMSE, MAE, and R^2 were 0.2787, 0.1762, and 0.9289 for the training set, and 0.3204, 0.2628, and 0.9202 for the test set, respectively. These statistical indicators demonstrate a good predictive capability of the SVM model. For the established SVM model, prediction results are also provided in Table S1, and model parameters are summarized in Tables S5 and S6 (Supporting Information).

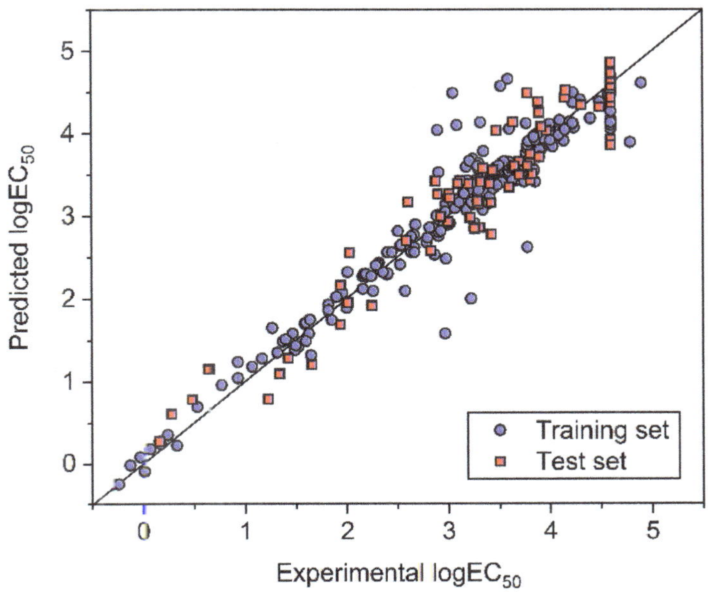

Figure 6. Comparison between experimental and SVM-predicted $logEC_{50}$ for ILs in the training and test datasets.

4. Model Comparison

Both the FNN and SVM models showed good predictive performances, as proven by the statistical results. Sorted in ascending order, the absolute errors between ML-predicted and experimental $logEC_{50}$ for the 71 ILs in the test dataset are shown in Figure 7. As indicated, the SVM model exhibited smaller absolute errors, revealing its higher predictive accuracy for IL toxicity.

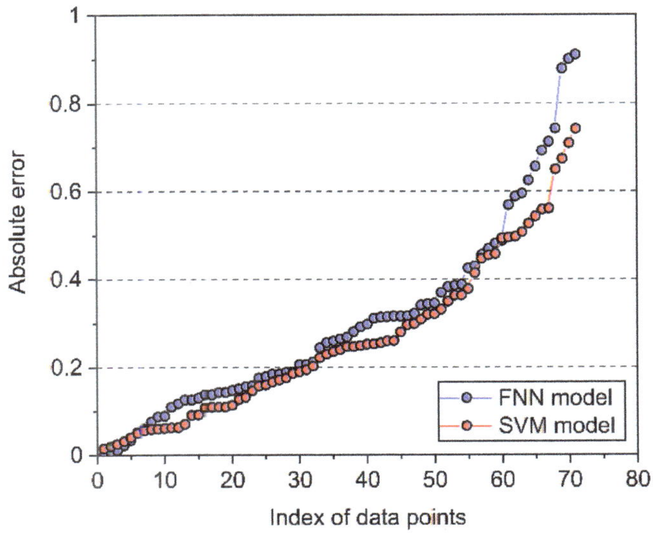

Figure 7. Absolute errors between model-predicted and experimental $logEC_{50}$ for the 71 ILs in the test dataset.

The two ML models established in this work were compared with the reported models developed by Sosnowska et al. [18] and Wu et al. [19]. As shown in Table 1, the FNN and SVM models presented much lower RMSE and MAE as well as a much higher R^2 value than the two previous models, indicating their higher predictive accuracy. Moreover, they have improved applicability due to the extended dataset used in the model development. Comparing the two models developed in this work, it was found that the SVM model presented relatively lower RMSE and MAE as well as a higher R^2 than the FNN model. Therefore, the SVM model can generally provide more accurate predictions on the toxicity of ILs against IPC-81, which confirms the conclusion obtained from Figure 7.

Table 1. Statistical comparisons of different models in predicting the toxicity of ILs. MLR: multiple linear regression.

Model	Sample Number	RMSE	MAE	R^2
MLR model (Sosnowska et al. [18])	304	0.51	-	0.77
MLR model (Wu et al. [19])	304	0.43	0.34	-
FNN model (this work)	355	0.3089	0.2294	0.9157
SVM model (this work)	355	0.2875	0.1935	0.9270

5. Conclusions

In this work, a dataset of 355 IL experimental $\log EC_{50}$ values was constructed to quantify the toxicity of ILs against IPC-81. Structural descriptors were generated with a feature extraction algorithm based on the SMILES strings of the ILs. Two ML frameworks (FNN and SVM) were built, and their structural parameters were optimized with five-fold cross validation. After determining the best set of structural parameters, the ML models were regressed using the training dataset, and the established models were evaluated with the test data. It was observed that the SVM model performed slightly better than the FNN model. However, compared with previously reported models, both models presented better predictive performance and improved applicability. Considering the satisfactory predictions on IL toxicity, the established ML models can be incorporated into computer-aided molecular design (CAMD) frameworks in order to identify suitable ILs that show low toxicity while meeting other requirements defined by the specific applications.

Supplementary Materials: The following are available online at https://www.mdpi.com/2227-9717/9/1/65/s1: Table S1: Details of the employed dataset and ML prediction results. Table S2: Structural descriptors of ILs for ML model development. Table S3: Occurrence frequencies of the descriptors in IL molecules. Table S4: Weight and bias values of the FNN model. Table S5: Support vectors of the SVM model. Table S6: Fitting parameters of the SVM model.

Author Contributions: Conceptualization: T.Z.; Methodology: T.Z. and Z.W.; Investigation: Z.W.; Writing—Original Draft Preparation: Z.W.; Writing—Review and Editing: Z.S. and T.Z. All authors have read and agreed to the published version of the manuscript.

Funding: This research received no external funding.

Institutional Review Board Statement: Not applicable.

Informed Consent Statement: Not applicable.

Data Availability Statement: The data presented in this study are available on request from the corresponding author.

Acknowledgments: Zihao Wang acknowledges the support from the International Max Planck Research School (IMPRS) for Advanced Methods in Process and Systems Engineering, Magdeburg, Germany.

Conflicts of Interest: The authors declare no conflict of interest.

References

1. Watanabe, M.; Thomas, M.L.; Zhang, S.; Ueno, K.; Yasuda, T.; Dokko, K. Application of ionic liquids to energy storage and conversion materials and devices. *Chem. Rev.* **2017**, *117*, 7190–7239. [CrossRef] [PubMed]
2. Zhang, X.; Zhang, X.; Dong, H.; Zhao, Z.; Zhang, S.; Huang, Y. Carbon capture with ionic liquids: Overview and progress. *Energy Environ. Sci.* **2012**, *5*, 6668–6681. [CrossRef]
3. Song, Z.; Hu, X.; Zhou, Y.; Zhou, T.; Qi, Z.; Sundmacher, K. Rational design of double salt ionic liquids as extraction solvents: Separation of thiophene/n-octane as example. *AIChE J.* **2019**, *65*, e16625. [CrossRef]
4. Song, Z.; Zhou, T.; Qi, Z.; Sundmacher, K. Systematic method for screening ionic liquids as extraction solvents exemplified by an extractive desulfurization process. *ACS Sustain. Chem. Eng.* **2017**, *5*, 3382–3389. [CrossRef]
5. Gharagheizi, F.; Sattari, M.; Ilani-Kashkouli, P.; Mohammadi, A.H.; Ramjugernath, D.; Richon, D. A "non-linear" quantitative structure–property relationship for the prediction of electrical conductivity of ionic liquids. *Chem. Eng. Sci.* **2013**, *101*, 478–485. [CrossRef]
6. Gharagheizi, F.; Sattari, M.; Ilani-Kashkouli, P.; Mohammadi, A.H.; Ramjugernath, D.; Richon, D. Quantitative structure—property relationship for thermal decomposition temperature of ionic liquids. *Chem. Eng. Sci.* **2012**, *84*, 557–563. [CrossRef]
7. Huang, Y.; Dong, H.; Zhang, X.; Li, C.; Zhang, S. A new fragment contribution-corresponding states method for physicochemical properties prediction of ionic liquids. *AIChE J.* **2013**, *59*, 1348–1359. [CrossRef]
8. Zhou, T.; Chen, L.; Ye, Y.; Chen, L.; Qi, Z.; Freund, H.; Sundmacher, K. An overview of mutual solubility of ionic liquids and water predicted by COSMO-RS. *Ind. Eng. Chem. Res.* **2012**, *51*, 6256–6264. [CrossRef]
9. Song, Z.; Zhang, C.; Qi, Z.; Zhou, T.; Sundmacher, K. Computer-aided design of ionic liquids as solvents for extractive desulfurization. *AIChE J.* **2018**, *64*, 1013–1025. [CrossRef]
10. Chong, F.K.; Eljack, F.T.; Atilhan, M.; Foo, D.C.; Chemmangattuvalappil, N.G. A systematic visual methodology to design ionic liquids and ionic liquid mixtures: Green solvent alternative for carbon capture. *Comput. Chem. Eng.* **2016**, *91*, 219–232. [CrossRef]
11. Chong, F.K.; Foo, D.C.Y.; Eljack, F.T.; Atilhan, M.; Chemmangattuvalappil, N.G. A systematic approach to design task-specific ionic liquids and their optimal operating conditions. *Mol. Syst. Des. Eng.* **2016**, *1*, 109–121. [CrossRef]
12. Zhou, T.; Shi, H.; Ding, X.; Zhou, Y. Thermodynamic modeling and rational design of ionic liquids for pre-combustion carbon capture. *Chem. Eng. Sci.* **2021**, *229*, 116076. [CrossRef]
13. Zhou, T.; McBride, K.; Linke, S.; Song, Z.; Sundmacher, K. Computer-aided solvent selection and design for efficient chemical processes. *Curr. Opin. Chem. Eng.* **2020**, *27*, 35–44. [CrossRef]
14. Shi, H.; Zhang, X.; Zhou, T.; Sundmacher, K. Model-based optimal design of phase change ionic liquids for efficient thermal energy storage. *Green Energy Environ.* **2021**. [CrossRef]
15. Ranke, J.; Stolte, S.; Störmann, R.; Arning, J.; Jastorff, B. Design of sustainable chemical products the example of ionic liquids. *Chem. Rev.* **2007**, *107*, 2183–2206. [CrossRef]
16. Stolte, S.; Matzke, M.; Arning, J.; Böschen, A.; Pitner, W.R.; Welz-Biermann, U.; Jastorff, B.; Ranke, J. Effects of different head groups and functionalised side chains on the aquatic toxicity of ionic liquids. *Green Chem.* **2007**, *9*, 1170–1179. [CrossRef]
17. Yan, F.; Xia, S.; Wang, Q.; Ma, P. Predicting the toxicity of ionic liquids in leukemia rat cell line by the quantitative structure–activity relationship method using topological indexes. *Ind. Eng. Chem. Res.* **2012**, *51*, 13897–13901. [CrossRef]
18. Sosnowska, A.; Grzonkowska, M.; Puzyn, T. Global versus local QSAR models for predicting ionic liquids toxicity against IPC-81 leukemia rat cell line: The predictive ability. *J. Mol. Liq.* **2017**, *231*, 333–340. [CrossRef]
19. Wu, T.; Li, W.; Chen, M.; Zhou, Y.; Zhang, Q. Estimation of Ionic Liquids Toxicity against Leukemia Rat Cell Line IPC-81 based on the Empirical-like Models using Intuitive and Explainable Fingerprint Descriptors. *Mol. Inf.* **2020**, *39*, 2000102. [CrossRef]
20. Varnek, A.; Baskin, I. Machine learning methods for property prediction in chemoinformatics: Quo vadis? *J. Chem. Inf. Model.* **2012**, *52*, 1413–1437. [CrossRef]
21. Liu, Q.; Zhang, L.; Tang, K.; Liu, L.; Du, J.; Meng, Q.; Gani, R. A machine learning-based atom contribution method for the prediction of charge density profiles and solvent design. *AIChE J.* **2020**, e17110. [CrossRef]
22. Zhang, L.; Mao, H.; Liu, L.; Du, J.; Gani, R. A machine learning based computer-aided molecular design/screening methodology for fragrance molecules. *Comput. Chem. Eng.* **2018**, *115*, 295–308. [CrossRef]
23. Su, Y.; Wang, Z.; Jin, S.; Shen, W.; Ren, J.; Eden, M.R. An architecture of deep learning in QSPR modeling for the prediction of critical properties using molecular signatures. *AIChE J.* **2019**, *65*, e16678. [CrossRef]
24. Wang, Z.; Su, Y.; Shen, W.; Jin, S.; Clark, J.H.; Ren, J.; Zhang, X. Predictive deep learning models for environmental properties: The direct calculation of octanol-water partition coefficients from molecular graphs. *Green Chem.* **2019**, *21*, 4555–4565. [CrossRef]
25. Zhou, T.; Jhamb, S.; Liang, X.; Sundmacher, K.; Gani, R. Prediction of acid dissociation constants of organic compounds using group contribution methods. *Chem. Eng. Sci.* **2018**, *183*, 95–105. [CrossRef]
26. Venkatraman, V.; Evjen, S.; Knuutila, H.K.; Fiksdahl, A.; Alsberg, B.K. Predicting ionic liquid melting points using machine learning. *J. Mol. Liq.* **2018**, *264*, 318–326. [CrossRef]
27. Song, Z.; Shi, H.; Zhang, X.; Zhou, T. Prediction of CO_2 solubility in ionic liquids using machine learning methods. *Chem. Eng. Sci.* **2020**, *223*, 115752. [CrossRef]
28. Zhao, Y.; Zhang, X.; Deng, L.; Zhang, S. Prediction of viscosity of imidazolium-based ionic liquids using MLR and SVM algorithms. *Comput. Chem. Eng.* **2016**, *92*, 37–42. [CrossRef]

29. Zhao, Y.; Zhao, J.; Huang, Y.; Zhou, Q.; Zhang, X.; Zhang, S. Toxicity of ionic liquids: Database and prediction via quantitative structure-activity relationship method. *J. Hazard. Mater.* **2014**, *278*, 320–329. [CrossRef]
30. Kim, S.; Chen, J.; Cheng, T.; Gindulyte, A.; He, J.; He, S.; Li, Q.; Shoemaker, B.A.; Thiessen, P.A.; Yu, B.; et al. PubChem 2019 update: Improved access to chemical data. *Nucleic Acids Res.* **2019**, *47*, D1102–D1109. [CrossRef]
31. Wang, Z.; Su, Y.; Jin, S.; Shen, W.; Ren, J.; Zhang, X.; Clark, J.H. A novel unambiguous strategy of molecular feature extraction in machine learning assisted predictive models for environmental properties. *Green Chem.* **2020**, *22*, 3867–3876. [CrossRef]
32. RDKit: Open-Source Cheminformatics Software. Available online: https://www.rdkit.org/ (accessed on 14 October 2020).
33. PyTorch. Available online: https://pytorch.org/ (accessed on 14 October 2020).
34. Kingma, D.P.; Ba, J. Adam: A method for stochastic optimization. *arXiv* **2014**, arXiv:1412.6980.
35. Rumelhart, D.E.; Hinton, G.E.; Williams, R.J. Learning representations by back-propagating errors. *Nature* **1986**, *323*, 533–536. [CrossRef]
36. Pedregosa, F.; Varoquaux, G.; Gramfort, A.; Michel, V.; Thirion, B.; Grisel, O.; Blondel, M.; Prettenhofer, P.; Weiss, R.; Dubourg, V.; et al. Scikit-learn: Machine learning in Python. *J. Mach. Learn. Res.* **2011**, *12*, 2825–2830.

processes

MDPI

Article

An Integrated Approach to the Design of Centralized and Decentralized Biorefineries with Environmental, Safety, and Economic Objectives

Antioco López-Molina [1], Debalina Sengupta [2], Claire Shi [3], Eman Aldamigh [2], Maha Alandejani [4] and Mahmoud M. El-Halwagi [2,5,*]

1 División Académica Multidisciplinaria de Jalpa de Mendez, Universidad Juárez Autonoma de Tabasco, Jalpa de Mendéz, Tab. 86205, Mexico; antiocolm83@gmail.com
2 Gas and Fuels Research Center, Texas A&M Engineering Experiment Station, College Station, TX 77843, USA; debalinasengupta@tamu.edu (D.S.); emar.aldamigh@tamu.edu (E.A.)
3 International School of Beijing, Shunyi District, Beijing 101318, China; Claire.Shi@student.isb.bj.edu.cn
4 Economics Department, Faculty of Economics and Administration, King Abdulaziz University, Jeddah 21589, Saudi Arabia; malandejani@kau.edu.sa
5 Department of Chemical Engineering, Texas A&M University, College Station, TX 77843, USA
* Correspondence: El-Halwagi@tamu.edu

Received: 3 November 2020; Accepted: 17 December 2020; Published: 20 December 2020

Abstract: Biorefineries provide economic, environmental, and social benefits towards sustainable development. Because of the relatively small size of typical biorefineries compared to oil and gas processes, it is necessary to evaluate the options of decentralized (or distributed) plants that are constructed near the biomass resources and product markets versus centralized (or consolidated) facilities that collect biomass from different regions and distribute the products to the markets, benefiting from the economy of scale but suffering from the additional transportation costs. The problem is further compounded when, in addition to the economic factors, environmental and safety aspects are considered. This work presents an integrated approach to the design of biorefining facilities while considering the centralized and decentralized options and the economic, environmental, and safety objectives. A superstructure representation is constructed to embed the various options of interest. A mathematical programming formulation is developed to transform the problem into an optimization problem. A new correlation is developed to estimate the capital cost of biorefineries and to facilitate the inclusion of the economic functions in the optimization program without committing to the type of technology or the size of the plant. A new metric called Total Process Risk is also introduced to evaluate the relative risk of the process. Life cycle analysis is applied to evaluate environmental emissions. The environmental and safety objectives are used to establish tradeoffs with the economic objectives. A case study is solved to illustrate the value and applicability of the proposed approach.

Keywords: biorefining; sustainable design; distributed manufacturing; centralized facilities; integration; CAPEX correlation

1. Introduction

A biorefinery is a processing facility which uses physical, chemical, biological, and/or biochemical processes to convert biomass into value-added products and energy. In recent years, biorefineries have garnered much interest and research as the energy demand and climate crisis become increasingly pressing issues. Biorefineries are essential for the sustainable development of energy and chemical resources while reducing greenhouse gas (GHG) emissions. Nonetheless, biorefinery sizes tend to be smaller than fossil-based refineries because of the relatively distributed nature of biomass compared to the large-scale production of fossil materials from specific regions [1,2]

The sources of biomass may be broadly classified as virgin biomass or waste biomass. Virgin biomass includes any new sources that are cultivated or otherwise grown for the purpose of using in the biorefineries. Examples of virgin biomass include sugarcane and corn for bioethanol, and soybeans and algae for biodiesel. Waste biomass may be derived from various sources such as municipal solid waste (MSW), agricultural waste, forestry-product and industrial wastes, and waste cooking oils and grease from restaurants. A biorefinery may be designed to handle multiple feedstocks [3]. In some cases, preliminary sorting is needed to separate the organic portions of the waste. For instance, sorting and mechanical methods are used to obtain refuse-derived fuel (RDF) from MSW.

Although the types of biomass are varied, the technologies to convert them to chemicals and fuels may be largely classified into the following main categories: biochemical or thermochemical platforms. Biochemical platforms include fermentation (aerobic or anaerobic) and transesterification. Thermochemical platforms include the utilization of heat and pressure in various combinations and processes like liquefaction, gasification, pyrolysis, and combustion are used to obtain products [4–6].

The biorefinery industry is complex and multifaceted and depends on many factors, including the scale, location, and its association with economic factors, logistic factors, market factors and industrial practices [7]. Furthermore, economic factors that affect a biorefinery as an industrial firm can be categorized into internal and external factors, the former are related to microeconomic or firm-level elements, such as capital, labor, operating cost, profit, and related risk. The external factors are related to macroeconomic and institutional-level determinants, such as real economic growth subsides, government intervention, institutional quality, and regulatory quality. From the microeconomic perspective, several measurements can be considered when evaluating the biorefinery process designs and performances. In order to acquire proper insights and accomplish significant outputs in connection with economic factors and optimal designs, it is crucial to estimate the economic feasibility of the process design of a new biorefinery plant. This is due to the high capital investment and operating costs that are commonly associated with biorefinery technologies [7–9]

Subsidies, which are an aspect of government economic policy, can be used as an instrument to stimulate renewable energy production [10]. Biorefinery projects may need government intervention and subsides to assist in technology development and profitable operations. Subsidies are an element of government expenditure and may be financed through public funds via several channels, including taxation and tariffs [11]. However, the subsidies implemented for renewable products have certain drawbacks. For example, subsidies that cannot correct market failures can result in an early exit from the industry; to illustrate, some energy products are potentially susceptive of market failure, as the prices of these products are determined by free market and international markets conditions, which can control the prices; that is a potential source of market failure that requires government intervention through adapting some new policies, such as corrective energy taxes [12,13]. Another form of subsidy that may be used in renewable energy production, particularly in the renewable energy certification system, is a hybrid of fees and subsidies. Examples of this include fee-rebate (or "Feebate"), Renewable and Green Certificate Market, and Renewable Identification Numbers (RIN) [14–16].

In order to protect the environment, mitigate climate change globally, and narrow the economic gap between developing and developed countries, several international institutions have established the Sustainable Development Goals (SDGs). The aim was to focus on innovation and sustainability in all forms and sectors, including industrial sectors, energy, ecosystem, climate change, and protecting natural species and the environment [17]. In addition, the SDGs have strongly recommended the development of bioenergy, biorefinery, and bioeconomy (UN-DESA/DSDG, 2018). Thus, a biorefinery has a significant role to play in the implementation of the goals within SDGs: (i) fostering resilience against climate change, (ii) enhancing the environment for green growth, (iii) creating opportunities for sustainable production and consumption of renewable resources, and (iv) conserving natural resources for future generations [18]. These goals will help create development opportunities that support local and global markets, estimated to be more than 1 trillion USD [18].

Important contributions have been made in the design and optimization of biorefineries. For recent reviews of the topic, the reader is referred to literature on the subject [5–8,19–23]. Relevant to the scope of this paper, several important research contributions have been made in the areas of optimal design of biorefineries and the assessment of their technical, economic, environmental, and safety objectives. To compare the options of standalone, centralized drop-in, and distributed drop-in (that uses pipelines for connection with processing facilities) for thermochemical biorefineries, Alamia et al. [24] calculated the efficiencies of the various options. Since distributed manufacturing offers unique benefits for environmental impact, Lan et al. [25] provided a life cycle analysis for decentralized preprocessing systems associated with biorefineries using pyrolysis. The complexity of designing integrated biorefineries and associated supply chains calls for the use of powerful optimization frameworks. Roni et al. [26] developed an optimization approach for the distributed supply chains of biorefineries with multiple feedstocks. Psycha et al. [27] developed a design approach for algae biorefineries using multiple feedstocks. To simplify the preliminary synthesis of integrated biorefineries, Bao et al. [28] introduced a shortcut process–synthesis method using high-level process technology details. Another shortcut approach developed by Tay et al. [29,30] used thermodynamic equilibrium models to design biorefineries. Because of the significant potential of integrated biorefineries towards sustainable development, sustainability metrics have been considered in several research efforts for the design of biorefineries. Meramo-Hurtado and González-Delga [31] used sustainability parameters to guide a hierarchical approach to designing biorefineries. Parada et al. [32] analyzed the incorporation of sustainability metrics and stakeholder perspectives in designing biorefineries. Andiappan et al. [33] developed a process synthesis approach for biorefineries with economic and environmental objectives. Traditionally, safety has not been considered as a primary objective during conceptual design of biorefineries. Typically, a biorefinery is first designed and then the safety aspects are subsequently assessed. There is a growing recognition of the importance of including safety during conceptual design [34,35]. Potential risks include fire, explosion, and release of toxic emissions. Different approaches for risk assessment may be used depending on the intended objectives, the stage of process design, and the available data and details. El-Halwagi et al. [8] presented a multiobjective optimization approach to the design of biorefineries with economic and safety objectives. Bowling et al. [36] addressed the problem of facility location of biorefineries as part of the integrated supply chains. Piñas et al. [37] carried out an economic assessment of centralized and decentralized biogas plants. Ng et al. [21], Tay et al. [38], and Ponce-Ortega et al. [39] used disjunctive fuzzy optimization for planning and synthesis of biorefineries while considering industrial symbiosis. Ng et al. [22] introduced a framework for optimizing mixture design in biorefining facilities. Sun and Fan [40] reviewed optimization methods for the supply chains associated with biorefineries. Notwithstanding the value of the aforementioned contributions, there are no published papers (to our knowledge) that provide a unified optimization-based approach to the screening, selection, and design of a combination of centralized and decentralized biorefineries while accounting for the technical, economic, environmental, and safety objectives.

Unlike fossil fuels that are produced in large quantities in specific regions, biomass is generated in a rather fragmented and nonconsolidated manner. Therefore, the sizes of biorefineries range from large, centralized facilities to smaller, decentralized facilities. Large processing facilities that use a single feedstock are able to achieve maximum economy of scale for capital expenditure (CAPEX) and produce market competitive products. Nonetheless, transportation of raw materials from distant areas increase the operating expenditure (OPEX) and may lead to detrimental environmental impact. On the other hand, small-scale biorefineries lose the advantageous economy of scale but reduce the transportation cost of biomass because they mostly use available raw materials from adjacent areas. Therefore, important decisions must be made on whether biomass should be collected from various regions and transported to a large facility (***centralized processing***) or if several biorefineries should be installed to treat biomass within a certain region (***decentralized or distributed processing***). Although economy of scale does not favor decentralized/distributed manufacturing, there are several advantages offered by decentralized biorefineries, such as: (1) lower costs and emissions for transporting biomass to the facility and products to the market; (2) higher integration opportunities with other processing facilities (e.g., refining, gas conversion); (3) enhanced resilience especially to natural disasters by virtue of geographical distribution; and (4) increased flexibility in handling multiple feedstocks and generating multiple products. There is also a critical need for further research to understand the economic, environmental, and safety tradeoffs for centralized versus decentralized biorefineries. This paper presents a systematic approach for the comparison of centralized and decentralized biorefineries. A superstructure-based optimization formulation is developed to reconcile the various objectives while making important decisions on the plant capacity, technology, feedstock source and distribution, and product rate and distribution. A methanol-from-RDF case study is solved to illustrate the merits of the proposed approach.

2. Problem Statement

Given are:

- A set of cities/regions: $\{i|i = 1, 2, \ldots, N_{Cities}\}$. Each city/region has a supply of available biomass with known flowrate (F_i), composition of $N_{components}$, and price ($Cost_i$, USD/tonne).
- A proposed location for a centralized facility, which is given the index $i = N_{Cities+1}$.
- A set of monetization technologies $\{j|j = 1, 2, \ldots, N_{Tech}\}$ that may be used to convert the biomass to a set of value-added chemicals and fuels $\{p|p = 1, 2, \ldots, N_{Products}\}$. The selling price of each product is referred to as C_p.
- A set of transportation options for biomass and for products with known cost (USD/tonne·mile).
- Market demand and selling price for each product.

It is desired to identify an optimal strategy for monetizing the biomass by answering the following questions:

- Should the biomass be processed in centralized facilities with industrial symbiosis or in decentralized facilities?
- Which technologies should be used?
- What is the capacity of each facility?
- What are the economic, environmental, and safety roles of transporting biomass to the biorefineries and the products to consumers?
- How should the economic, environmental, and safety objectives for the integrated systems be evaluated and reconciled?

3. Methodology

The main steps in the proposed methodology are shown in the flowchart of Figure 1. The methodology is streamlined via the following tasks:

- Development of superstructure and optimization formulation;
- Development of a correlation for capital cost to be used in the economic optimization;
- Life cycle analysis of the proposed pathways;
- Safety analysis of the proposed pathways;
- Reconciliation of economic, environmental, and safety objectives.

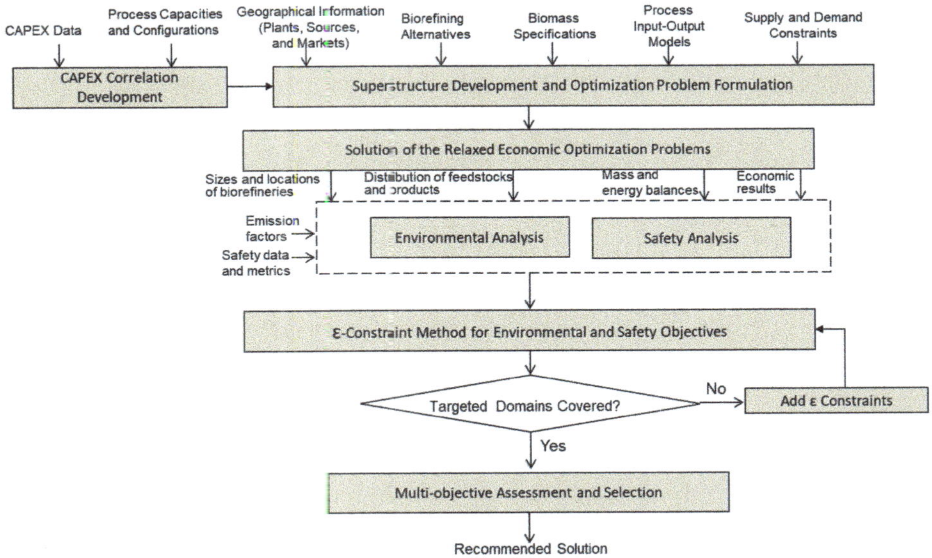

Figure 1. Flowchart summarizing the methodology.

First, the various configurational alternatives are embedded through a superstructure representation that represents the basis for formulating the optimization program. In order to explicitly include cost optimization within the optimization formulation, a correlation is developed for estimating capital cost of the proposed biorefineries. Three objectives are considered using economic, environmental, and safety metrics. The ε-constraintmethod is used to solve the multiobjective optimization problem. The economic metric (e.g., maximizing return on investment) is used as the objective function while the safety and environmental objectives are placed in the constraints with certain bounds that are repeatedly altered to generate the tradeoff results. First, the relaxed optimization problem is solved for maximizing profit with no bounds on the safety and environmental metrics. Next, the safety and environmental bounds are added and the optimization program is solved to generate alternate configurations with tradeoffs that constitute the basis for decision making.

The following sections summarize the key steps used in setting up the optimization formulation and the solution approach.

3.1. Development of Superstructure and Optimization Formulation

The first step is to create a superstructure that embeds the options of interest. Figure 2 shows the superstructure with centralized and decentralized options. A source–interception–sink representation [41] is used. Sources represent the biomass available from each region. Sinks represent the market demands in the various regions. The centralized and decentralized biorefineries represent the interceptors. Biomass may be processed within the same city in a decentralized facility or hauled to a larger centralized plant that accepts feedstock from multiple regions. The products may also be used locally or transported for sales in other regions. The flowrates of biomass and products assigned within the superstructure (represented by dashed arrows) as well as the throughput through each biorefinery are unknown and to be determined via the solution of the optimization formulation. A zero value indicates the absence of an option.

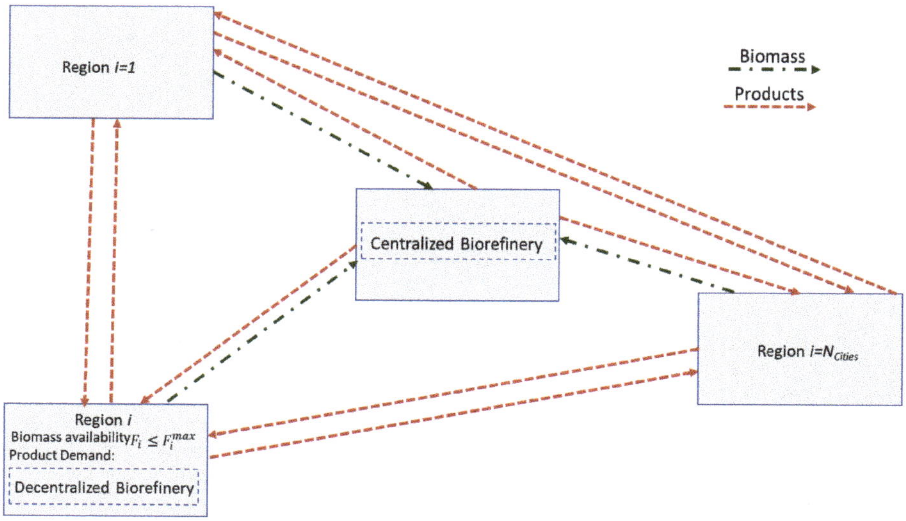

Figure 2. Superstructure of centralized and decentralized biorefining options.

Using the aforementioned superstructure, the optimization formulation is given below:

$$\text{Maximize Profitability (e.g., return on investment (ROI))} \tag{1}$$

Subject to:

$$\text{Annual Sales} = \sum_{i=1}^{N_{Cities+1}} \sum_{j=1}^{N_{Tech}} \sum_{p=1}^{N_{Products}} C_p * R_{i,j,p} \tag{2}$$

where $R_{i,j,p}$ is the production capacity of product p in the plant in city i using technology j (which is an optimization variable).

To ensure a practical capacity of each plant between a lower bound (LB) and an upper bound (UB), the following constraint is used:

$$LB * I_{i,j,p} \le R_{i,j,p} \le UB * I_{i,j,p} \tag{3}$$

where $I_{i,j,p}$ is a binary integer variable that takes the value of 1 when product p is produced in region/city i using technology j. Otherwise, it is zero.

To limit the number of facilities per city to a maximum number, the following constraint is used:

$$\sum_{j=1}^{N_{Tech}} \sum_{p=1}^{N_{Products}} I_{i,j,p} \le NPlants_i^{max} \; \forall i \tag{4}$$

The fixed capital investment (FCI) is given by:

$$FCI = \sum_{i=1}^{N_{Cities+1}} \sum_{j=1}^{N_{Tech}} \sum_{p=1}^{N_{Products}} FCI_{i,j,p} \tag{5}$$

The total capital investment (TCI) is given by

$$\text{TCI} = \text{FCI} + \text{WCI} \tag{6}$$

where WCI is the working capital investment. In this work, WCI is taken to be 15/85 of the FCI (El-Halwagi, 2017a).

The annual net profit (ANP) is expressed as (El-Halwagi, 2017a):

$$\text{ANP} = (\text{Annual Sales} - \text{Annual OPEX} - \text{Annualized FCI}) * (1 - \text{Tax Rate}) + \text{Annualized FCI} \quad (7)$$

The ε-constraintmethod is used to account for the environmental and safety objectives:

$$EM \leq EM^{Max} \quad (8)$$

$$RM \leq RM^{Max} \quad (9)$$

where EM and RM are the environmental and risk metrics, respectively, and Max designates an upper bound on the metric. The metrics may be selected from a wide variety of options [34,35,41]. The case study in this paper shows an example of the calculation and use of such metrics.

The process yield is:

$$G_{i,j,p} = \varnothing\left(F_{i,j,p}\right) \quad (10)$$

where ∅ is the yield function that depends on the feed characteristics and technology.

The availability of feedstock is:

$$F_i \leq F_i^{max}\ \forall i \quad (11)$$

The market limitations are:

$$G_{i,p} \leq G_{i,p}^{max}\ \forall i \quad (12)$$

The flowrate of the biomass received in city i, for technology j, to produce product p is obtained from shipments coming from all cities. Therefore,

$$F_{i\,j,p} = \sum_{i'=1}^{N_{Cities+1}} F_{i,i',j,p} \quad (13)$$

where $F_{i,i',j,p}$ is the biomass flowrate assigned from city i to city i' and technology j to produce product p.

Therefore, the sum of all shipments of biomass from a city is given by the shipments distributed from sources to sinks:

$$F_i = \sum_{j=1}^{N_{Tech}} \sum_{p=1}^{N_{Products}} F_{i,j,p}\ \forall i \quad (14)$$

The net production of product p is:

$$G_{i,p} = \sum_{j=1}^{N_{Tech}} G_{i,j,p} - \sum_{i'=1}^{N_{Cities+1}} G_{i,i',p}^{Trasported}\ \forall i, p \quad (15)$$

The first term on the right-hand side represents that total amount of product p generated by all technologies in city i. Product p may also be shipped to or from city i. The second term represents the net amount of product p shipped out of city i. The term $G_{i,i',p}^{Trasported}$ has a positive value when it is shipped out of city i and a negative value when it is shipped to city i.

3.2. Capital Cost Estimation

A convenient approach to including a parametric function for the estimation of CAPEX in the optimization formulation is to use a generic formulation that works for various biorefining technologies, feedstocks, and products. Because of the nature of this work intended to generate high-level directions for decision making, order-of-magnitude cost estimates are appropriate. Towards this end, we have carried out data analytics for the information extracted from 40 biorefineries. A functional form similar to the correlation developed by Zhang and El-Halwagi [42] was used. The correlation relates the FCI to two main factors: process throughput (feed flowrate of biomass) and number of functional steps "N" (which represent primary operation such as separation, reaction, and waste treatment). The data sources are given in the Supplementary Materials.

The resulting correlation is expressed as:

$$FCI\ (\text{in USDMM}) = 0.16 * N * (\text{Flowrate of biomass feed in 1000 tonnes/year}) * 0.84 \quad (16)$$

As indicated earlier, for the intended purposes of high-level decision making, an order-of-magnitudes cost estimation has the proper level of accuracy [41,42].

3.3. Life Cycle for RDF to Methanol Process

The centralized vs. decentralized processing for the RDF to methanol process needs an account of emissions at different stages of the life cycle. A life cycle assessment framework [43] is used to compute these emissions. For the process, the system boundaries are considered to be at the processing plant gate, where the RDF is brought. For the transportation emissions, the latest U.S. Environmental Protection Agency (EPA) emission factors for transportation are considered from the Emission Factors for Greenhouse Gas Inventories [44].

3.4. Safety Analysis

Two types of risks are evaluated for the process and for transportation. A relative risk approach is used to compare and rank the risks of the various alternatives. For process risk, hazard identification is carried out using the Hazardous Process Stream Index (HPSI) [45]. The HPSI is proposed to define and compare the level of hazard of each stream. The estimation is based on five normalized indicators, which consider pressure, density, molar flowrate, heat of combustion, and flash point, as follows:

$$I_P = \frac{\text{pressure value of individual stream}}{\text{average pressure for all streams}} \quad (17)$$

$$I_\rho = \frac{\text{density value of individual stream}}{\text{average density for all streams}} \quad (18)$$

$$I_{MF} = \frac{\text{molar flow value of individual stream}}{\text{average molar flow for all streams}} \quad (19)$$

$$I_{\Delta Hc} = \frac{\text{heat of combustion of individual stream}}{\text{average heat of combustion for all streams}} \quad (20)$$

$$I_{Fp} = \frac{\text{flash point score of individual stream}}{\text{average flash point score for all streams}} \quad (21)$$

The result of *HPSI* is calculated through the following expression:

$$HPSI = \left(\frac{I_P \cdot I_{MF} \cdot I_{\Delta Hc} \cdot I_{FP}}{I_\rho}\right) \cdot W \quad (22)$$

This index accounts for the relationship between the five dimensionless indicators of each process stream, and W is used as a scaling factor. As such, an increase in the capacity implies an increase in flowrate and inherently a more significant mass release in the case of loss of containment and its consequences. Therefore, W is defined as the ratio of the production capacity of the process with capacity i (CP_i) to the reference process capacity (CP_{base}). It is worth noting that the proposed index takes into account the impact of process capacity. For the same process technology and flowsheet, the smallest-scale process risk is considered the reference if the production capacity is different.

$$W = \frac{CPi}{CP_{base}} \quad (23)$$

The process line with the highest value of HPSI poses the highest risk levels, reflecting the severity of the process stream in the loss of containment case, leading to a fire or explosion.

The relative process risk level is estimated from the HPSI value. A new normalization approach is proposed to enable the identification of relative risk using the following equation:

$$R_i = \frac{HPSI_i - HPSI_{min}}{HPSI_{max} - HPSI_{min}} \quad (24)$$

where $HPSI_i$ is the individual index value of the process streams, $HPSI_{min}$ is the smallest index value of the process streams, $HPSI_{max}$ is the maximum index value of the process streams, and R_i is the individual risk of the process streams.

Once the individual risk has been estimated, the total relative process risk (R_T) is estimated by the arithmetic average of the individual risks as follows:

$$R_T = \frac{\sum_i^n R_i}{PS} \quad (25)$$

where PS is the total number of process streams. The risk scale is defined according to Table 1.

Table 1. Risk scale based on the Hazardous Process Stream Index (HPSI).

Total Process Risk (R_T)	Risk Level
0.0–0.2	Very Low Risk
0.2–0.4	Low Risk
0.4–0.6	Medium Risk
0.6–0.8	High Risk
0.8–1.0	Very High Risk

If different process technologies are compared, HPSI must be assessed for each process using the same production capacity and then the assessment is carried out for the total relative process risk.

It is worth noting that alternate approaches for risk assessment may be used. For instance, an alternative approach to the aforementioned HPSI to determine the risk of the process at a different production scale is to consider the sum of the individual risks instead of the average used in this work. To classify risk based on summation, a risk scale may be defined to categorize/classify the different processes based on the sum.

For the methanol transportation risk analysis, the approach proposed by Zhang et al. (2018) is adopted. The following is a summary of the approach. Two transportation modes are considered: highway and railroad. Three variables were defined to assess the risk factor of methanol transportation using the two alternatives: (1) value loss per accident (C_t), (2) number of trips (N_t) per year, and (3) overall probability of accident as a function of distance (d) when methanol is moved among cities ($P_{i,j,d}$). The number of trips is calculated, dividing the maximum demand of each city by the load capacity of each transport. The overall probability of an accident during methanol transport is estimated by the product of general accident rate for transportation mode and the cumulative probability as a function of the natural logarithm of distance for methanol highway and railroad incidents. The product of these three variables provides the risk factor of each transport option

4. Case Study: Centralized vs. Decentralized Conversion of MSW to Methanol

Consider the case of three cities (A, B, and C) with known biomass type in the form of refuse derived fuels (RDF) obtained from MSW at a cost of USD55/tonne. Gasification-based technology is used to convert RDF to methanol. Detailed description of this process consists of three main stages: gasification, purification, and methanol synthesis. Description of the technical, design, and operational details are given in literature [46–51]. Simulation results for the streams, operating conditions, and design specifications are taken from literature [52–54]. Based on the data provided by Iaquaniello et al. [55], key process information (on the use of electric energy, water, methanol yield, and assumed costs of electric energy, water, and methanol) are shown in Table 2.

Table 2. Key process information.

RDF (Refuse-Derived Fuel)	Electric Energy Needed	Cost of Electric Energy (Assuming USD0.05/kWh)	Demineralized Water Needed	Cost of Water (Assuming USD3/Tonne)	Produced Methanol	Value of Methanol (Assuming USD350/Tonne)
1 tonne	500 kWh	25	0.15 tonne	USD0.45	0.4 tonne	USD140

The flowrate of RDF available in each city and the relative locations of the cities and the proposed centralized facility are show by Figure 3. A decision needs to be made on whether to build one centralized facility or multiple decentralized (smaller) facilities. The tradeoffs include cost, transportation, environmental impact, and risk.

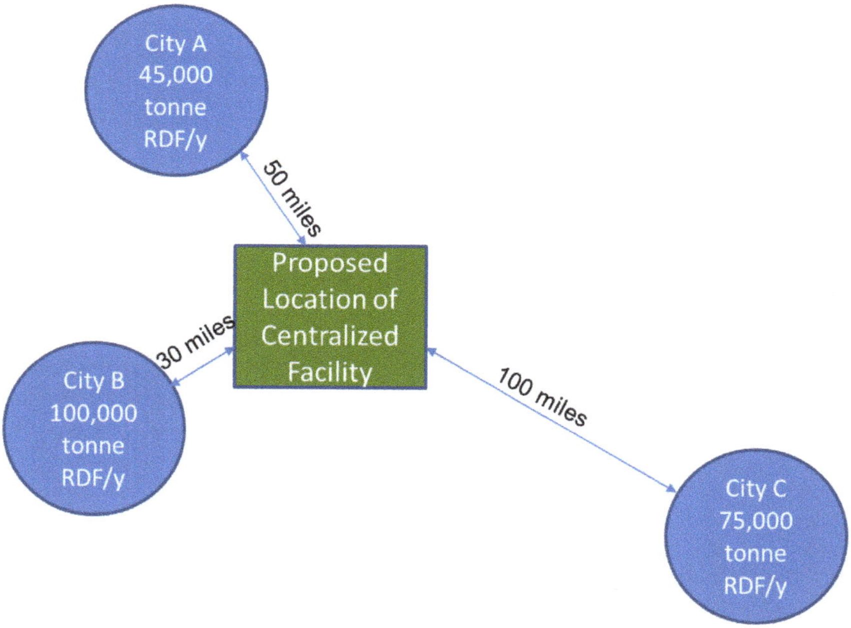

Figure 3. Biomass availability and relative locations for the case study.

The maximum demands for methanol in cities A, B, and C are 0, 60,000, and 25,000 tonne/year, respectively. The transportation costs were based on data provided by Zhang et al. [56] and are USD0.011/(tonne RDF·mile) and USD0.015/(tonne methanol·mile).

Life Cycle Assessment for RDF to Methanol Process

The proposed approach and optimization formulation are generally applicable for any environmental metrics. For this case study, only greenhouse gas emissions and their global warming impacts are considered to illustrate the applicability of the proposed approach and to discuss the tradeoffs. For transportation, two methods are considered for service: railroad and diesel-fueled heavy trucks. When a decentralized facility is co-located within the same city sourcing RDF, the transportation distance is assumed to be negligible.

The data of the RDF-to-methanol process described by Salladini et al. [48] are adapted to calculate the process emissions and to create the diagram shown by Figure 4 for a functional unit of one tonne of methanol produced. A linear proportionality consumption of overall inputs and outputs is assumed.

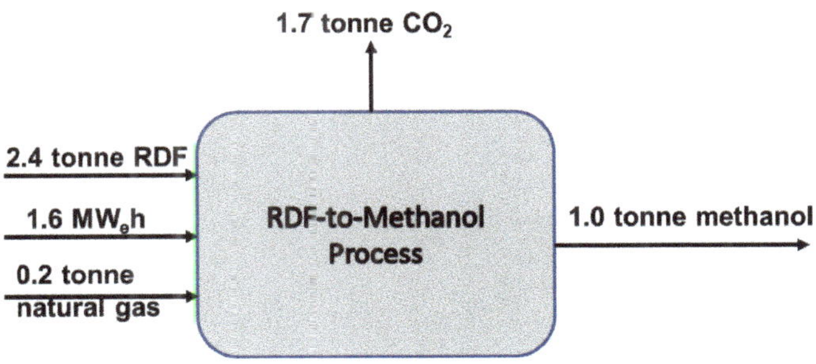

Figure 4. Key process inputs and outputs for a basis of one tonne methanol.

The emissions for the upstream production of electricity (Table 3) and natural gas (Table 4) are computed from emission factors provided by the Emission Factors for Greenhouse Gas Inventories (EPA, 2020). Assuming a natural gas power plant, the average emission factors for electricity in the United States are used.

Table 3. Non-baseload emission factors.

Item	CO_2 Factor Tonne/MWh	CH_4 Factor Tonne/MWh	N_2O Factor Tonne/MWh
US Average Energy Mix	0.65	5.30×10^{-5}	0.77×10^{-5}

Table 4. Steam and heat emission factors assuming natural gas is entirely used at 80% efficiency.

Item	MMBtu per Tonne of Natural Gas (LNG Factor Used)	CO_2 Factor (kg CO_2/MMBtu)	CH_4 Factor (g CH_4/MMBtu)	N_2O Factor (g N_2O/MMBtu)
Steam and Heat (from Natural Gas)	51.7	53.06	1.0	0.10

The ton-mile factors for medium and heavy-duty trucks are used for the transportation of methanol from the centralized facility to the cities A, B, and C. A tonne-mile is one tonne of freight carried one mile, as a unit of traffic.

For the HPSI evaluation, the stream data were taken from literature [48,52–54]. The smallest plant (25,000 tonne/year) was taken as the reference.

5. Results and Discussion

The optimization formulation is a mixed-integer nonlinear program (MINLP) with 68 variables. The solution of the relaxed problem with no bounds on the environmental or safety objectives (using the Global Solver of LINGO software) is shown by Figure 5. The global solution of the relaxed optimization problem features a centralized facility and achieves a return on investment (ROI) of 11.8%/year. The GHG emissions of the relaxed problem is 291 × 10^3 tonnes CO_{2eq}/year and the process risk index, HPSI, is 0.75 (high risk). As described in the methodology, the ε-constraintmethod was used to establish the tradeoffs among the various objectives. Using bounds of 285 × 10^3 tonnes CO_{2eq}/year on the GHG emissions and 0.5 on HPSI (medium risk), another solution (shown by Figure 6) is generated, which is economically suboptimal but superior for the environmental and safety objectives. It has a common facility for cities (A and B) and a local facility in city C. The ROI for the common facility matches that of the global solution, but there is a decrease in the ROI for the facility in city C to 10.1%/year. Table 5 shows the GHG emissions and safety for the optimal and suboptimal solutions.

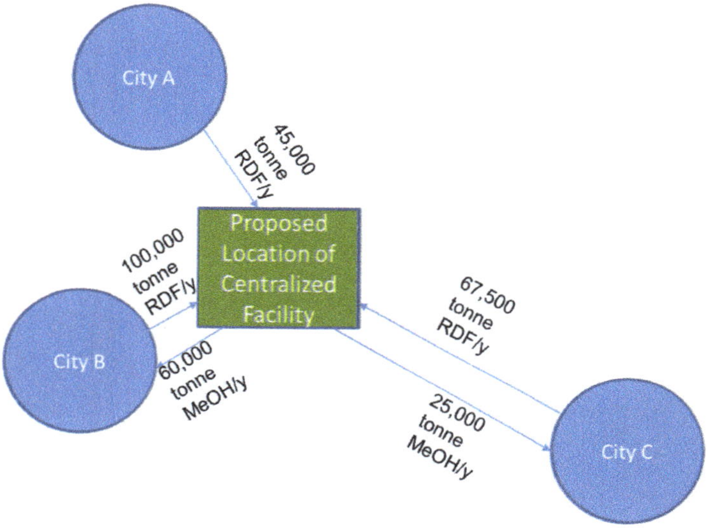

Figure 5. Economically optimum solution with a centralized facility (ROI is return on investment).

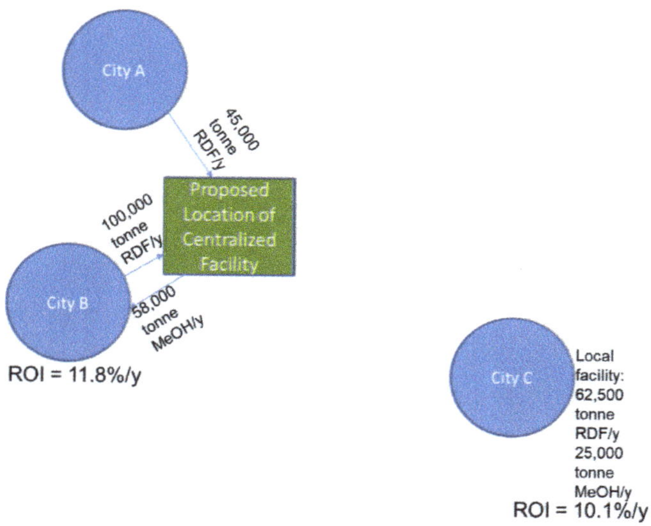

Figure 6. Suboptimal solution with centralized and decentralized facilities.

Table 5. Risk results for methanol transportation.

Solution	Risk Factor by City	Highway	Railroad
Optimal	Transporting methanol to city B	207.27	128.08
	Transporting methanol to city C	426.87	673.17
	Total risk factor from centralized facility	634.14	801.24
Suboptimal	Transporting methanol to city B	200.39	297.13

The results of the transportation risk analysis resulting from spillage/noncontainment due to accidents are shown by Table 5. The risk for transporting biomass was assumed to be negligible compared to the risk of transporting methanol (which involves loss of containment, toxicity, flammability, and explosion issues). For the optimal solution, transporting methanol to cities B and C by highway is safer than by railroad. The total risk factor using a highway gives a value of 634.14, while the total risk factor using a railroad is 801.24. The differences in risk are attributed to distance. Therefore, the lower risk factor for transporting methanol from a centralized facility is obtained by using the railroad to supply methanol to city B and by using the highway for city C, giving a value of 554.95. In the suboptimal solution, the use of highway as a transport medium is safer than railroad. Table 6 summarizes the carbon footprint and risk for the economically optimal and the suboptimal solutions. Although the economically optimal solution offers a higher ROI, it involves higher carbon footprint and higher risk. The decision makers should balance these conflicting objectives and select a strategy that reconciles the importance of the economic, environmental, and safety objectives. Alternatively, the decision makers may use an economic platform to incorporate sustainability, risk, and resilience into the ROI calculations [57–62].

Table 6. Greenhouse gas (GHG) emissions and risk for the optimal and suboptimal solutions.

Solution	Total GHG Emissions (Tonnes CO_{2eq}/year)	Process Risk (HPSI)	Lower Transportation Risk Factor
Optimal with Maximum ROI (11.8%)	290,877 Total (104,061 from electricity, 39,659 from natural gas, 144,500 from process emissions, 2269 from raw material transport, 388 from product transport)	0.75 (High Risk)	554.9 (using the highway to transport methanol to city B and using railroad to transport methanol to city C)
Suboptimal Solution	282,760 Total (101,613 from electricity, 38,726 from natural gas, 141,100 from process emissions, 993 from raw material transport, 329 from product transport)	0.50 (Medium Risk)	200.4 (using the highway to transport methanol to city B)

6. Conclusions and Recommendations for Future Research

This paper introduces a systematic approach to the design and comparison of centralized versus decentralized biorefining options. The paper provides the following new contributions: (1) a superstructure representation embedding all configurations of interest; (2) an optimization formulation with economic, environmental, and safety objectives that are solved using the ε-constraintmethod to establish the tradeoffs among the multiple objectives; and (3) incorporation of transportation risk (in addition to process risk using a new metric quantifying total process risk). Furthermore, a CAPEX cost correlation is developed for order-of-magnitude estimation purposes. This correlation offers the advantages of using few data (feed flowrate and number of functional steps) and enabling the explicit incorporation of cost functions in the optimization formulation without committing to the type of technology or the size of the plant. The limitation of this correlation is that its level of accuracy is an order-of-magnitude estimate and should therefore be used with caution, with the results checked using more detailed cost estimation methods once a detailed design is available. Alternatively, if more accurate cost correlations are available, they can be readily used in the optimization formulation, which does not depend on a specific correlation. A case study is solved to address centralized versus decentralized options for converting RDF to methanol. The centralized option showed better profitability but higher levels of carbon footprint and risk.

This work constitutes the basis for several future research directions. Specifically, the following topics are recommended for future research:

- Inclusion of multiple environmental metrics—while the case study in this paper focused on GHG emissions, other environmental metrics may be used, such as land use change, water usage and discharge, pollutant discharge, acidification, and eutrophication;
- Consideration of additional objectives such as resilience [57], reliability, availability, and maintainability [62];
- Integration of centralized and decentralized biorefineries with other infrastructures through the concept of industrial symbiosis [63,64];
- Consideration of feedstock variability and uncertainty [60,65].

Supplementary Materials: The following are available online at http://www.mdpi.com/2227-9717/8/12/1682/s1, Figure S1: Data used in developing FCI correlation, Table S1: Summary of extracted data.

Author Contributions: All authors contributed to conceptualization, methodology, validation, formal analysis, investigation, data curation, writing—original draft preparation, writing—review and editing, and visualization. All authors have read and agreed to the published version of the manuscript.

Funding: This research received no external funding.

Conflicts of Interest: The authors declare no conflict of interest.

Nomenclature

C_p	Selling price of product *p*
$Cost_i$	Cost of biomass in city *i*
CP_i	Process production capacity
CP_{base}	Reference process capacity
EM	Environmental metric
F_i	Flowrate of biomass available in city *i*
F_i^{max}	Maximum flowrate of biomass available in city *i*
$F_{i,j,p}$	Flowrate of biomass assigned to city *i* and technology *j* to produce product *p*
$F_{i,i',j,p}$	Flowrate of biomass assigned from city *i* to city *i′* and technology *j* to produce product *p*
FCI	Fixed capital investment
$G_{i,\,p}$	Net production of product *p* in the plant in city *i*
$G_{i,j,p}$	Production capacity of product *p* in the plant in city *i* using technology *j*
$G_{i,i',p}^{Trasported}$	Net amount of product *p* shipped out of city *i*
HPSI	Hazardous process stream index
$HPSI_{min}$	Smallest HPSI value of reference process
$HPSI_{max}$	Highest HPSI value of reference process
i	Index for cities/regions
$I_{i,j,p}$	Binary integer variable that takes the value of 1 when product *p* is produced in city *i* using technology *j*.
I_{Fp}	Flash point indicator
$I_{\Delta Hc}$	Heat of combustion indicator
I_{MF}	Molar flow indicator
I_P	Pressure indicator
I_ρ	Density indicator
j	Index for technologies
N_{Cities}	Total number of cities/regions
N_{Tech}	Total number of technologies
P	Index for products
R_i	Individual risk of process streams
R_T	Total process risk

RM	Risk metric
TCI	Total capital investment
W	HPSE scaling factor
WCI	Working capital investment
Greek	
∅	A process yield function that relates products to reactants

References

1. Santibañez-Aguilar, J.E.; González-Campos, J.B.; Ponce-Ortega, J.M.; Serna-González, M.; El-Halwagi, M.M. Optimal planning and site selection for distributed multiproduct biorefineries involving economic, environmental and social objectives. *J. Clean. Prod.* **2014**, *65*, 270–294. [CrossRef]
2. Ubando, A.T.; Felix, C.B.; Chen, W.-H. Biorefineries in circular bioeconomy: A comprehensive review. *Bioresour. Technol.* **2020**, *299*, 122585. [CrossRef] [PubMed]
3. Pyrgakis, K.; Kokossis, A.C.; Kokossis, A.C. A Total Site Synthesis approach for the selection, integration and planning of multiple-feedstock biorefineries. *Comput. Chem. Eng.* **2019**, *122*, 326–355. [CrossRef]
4. Ng, D.K.; Tan, R.R.; Foo, D.C.; El-Halwagi, M.M. *Process Design Strategies for Biomass Conversion Systems*; John Wiley & Sons: Hoboken, NJ, USA, 2015.
5. Sengupta, D.; Pike, R.W. *Chemicals from Biomass: Integrating Bioprocesses into Chemical Production Complexes for Sustainable Development*; CRC Press: Boca Raton, FL, USA, 2012.
6. Stuart, P.; El-Halwagi, M.M. (Eds.) *Integrated Biorefineries: Design, Analysis, and Optimization*; Taylor and Francis/CRC: Boca Raton, FL, USA, 2013.
7. Sadhukhan, J.; Kok, S.N.; Martinez, E. *Biorefineries and Chemical Processes. Design, Integration and Sustainability Analysis*; John Wiley & Sons: West Sussex, UK, 2014.
8. El-Halwagi, A.M.; Rosas, C.; Ponce-Ortega, J.M.; Jiménez-Gutiérrez, A.; Mannan, M.S.; El-Halwagi, M.M. Multiobjective optimization of biorefineries with economic and safety objectives. *AIChE J.* **2013**, *59*, 2427–2434. [CrossRef]
9. Conrad, K.; Wang, J. On the Design of Incentive Mechanisms in Environmental Policy. In *Economic Incentives and Environmental Policies*; Springer Science and Business Media LLC: Berlin, Germany, 1994; pp. 69–85.
10. Cruce, J.R.; Quinn, J.C. Economic viability of multiple algal biorefining pathways and the impact of public policies. *Appl. Energy* **2019**, *233*, 735–746. [CrossRef]
11. Lohman, L.D.S. Economic Incentives in Environmental Policy: Why are They White Ravens? In *Economic Incentives and Environmental Policies*; Springer Science and Business Media LLC: Berlin, Germany, 1994; pp. 55–67.
12. Parry, I.W.; Heine, M.D.; Lis, E.; Li, S. *Getting Energy Prices Right: From Principle to Practice*; International Monetary Fund: Washington, DC, USA, 2014.
13. Clements, M.B.J.; Coady, D.; Fabrizio, M.S.; Gupta, M.S.; Alleyne, M.T.S.C.; Sdralevich, M.C.A. *Energy Subsidy Reform: Lessons and Implications*; International Monetary Fund: Washington, DC, USA, 2014.
14. Greene, D.L.; Patterson, P.D.; Singh, M.; Li, J. Feebates, rebates and gas-guzzler taxes: A study of incentives for increased fuel economy. *Energy Policy* **2005**, *33*, 757–775. [CrossRef]
15. Aune, F.R.; Dalen, H.M.; Hagem, C. Implementing the EU renewable target through green certificate markets. *Energy Econ.* **2012**, *34*, 992–1000. [CrossRef]
16. Brown, T.R.; Thilakaratne, R.; Brown, R.C.; Hu, G. Regional differences in the economic feasibility of advanced biorefineries: Fast pyrolysis and hydroprocessing. *Energy Policy* **2013**, *57*, 234–243. [CrossRef]
17. SDGs Kenya Forum: Vision 2030s Medium Term Plan as a Framework for Implementation of the Sustainable Development Goals, SDGs Kenya Forum for Sustainable Development. Available online: http://www.developlocal.org/wp-content/uploads/2016/04/ImplementingTheSDGs.pdf (accessed on 1 August 2020).
18. Sadhukhan, J.; Martinez-Hernandez, E.; Murphy, R.J.; Ng, D.K.; Hassim, M.H.; Ng, K.S.; Kin, W.Y.; Jaye, I.F.M.; Hang, M.Y.L.P.; Andiappan, V. Role of bioenergy, biorefinery and bioeconomy in sustainable development: Strategic pathways for Malaysia. *Renew. Sustain. Energy Rev.* **2018**, *81*, 1966–1987. [CrossRef]
19. Sánchez, E.N.; Ojeda, K.; Elhalwagi, M.M.; Kafarov, V. Biodiesel from microalgae oil production in two sequential esterification/transesterification reactors: Pinch analysis of heat integration. *Chem. Eng. J.* **2011**, *176–177*, 211–216. [CrossRef]

20. Dickinson, S.; Mientus, M.; Frey, D.; Amini-Hajibashi, A.; Ozturk, S.; Shaikh, F.; Sengupta, D.; El-Halwagi, M.M. A review of biodiesel production from microalgae. *Clean Technol. Environ. Policy* **2016**, *19*, 637–668. [CrossRef]
21. Ng, R.T.L.; Ng, D.K.; Tan, R.R.; El-Halwagi, M.M. Disjunctive fuzzy optimisation for planning and synthesis of bioenergy-based industrial symbiosis system. *J. Environ. Chem. Eng.* **2014**, *2*, 652–664. [CrossRef]
22. Ng, L.Y.; Andiappan, V.; Chemmangattuvalappil, N.G.; Ng, D.K. A systematic methodology for optimal mixture design in an integrated biorefinery. *Comput. Chem. Eng.* **2015**, *81*, 288–309. [CrossRef]
23. Liew, W.H.; Hassim, M.H.; Ng, D.K.S. Review of evolution, technology and sustainability assessments of biofuel production. *J. Clean. Prod.* **2014**, *71*, 11–29. [CrossRef]
24. Alamia, A.; Gardarsdottir, S.O.; Larsson, A.; Normann, F.; Thunman, H. Efficiency Comparison of Large-Scale Standalone, Centralized, and Distributed Thermochemical Biorefineries. *Energy Technol.* **2017**, *5*, 1435–1448. [CrossRef]
25. Lan, K.; Ou, L.; Park, S.; Kelley, S.S.; Yao, Y. Life Cycle Analysis of Decentralized Preprocessing Systems for Fast Pyrolysis Biorefineries with Blended Feedstocks in the Southeastern United States. *Energy Technol.* **2020**, *8*. [CrossRef]
26. Roni, M.S.; Thompson, D.N.; Hartley, D.S. Distributed biomass supply chain cost optimization to evaluate multiple feedstocks for a biorefinery. *Appl. Energy* **2019**, *254*, 113660. [CrossRef]
27. Psycha, M.; Kapnisi, M.; Kokossis, A.C. Extended Value Chain Synthesis towards the Design of Multi-Feedstock Algae Biorefineries. In *Computer Aided Chemical Engineering;* Elsevier BV: Amsterdam, The Netherlands, 2018; Volume 43, pp. 645–650.
28. Bao, B.; Ng, D.K.; Tay, D.H.; Jiménez-Gutiérrez, A.; El-Halwagi, M.M. A shortcut method for the preliminary synthesis of process-technology pathways: An optimization approach and application for the conceptual design of integrated biorefineries. *Comput. Chem. Eng.* **2011**, *35*, 1374–1383. [CrossRef]
29. Tay, D.H.S.; Ng, D.K.; Kheireddine, H.; El-Halwagi, M.M. Synthesis of an integrated biorefinery via the C–H–O ternary diagram. *Clean Technol. Environ. Policy* **2011**, *13*, 567–579. [CrossRef]
30. Tay, D.H.S.; Kheireddine, H.; Ng, D.K.S.; El-Halwagi, M.M.; Tan, R.R. Conceptual Synthesis of Gasification-Based Biorefineries Using Thermodynamic Equilibrium Optimization Models. *Ind. Eng. Chem. Res.* **2011**, *50*, 10681–10695. [CrossRef]
31. Meramo-Hurtado, S.-I.; González-Delgado, Á.D. Biorefinery synthesis and design using sustainability parameters and hierarchical/3D multi-objective optimization. *J. Clean. Prod.* **2019**, *240*, 118134. [CrossRef]
32. Parada, M.P.; Osseweijer, P.; Posada, J. Sustainable biorefineries, an analysis of practices for incorporating sustainability in biorefinery design. *Ind. Crops Prod.* **2017**, *106*, 105–123. [CrossRef]
33. Andiappan, V.; Ko, A.S.Y.; Lau, V.W.S.; Ng, L.Y.; Ng, R.T.L.; Chemmangattuvalappil, N.G.; Ng, D.K. Synthesis of sustainable integrated biorefinery via reaction pathway synthesis: Economic, incremental environmental burden and energy assessment with multiobjective optimization. *AIChE J.* **2014**, *61*, 132–146. [CrossRef]
34. Roy, N.; Eljack, F.; Jiménez-Gutiérrez, A.; Zhang, B.; Thiruvenkataswamy, P.; El-Halwagi, M.; Mannan, M.S. A review of safety indices for process design. *Curr. Opin. Chem. Eng.* **2016**, *14*, 42–48. [CrossRef]
35. Park, S.; Xu, S.; Rogers, W.; Pasman, H.; El-Halwagi, M.M. Incorporating inherent safety during the conceptual process design stage: A literature review. *J. Loss Prev. Process. Ind.* **2020**, *63*, 104040. [CrossRef]
36. Bowling, I.M.; Ponce-Ortega, J.M.; El-Halwagi, M.M. Facility Location and Supply Chain Optimization for a Biorefinery. *Ind. Eng. Chem. Res.* **2011**, *50*, 6276–6286. [CrossRef]
37. Piñas, J.A.V.; Venturini, O.; Lora, E.E.S.; Del Olmo, O.A.; Roalcaba, O.D.C. An economic holistic feasibility assessment of centralized and decentralized biogas plants with mono-digestion and co-digestion systems. *Renew. Energy* **2019**, *139*, 40–51. [CrossRef]
38. Tay, D.H.; Ng, D.K.; Sammons, N.E., Jr.; Eden, M.R. Fuzzy optimization approach for the synthesis of a sustainable integrated biorefinery. *Ind. Eng. Chem. Res.* **2011**, *50*, 1652–1665. [CrossRef]
39. Ponce-Ortega, J.M.; Pham, V.; El-Halwagi, M.M.; El-Baz, A.A. A Disjunctive Programming Formulation for the Optimal Design of Biorefinery Configurations. *Ind. Eng. Chem. Res.* **2012**, *51*, 3381–3400. [CrossRef]
40. Sun, O.; Fan, N. A Review on Optimization Methods for Biomass Supply Chain: Models and Algorithms, Sustainable Issues, and Challenges and Opportunities. *Process. Integr. Optim. Sustain.* **2020**, *4*, 203–226. [CrossRef]
41. El-Halwagi, M.M. *Sustainable Design through Process Integration: Fundamentals and Applications to Industrial Pollution Prevention, Resource Conservation, and Profitability Enhancement*, 2nd ed.; Elsevier: Amsterdam, The Netherlands, 2017.

42. Zhang, C.; El-Halwagi, M.M. Estimate the Capital Cost of Shale-Gas Monetization Projects. *Chem. Eng. Prog.* **2017**, *113*, 28–32.
43. Chouinard-Dussault, P.; Bradt, L.; Ponce-Ortega, J.M.; El-Halwagi, M.M. Incorporation of process integration into life cycle analysis for the production of biofuels. *Clean Technol. Environ. Policy* **2010**, *13*, 673–685. [CrossRef]
44. EPA (Environmental Protection Agency). Emission Factors for Greenhouse Gas Inventories. 2020. Available online: https://www.epa.gov/sites/production/files/2020-04/documents/ghg-emission-factors-hub.pdf (accessed on 4 June 2020).
45. López-Molina, A.; Huerta-Martínez, Á.A.; Gómez-Castro, F.I.; Conde-Mejía, C. A novel approach to identify hazards in non-conventional/intensified biofuels production processes. *Chem. Eng. Process. Process. Intensif.* **2020**, *157*, 108139. [CrossRef]
46. Park, J.; Cho, J.; Lee, Y.; Park, M.-J.; Lee, W.B. Practical Microkinetic Modeling Approach for Methanol Synthesis from Syngas over a Cu-Based Catalyst. *Ind. Eng. Chem. Res.* **2019**, *58*, 8663–8673. [CrossRef]
47. Ramzan, N.; Ashraf, A.; Naveed, S.; Malik, A. Simulation of hybrid biomass gasification using Aspen plus: A comparative performance analysis for food, municipal solid and poultry waste. *Biomass-Bioenergy* **2011**, *35*, 3962–3969. [CrossRef]
48. Salladini, A.; Agostini, E.; Borgogna, A.; Spadacini, L.; Annesini, M.; Iaquaniello, G. Analysis on High Temperature Gasification for Conversion of RDF into Bio-Methanol. *Gasif. Low-grade Feedstock* **2018**, *143*. [CrossRef]
49. Julián-Durán, L.; Ortiz-Espinoza, A.P.; El-Halwagi, M.M.; Jiménez-Gutiérrez, A. Techno-economic assessment and environmental impact of shale gas alternatives to methanol. *ACS Sustain. Chem. Eng.* **2014**, *2*, 2338–2344. [CrossRef]
50. Ehlinger, V.M.; Gabriel, K.J.; Noureldin, M.M.B.; El-Halwagi, M.M. Process Design and Integration of Shale Gas to Methanol. *ACS Sustain. Chem. Eng.* **2014**, *2*, 30–37. [CrossRef]
51. Alsuhaibani, A.S.; Afzal, S.; Challiwala, M.; Elbashir, N.O.; El-Halwagi, M.M. The impact of the development of catalyst and reaction system of the methanol synthesis stage on the overall profitability of the entire plant: A techno-economic study. *Catal. Today* **2020**, *343*, 191–198. [CrossRef]
52. Shehzad, A.; Bashir, M.J.; Sethupathi, S. System analysis for synthesis gas (syngas) production in Pakistan from municipal solid waste gasification using a circulating fluidized bed gasifier. *Renew. Sustain. Energy Rev.* **2016**, *60*, 1302–1311. [CrossRef]
53. Leo, D. Process Modelling and Simulation of a Methanol Synthesis Plant Using Syngas Streams Obtained from Biomass. Master's Thesis, Politecnico di Milano, Milan, Italy, 2018. Available online: https://www.politesi.polimi.it/bitstream/10589/142748/3/2018_10_Leo.pdf (accessed on 4 June 2020).
54. Lücking, L.E. Methanol Production from Syngas: Process Modelling and Design Utilising Biomass Gasification and Integrating Hydrogen Supply. Master's Thesis, Delft University of Technology, Delft, The Netherlands, 2017. Available online: http://repository.tudelft.nl/ (accessed on 4 June 2020).
55. Iaquaniello, G.; Centi, G.; Salladini, A.; Palo, E.; Perathoner, S.; Spadaccini, L. Waste-to-methanol: Process and economics assessment. *Bioresour. Technol.* **2017**, *243*, 611–619. [CrossRef] [PubMed]
56. Zhang, C.; Nguyen, C.; Eljack, F.; Linke, P.; El-Halwagi, M.M. Integration of Safety in the Optimization of Transporting Hazardous Materials. *Process. Integr. Optim. Sustain.* **2018**, *2*, 435–446. [CrossRef]
57. El-Halwagi, M.M. A return on investment metric for incorporating sustainability in process integration and improvement projects. *Clean Technol. Environ. Policy* **2017**, *19*, 611–617. [CrossRef]
58. El-Halwagi, M.M.; Sengupta, D.; Pistikopoulos, E.N.; Sammons, J.; Eljack, F.; Kazi, M.-K. Disaster-Resilient Design of Manufacturing Facilities Through Process Integration: Principal Strategies, Perspectives, and Research Challenges. *Front. Sustain. Food Syst.* **2020**, *1*. [CrossRef]
59. Guillen-Cuevas, K.; Ortiz-Espinoza, A.P.; Ozinan, E.; Jiménez-Gutiérrez, A.; Kazantzis, N.K.; El-Halwagi, M.M. Incorporation of Safety and Sustainability in Conceptual Design via a Return on Investment Metric. *ACS Sustain. Chem. Eng.* **2018**, *6*, 1411–1416. [CrossRef]
60. Ortiz-Espinoza, A.P.; Jiménez-Gutiérrez, A.; El-Halwagi, M.M.; Kazantzis, N.K.; Kazantzi, V. Comparison of safety indexes for chemical processes under uncertainty. *Process. Saf. Environ. Prot.* **2021**, *148*, 225–236. [CrossRef]
61. Moreno-Sader, K.; Jain, P.; Tenorio, L.C.B.; Mannan, M.S.; El-Halwagi, M.M. Integrated Approach of Safety, Sustainability, Reliability, and Resilience Analysis via a Return on Investment Metric. *ACS Sustain. Chem. Eng.* **2019**, *7*, 19522–19536. [CrossRef]

62. Al-Douri, A.; Kazantzi, V.; Eljack, F.T.; Mannan, M.S.; El-Halwagi, M.M. Mitigation of operational failures via an economic framework of reliability, availability, and maintainability (RAM) during conceptual design. *J. Loss Prev. Process. Ind.* **2020**, *67*, 104261. [CrossRef]
63. El-Halwagi, M.M. A Shortcut Approach to the Multi-scale Atomic Targeting and Design of C–H–O Symbiosis Networks. *Process. Integr. Optim. Sustain.* **2017**, *1*, 3–13. [CrossRef]
64. Noureldin, M.M.; El-Halwagi, M.M. Synthesis of C-H-O Symbiosis Networks. *AIChE J.* **2015**, *61*, 1242–1262. [CrossRef]
65. Mukherjee, R.; Asani, R.R.; Boppana, N.; El-Halwagi, M.M. Performance evaluation of shale gas processing and NGL recovery plant under uncertainty of the feed composition. *J. Nat. Gas Sci. Eng.* **2020**, *83*, 103517. [CrossRef]

processes

MDPI

Article

Effect of Hydrogen Bond Donors and Acceptors on CO_2 Absorption by Deep Eutectic Solvents

Tausif Altamash [1], Abdulkarem Amhamed [1], Santiago Aparicio [2,*] and Mert Atilhan [3,*]

[1] Qatar Environment & Energy Research Institute, Hamad Bin Khalifa University, Doha 34110, Qatar; taltamash@hbku.edu.qa (T.A.); aamhamed@hbku.edu.qa (A.A.)

[2] Department of Chemistry, University of Burgos, 09001 Burgos, Spain

[3] Department of Chemical and Paper engineering, Western Michigan University, Kalamazoo, MI 49008, USA

* Correspondence: sapar@ubu.es (S.A.); mert.atilhan@wmich.edu (M.A.)

Received: 7 November 2020; Accepted: 22 November 2020; Published: 25 November 2020

Abstract: The effects of a hydrogen bond acceptor and hydrogen bond donor on carbon dioxide absorption via natural deep eutectic solvents were studied in this work. Naturally occurring non-toxic deep eutectic solvent constituents were considered; choline chloride, b-alanine, and betaine were selected as hydrogen bond acceptors; lactic acid, malic acid, and fructose were selected as hydrogen bond donors. Experimental gas absorption data were collected via experimental methods that uses gravimetric principles. Carbon dioxide capture data for an isolated hydrogen bond donor and hydrogen bond acceptor, as well as natural deep eutectic solvents, were collected. In addition to experimental data, a theoretical study using Density Functional Theory was carried out to analyze the properties of these fluids from the nanoscopic viewpoint and their relationship with the macroscopic behavior of the system, and its ability for carbon dioxide absorption. The combined experimental and theoretical reported approach work leads to valuable discussions on what is the effect of each hydrogen bond donor or acceptor, as well as how they influence the strength and stability of the carbon dioxide absorption in deep eutectic solvents. Theoretical calculations explained the experimental findings, and combined results showed the superiority of the hydrogen bond acceptor role in the gas absorption process, with deep eutectic solvents. Specifically, the cases in which choline chloride was used as hydrogen bond acceptor showed the highest absorption performance. Furthermore, it was observed that when malic acid was used as a hydrogen bond donor, it led to low carbon dioxide solubility performance in comparison to other studied deep eutectic solvents. The cases in which lactic acid was used as a hydrogen bond donor showed great absorption performance. In light of this work, more targeted, specific, deep eutectic solvents can be designed for effective and alternative carbon dioxide capture and management.

Keywords: absorption; carbon dioxide capture; deep eutectic solvents; density functional theory; hydrogen bond

1. Introduction

In recent years, the unprecedented amount of carbon dioxide (CO_2) emissions, resulting mainly from fossil fuel utilization/based activities, have impacted global warming [1,2] and climate change [3–5]. Conventional amine-based CO_2 mitigation techniques have been considered as effective CO_2 capture methods over the past several decades, despite some serious drawbacks, such as solvent loss, corrosion, degradation and, more importantly, high regeneration energy cost [6–12]. Hence, there is need for alternative solvent systems that can effectively scrub CO_2 in gaseous effluent streams, at both pre- and post- combustion processes, with minimum requirement of infrastructure retrofitting costs for existing CO_2 capture process units in plants. For this purpose, various materials have been developed over the past few decades. Porous adsorbents [13–17] as well as liquid solvents [18,19] have been considered in both academia and in the industry; however, due to low manufacturing costs and less requirements on new processing equipment, liquid systems have been considered more attractive. Novel solvent development for CO_2 capture purposes has been centered on ionic liquids [20–25] and liquid polymers [26–29]. In recent years, deep eutectic solvents (DES) [30–32] obtained from ionic liquids, as well as from natural products (natural deep eutectic solvents (NADES)) [33–35], have also been considered for CO_2 management, especially at low to moderate pressures. DES have important advantages over the other potential candidates as they are low-cost materials with appreciable renewability capability [36,37], low toxicity [32,38,39], have less environmental impact [40,41], and good solvent recovery ratio [42].

DES or NADES are formed by mixing the combination of hydrogen bond donors (HBD) and hydrogen bond acceptors (HBA) at various molar mixing ratios to form a low melting point eutectic mixture. Experimental studies show that DES and NADES samples yield to CO_2 solubility that can be compared to current state-of-the-art methods [43]. Several studies on gas solubility in DES with ionic liquids are being used as HBA [30,44–47]. In contrast with DES, only a handful of studies use NADES for CO_2 solubilities; these studies focus mainly on choline chloride as HBA mixed with glycerol/propanediol/malic acids [34] and monoethanolamine [48]. In our previous work, we studied gas absorption performances via novel NADES, which was obtained by considering choline chloride (ChCl), alanine (Al), and betaine (Be) as HBA, and lactic acid (La), malic acid (Ma), and phenylacetic acid (Paa) as HBD. We showed that these NADES perform very well for CO_2 capture, especially at moderate pressures up to 50 bars, and between 298.15 K and 323.15 K isotherms [49–51]. CO_2 sorption experiments showed maximum absorption performance between 3.5 and 5 mmol CO_2/gr of NADES in these experiments at the highest pressures. Considering the monoethanolamine solution (MEA) CO_2 capture performance as 1.8 mmol CO_2/g MEA (117 mg/g) at 24 bar and 313 K [52], CO_2 absorption via NADES have great potential for chemical processes. Despite these promising results, there is no systematic study on how HBA/HBD affect the gas sorption performance, what binding role HBA/HBD play during CO_2 capture, which HBA/HBD have a superior effect once exposed to CO_2 environment, and whether ionic liquid based HBA perform better than the amino acidic-based ones. In this work, we attempt to find answers to these questions, present new CO_2 capture data for HBA and HBD, and compare their performances against their NADES mixtures. Furthermore, detailed density functional theory (DFT) calculations were carried out to explain the details of the binding energies and infer on the CO_2 interaction mechanism of HBA/HBD/NADES separately.

2. Materials and Methods

The following were purchased from Sigma Aldrich: alanine (Al) with ≥98% purity (Chemical Abstracts Service (CAS) number 56-41-7) with melting point of 258 °C; betaine (Be) with ≥98% purity (CAS number 107-43-7) with melting point of 310 °C; choline chloride (ChCl) with ≥98.0% purity (CAS number 67-48-1) with melting point of 302 °C; DL-malic acid (Ma) with 99.0% purity (CAS number 6915-15-7); lactic acid (La) with 85% purity (CAS number 50-21-5); d-fructose (Fr) with ≥99% purity (CAS number 57-48-7); and citric acid (Ca) with ≥99.5% purity (CAS number 77-92-9). All materials with mentioned purities on the boxes were used without further treatment. Carbon dioxide (CO_2) gas with purity of ≥99.99% was obtained from Praxair. In order to form NADES samples, Al, Be, ChCl, were mixed with La, Ma, Ca, and Fr with 1 to 1 (1:1) molar mixing ratios by following vigorous stirring of the mixture until a clear-homogenous solution was obtained, in a glove box, in which humidity and ambient conditions were controlled. Table 1 shows the chemical structures of the studied HBA and HBD, which also form the NADES samples, Al:La, Al:Ma, Be:La, ChCl:Ma, ChCl:La, ChCl:Fr. All of the prepared NADES samples were observed to be liquid state at room temperature and at atmospheric pressure. Structures of the HBA and HBD are provided in Table 1.

Table 1. Structures of hydrogen bond acceptors (HBA) and hydrogen bond donors (HBD) compounds.

Hydrogen Bond Donor (HBD)	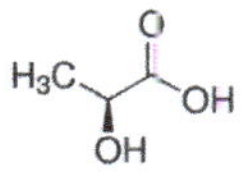 **Lactic Acid**	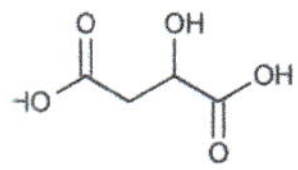 **Malic Acid**	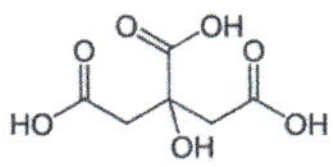 **Malic Acid**	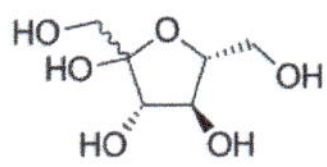 **Fructose**
Hydrogen Bond Acceptor (HBA)	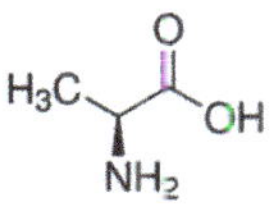 **Alanine**	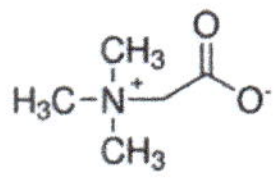 **Betaine**	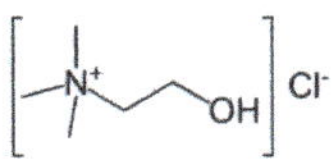 **Choline Chloride**	

2.1. Experimental

CO_2 sorption performance of NADES samples were reported in a previous study, in which the contactless magnetic suspension gravimetric method was used for high-pressure gas absorption-desorption measurements via Rubotherm apparatus. A magnetic suspension apparatus that is equipped with an automated syringe pump was used to conduct both low- and high-pressure gas experiments. In a typical procedure, a few ml of sample is placed in a measurement chamber and it is evacuated overnight at temperatures around 70 °C. Once the material preparation is complete, then solubility measurements take place in the apparatus, starting from the lower pressures toward higher pressures, with stepwise gradual pressure increments. It takes about 60 min to establish equilibrium thermal in between each pressure increment. Once the highest pressure is measured, then few pressure points are re-experimented during the desorption period, in order to observe whether hysteresis occurs during the reversed conditions. The described method was used to obtain HBA + CO_2 and HBD + CO_2 sorption performances in this work, and further details of the measurement technique, calibration, and overall uncertainty can be obtained elsewhere [53].

2.2. Theoretical

Initial structures for all of the considered HBA, HBD compounds, and NADES molecular clusters were built with the Avogadro program [54]; ORCA code [55] was used for DFT calculations along this study for B3LYP functional [56,57], coupled with van der Waals semi-empirical contribution from the DFT-D3 method by Grimme [58], and 6-311++G** basis set. The interaction energy (ΔE) for all of the considered structures were calculated by considering counterpoise correction for Basis Set Superposition Error (BSSE) [59]. The quantum Bader's atoms-in-molecules (AIM) theory [60] was used for interpretation of intermolecular interactions, and for topological analysis, which was obtained via Multiwfn program [61]. From this analysis, specifically interaction regions that were characterized by bond critical points (BCP, (3,−1) type, according to Bader's terminology). The corresponding values of electron density, ρ, and Laplacian (∇) of electron density, $\nabla^2\rho$ were obtained [62]. The properties of the inferred critical points may be related with the strength of the interactions [63,64]. Likewise, quantitative bond critical points and its implications on bonding strength were further analyzed via reduced density gradient analysis (RDG) for visual representation of strength and the nature of intermolecular forces through colored isosurfaces [65].

3. Results

CO_2 solubility experiments for NADES molecular structures were reproduced from previous work. In this study, experimental data on CO_2 capture performance via HBA and HBD were received successfully and illustrated graphically. For comparison purposes, 298.15 K isotherm was considered for qualitative and quantitative analysis. A total of 24 pressure points were collected for each HBA or HBD during CO_2 sorption experiments, in which 12 were for adsorption, and 12 for desorption measurements. Figure 1 shows the CO_2 capture performance for HBA and HBD prior to mixing to form the deep eutectic solvent at 298.15 K isotherm. At a low-pressure side ($p < 5$bars), there is no distinct effect observed on the sorption performance and the data are clustered around ~0.3 mmol CO_2/g sorbent. As the pressure is increased to moderate to high pressures, there is a clear separation between ChCl + CO_2, Be + CO_2 (~4.95 mmol CO_2/g) cases in comparison to Al + CO_2, La + CO_2, Ma + CO_2, Ca + CO_2 (~3.77 mmol CO_2/g) cases. However, Fr + CO_2 falls in between of the trend of these two groups at high pressures yielding 4.45 mmol CO_2/g performance.

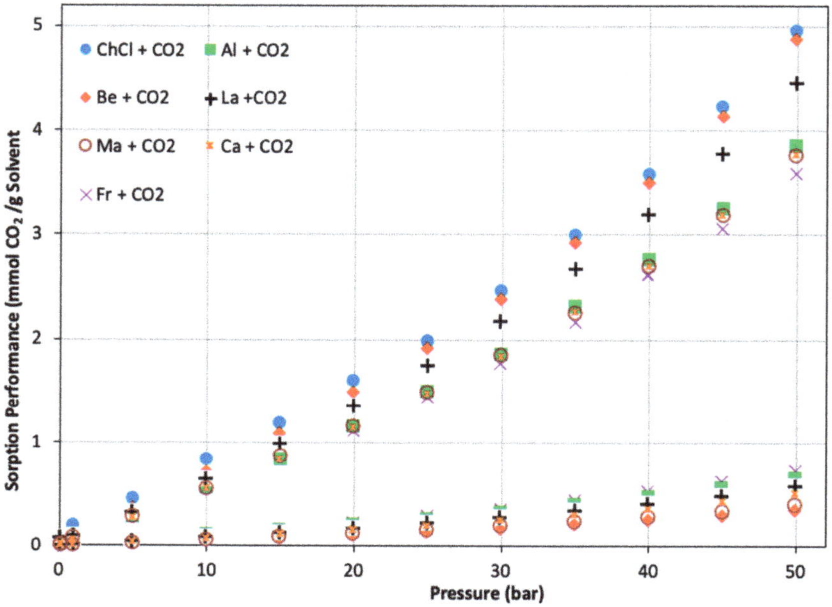

Figure 1. CO_2 capture performance for hydrogen bond acceptors and hydrogen bond donors prior to mixing to form deep eutectic solvent at 298.15 K isotherm.

Furthermore, absorption capacity of DES is provided in electronic supporting information for CO_2 loading amounts in mol/mol units. Figure S1 reveals significant trends on the effect of hydroxyl groups. As the amount of hydroxyl groups in the HBD increases, the absorption of capacity of the CO_2 increase. This is related to the intramolecular and intermolecular hydrogen-bonding forces between the HBA and the CO_2 molecule. For La, Ma, Fr, Ca there are 4, 3, 2, 1 hydroxyl groups in the HBD structure, and the absorption capacity in Figure S1 overlaps with this argument. On the other hand, for the case of HBA, as the volume of the of the group that is attached on the N atom in the HBA structure increases, it affects the distance between center of mass and the CO_2 molecule and, thus, develops a hydrogen bond easier. This leads to the hydrogen-bonding force of DES and improved solvent properties. This phenomenon is also observed in this work when the molecular structures of ChCl, Be, and Al are considered, and their CO_2 sorption capacities are examined in Figure S1. The development of hydrogen bonding is discussed in further detail, from the molecular point of view, in the following section of this manuscript.

CO_2 capture performance comparison of the HBA and HBD with respect to NADES has been studied and presented in Figure 2. It was expected for NADES (or DES) CO_2 capture performance to fall in between its former HBA and HBD compounds. However, interestingly, only Al:La + CO_2 showed this behavior with distinct performance separation between the HBA and HBD CO_2 capture data, and fall right in the middle of these two curves (Figure 2a). For the case of ChCl:La + CO_2, the NADES sorption coincide over the ChCl + CO_2 curve, which is observed to be underperforming in comparison to La + CO_2 case (Figure 2c). For Al:MA + CO_2 case, it has a similar trend with ChCl:La + CO_2 and the NADES + CO_2 trend overlaps again with the HBA trend (Figure 2d), with the exception of a slight departure of NADES towards higher performance at pressures higher than 35 bars. However, for the case of Al:La + CO_2, there is a distinct segregation of HBA+CO_2, HBD+CO_2 and NADES+CO_2 trends (Figure 2c). In this specific case, pressures above 10 bars, CO_2 capture performance trend was observed as La + CO_2 > NADES + CO_2 > Al + CO_2, or in other words HBD + CO_2 > NADES + CO_2 > HBA + CO_2. Maximum solubility performances were obtained via the highest achieved experimental pressure at 50 bars, reported in Table 2. For all cases, except for Be:La + CO_2 (Figure 2b), HBA experimental sorption performances was superior than that of HBD, which was also mentioned by D.O. Abranchesn et al. (that Be possess weak interaction with itself, but act as excellent HBA) [66]. ChCl+CO_2 showed the best performance with 4.96 mmol CO_2/g, whereas Al + CO_2 showed the worst capture performance with 3.86 mmol CO_2/g. In the case of Be:La + CO_2, the NADES + CO_2 profile was observed to be lower than its constituents (Be and La), which can be explained due to the negative excess volume that was created via mixing the HBA and HBD [67]. Likewise, in the case of Al:Ma + CO_2, the NADES profile was observed to be higher than its HBA and HBD, which is a sign of positive excess volume when Al and Ma was mixed to form NADES.

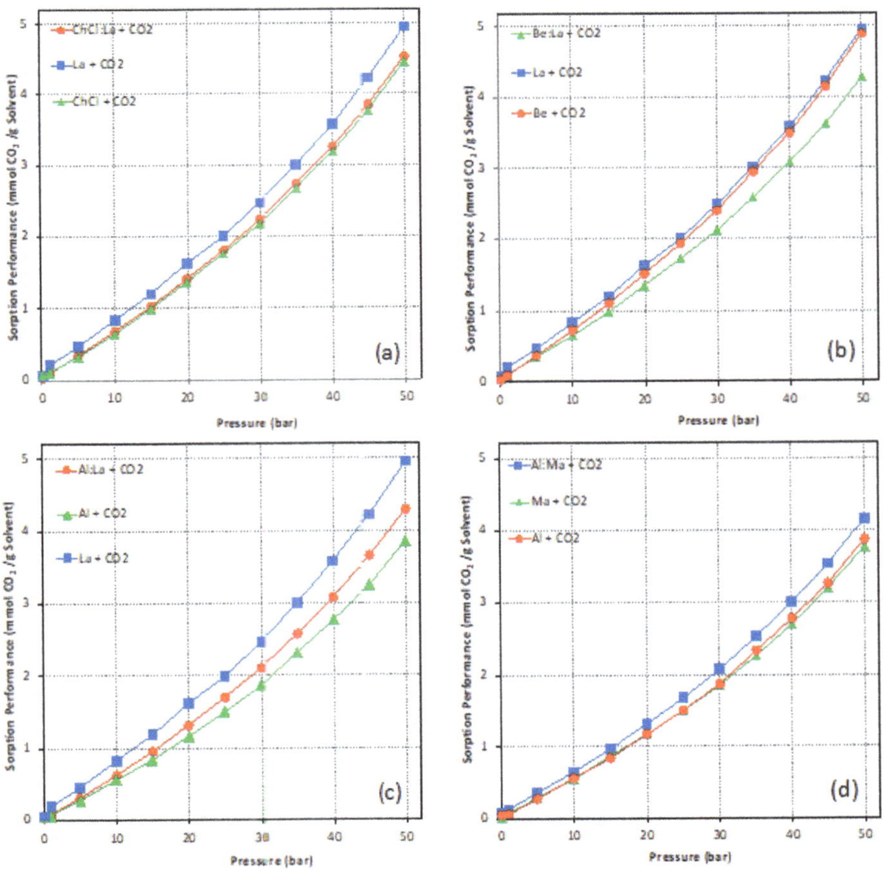

Figure 2. CO_2 capture performance of studied deep eutectic solvents (DES) systems and comparison to their constituents. (**a**) ChCl:La + CO_2, (**b**) Be:La + CO_2, (**c**) Al:La + CO_2, (**d**) Al:Ma + CO_2.

Table 2. Gas sorption data and binding energies ranking.

Structure	Sorption (mmol CO_2/g)	Sorption Comparison	Binding Energy (eV)	Binding Comparison
ChCl + CO_2	4.96		−0.137	
ChCl:La + CO_2	4.52	HBA > NADES > HBD	1.433	HBA > HBD
La + CO_2	4.46		−0.122	
Be + CO_2	4.88		−0.112	
Be:La + CO_2	4.26	HBA > HBD > NADES	1.292	HBD > HBA
La + CO_2	4.46		−0.122	
Al + CO_2	3.86		−0.119	
Al:La + CO_2	4.30	HBD > NADES > HBA	0.828	HBA ~ HBD
La + CO_2	4.46		−0.122	
Al + CO_2	3.86		−0.119	
Al:Ma + CO_2	4.14	NADES > HBA > HBD	N/A	HBA > HBD
Ma + CO_2	3.76		−0.104	
ChCl + CO_2	4.96		−0.137	
ChCl:Fr + CO_2	4.24	HBA > NADES > HBD	2.199	HBA > HBD
Fr + CO_2	3.58		−0.101	
ChCl + CO_2	4.96		−0.137	
ChCl:Ma + CO_2	4.22	HBA > NADES > HBD	1.514	HBA > HBD
Ma + CO_2	3.76		−0.104	

As clearly reported in Table 1, the CO_2 capture performance of NADES is lower in comparison to sole HBA and HBD cases, for most of the studied cases. When the melting point of the studied HBA and HBD are considered, they are solid in room temperatures, and even at elevated process temperatures, at which typical pre- and post-combustion CO_2 capture operations take place. However, the NADES compounds exist in liquid phase at mentioned temperatures, thus, making their processability much easier than their constituents. One of the main objectives of considering solvents for CO_2 capture is to be able to utilize existing infrastructures that were built with the consideration of current-state-of-the-art capture agents (e.g., monoethanolamine-based solvents), with minimum retrofitting requirements on the equipment infrastructure.

DFT simulations were carried out for the same structures for which the experimental findings were shared above. The main purpose of the utilization DFT simulations was to obtain insights on electronic configuration of the studied compounds and infer on the behavior of the interaction sites between the HBA/HBD/NADES structures and the CO_2 molecule. Macroscopic properties can be estimated through calculation intensive molecular dynamic simulations. DFT simulations were carried out for each NADES, HBA, and HBD with CO_2 gas presented around them at various spatial positions. Three different spatial positions were considered for potentially high CO_2 interactions sites for NADES and two positions were considered for HBA/HBD compounds, respectively. Obviously, a restricted amount of CO_2 molecules around the NADES structure would not represent the overall bulk phase conditions and the entire solubility phenomena. It should be noted that DFT is considered a tool to assess the interaction nature as well as characteristics of the studied compounds in case they are exposed to CO_2.

Once the DFT simulations were obtained and the interactions energies were corrected for the BSSE [59], the cases that gave the highest interaction energies were considered for further DFT analysis. In this work B3LYP with the 6-311++G** theory level was selected, based on its accuracy and reasonable computational time [49,50]. The summary of interaction energies for the selected cases are provided in Table 3 in eV units. Binding energies for each HBA, HBD, and NADES with CO_2 are provided in Table 4, along with comparisons of binding energies with respect to each other within each NADES group.

Table 3. Summary of energies of optimized structures.

Group	Structure	Energy (eV)
NADES	Al:La	−18,157.79
	Be:La	−20,295.39
	ChCl:Fr	−40,166.78
	ChCl:La	−30,819.46
	ChCl:Ma	−35,950.19
Gas	CO_2	−5,131.10
HBD	La	−9,348.54
	Fr	−18,695.51
	Ma	−14,479.15
HBA	Al	−8,807.68
	Be	−10,944.60
	ChCl	−21,468.77

Table 4. Optimized energies of HBA, HBD, and natural deep eutectic solvents (NADES) interacting with CO_2 at various spatial positions.

Group	Structure	Energy (eV)	Binding Energy (eV)	Average Binding Energy (eV)
NADES + CO_2	Al:La + CO2_p1	−23,288.051	0.840	0.828
	Al:La + CO2_p2	−23,288.072	0.819	
	Al:La + CO2_p3	−23,288.065	0.826	
	Be:La + CO2_p1	−25,425.242	1.248	1.292
	Be:La + CO2_p2	−25,425.200	1.289	
	Be:La + CO2_p3	−25,425.150	1.339	
	ChCl:Fr + CO2_p1	−45,295.714	2.167	2.199
	ChCl:Fr + CO2_p2	−45,295.679	2.202	
	ChCl:Fr + CO2_p3	−45,295.654	2.227	
	ChCl:La + CO2_p1	−35,949.184	1.370	1.433
	ChCl:La + CO2_p2	−35,949.103	1.451	
	ChCl:La + CO2_p3	−35,949.075	1.478	
	ChCl:Ma + CO2_p1	−41,079.907	1.379	1.514
	ChCl:Ma + CO2_p2	−41,079.733	1.553	
	ChCl:Ma + CO2_p3	−41,079.676	1.610	
HBA + CO_2	Al + CO2_p1	−13,938.915	−0.138	−0.119
	Al + CO2_p2	−13,938.876	−0.099	
	Be + CO2_p1	−16,075.871	−0.173	−0.112
	Be + CO2_p2	−16,075.749	−0.051	
	ChCl + CO2_P1	−26,600.028	−0.158	−0.137
	ChCl + CO2_P2	−26,599.997	−0.127	
	ChCl + CO2_P3	−26,599.996	−0.126	

Table 4. *Cont.*

Group	Structure	Energy (eV)	Binding Energy (eV)	Average Binding Energy (eV)
HBD + CO_2	Fr + CO2_p1	−23,826.696	−0.093	−0.101
	Fr + CO2_p2	−23,826.712	−0.108	
	La + CO2_p1	−14,479.781	−0.145	−0.122
	La + CO2_p2	−14,479.736	−0.099	
	Ma + CO2_p1	−19,610.352	−0.101	−0.104
	Ma + CO2_p2	−19,610.358	−0.106	

Once the binding energy results are deeply analyzed, HBA > HBD trend is observed for CO_2 interactions, which is similar to the experimental behavior CO_2 sorption performance in HBA/HBD. The only exception was observed in Be + CO_2 and La + CO_2 comparisons, for which the binding energy is higher for HBD + CO_2 than HBA + CO_2 case. There is no particular correlation between the maximum sorption amount and the binding energy for the NADES + CO_2 cases. On the other hand, the maximum sorption amounts and the binding energies for both HBD and HBA follows the same rank. In other words, it was observed that as the binding energy for CO_2 increases, the CO_2 sorption performance also increases for both HBD and HBA cases.

Furthermore, the interfacial properties of NADES/HBA/HBD and CO_2 can also be studied using molecular dynamics simulations. Garcia et al. showed how CO_2 establishes remarkable interactions with the bulk NADES structure through analyzing the evolution of the number of hydrogen bondings between the HBA and HBD [46]. Furthermore, strong affinity between DES and CO_2 molecules were quantified by the corresponding interaction energies. Results shared by Garcia et al. showed the large interaction energies for interaction of CO_2 molecules, with both the HBA and HBD, with slightly higher values between HBA and CO_2 in comparison to HBD–CO_2. Relatively small differences between the HBA and CO_2 and HBD–CO_2 suggests a mutual effect on CO_2 solubility [46].

On the other hand, a study by Wang et al. showed the effect of the type and molar mixing ratio of HBA and HBD for ionic liquid-based DES through analyzing the radial distribution functions. It was reported that the cation of the HBA plays a superior role on the hydrogen bond development between the DES and CO_2, yet again supporting the observed trend in this work [68].

Another molecular dynamics-based study that supports the CO_2 affinity is driven by HBA of the DES was reported elsewhere [69]. Besides analyzing the hydrogen bond evolution between the CO_2 molecule and HBA/HBD, they proved the superior effect of HBA on CO_2 affinity via displaying the higher degree of clustering of CO_2 molecules around the HBA through the spatial distribution function isosurfaces. All of these studies are in line with the arguments that are claimed in this work through DFT simulations on the superior binding effect of HBA on CO_2.

The reduced density gradients (RDG) isosurfaces were used to visualize the interaction of CO_2 with HBA and HBD, Figure 3. Strong van der Waals type interactions are recorded for CO_2 binding at relevant interaction sites at HBA and HBD, which is a sign of reversible process for CO_2 sorption since H-bonding has not been observed. Especially at ChCl, both -H and -Cl^- sites showed more intensified RDG volumes between the -O site of CO_2 (Figure 3c), which is also quantified with the highest binding energy with −0.137 eV (Table 2). Likewise, weakest interaction was observed with less intense isosurface between CO_2 and Ma (Figure 3e), and it is evident with the binding energy reported in Table 2 for Ma + CO_2.

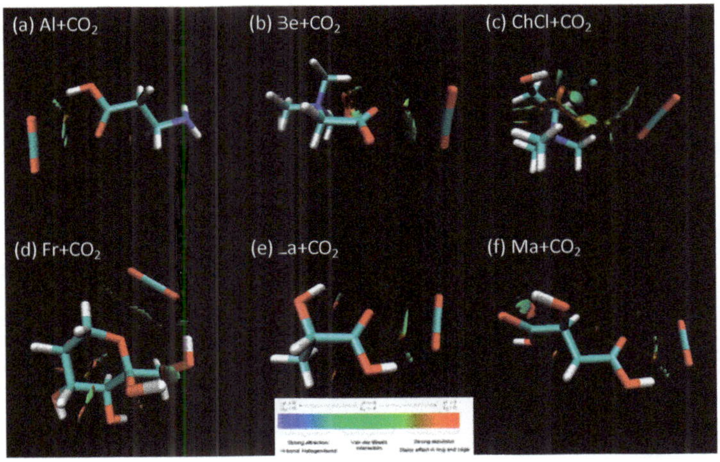

Figure 3. Reduced Density Gradient (RDG) isosurfaces for HBA + CO_2 and HBD + CO_2 cases. (**a**) Al + CO_2, (**b**) Be + CO_2, (**c**) ChCl + CO_2, (**d**) Fr + CO_2, (**e**) La + CO_2, (**f**) Ma + CO_2.

Quantum theory atom-in-a-molecule (QTAIM) analysis and RDG isosurfaces for NADES was presented for NADES + CO_2 cases in Figures 4 and 5, respectively. For most of the cases, CO_2 established critical binding with HBA of the NADES structure, except for the ChCl:Fr + CO_2 case. In Figure 4, it is evident that -O site of CO_2 is the most recurring interaction between the HBA -H site. The experimental values for the CO_2 solubility in NADES can be ranked as ChCl:La > Al:La ~ ChCl:Ma > Be:La > Al:La. Whereas the binding energies that are calculated via DFT simulations for CO_2 are ranked as ChCl:Fr > ChCl:La ~ ChCl:Ma > Be:La > Al:La. These two comparisons do not overlap with each other. A more logical comparison on overall solubility performance, with respect to binding energies, would be between the NADES constituent prior to the mixing, since they form excess volume pockets once they are mixed, which leads to excess CO_2 capture performance.

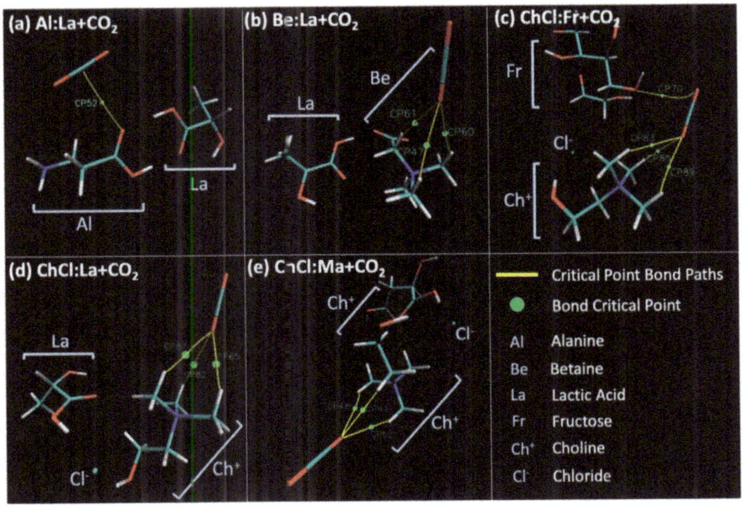

Figure 4. Density functional theory (DFT) figures for each structure final optimized geometry.

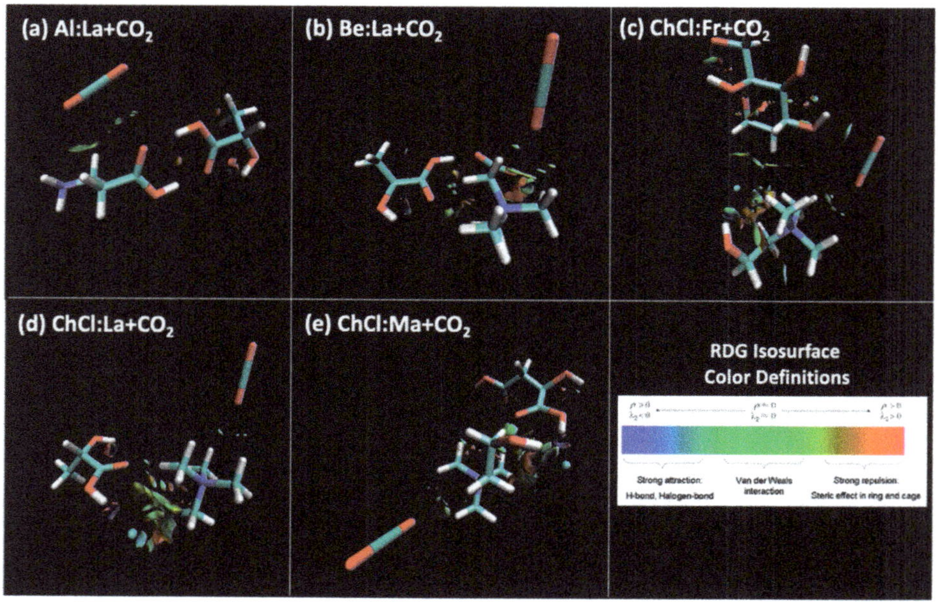

Figure 5. Reduced density gradient (RDG) isosurfaces for HBA + CO_2 and HBD + CO_2 cases.

4. Conclusions

In this study, the mechanism of CO_2 sorption amino acid based NADES, the effect of the hydrogen bond acceptor and hydrogen bond donor on carbon dioxide absorption via natural deep eutectic solvent systems is presented for the first time. These results will assist the scientific community while designing NADES for gas sorption applications. Both experimental and DFT simulation for NADES, HBA, and HBD were studied for their CO_2 sorption performances. Strong van der Waals forces were observed as binding characteristics of CO_2 with all of the studied cases, which supports the physisorption behavior of the HBA + CO_2, HBD + CO_2, as well as NADES + CO_2 experimental behavior that did not yield any adsorption-desorption hysteresis. Furthermore, it has been observed that HBA plays a leading role in CO_2 solubility for NADES cases. Polarity unbalance of Be and ChCl based NADES cases causes strong negative deviations from ideality when mixed with their organic HBD. This effect was also observed on the CO_2 absorption plots (Figure 2), at which the NADES solubility performances were pushed further down, yielding lower performance in comparison to both HBA and HBD. Furthermore, considering Be and ChCl cases, both the experimental sorption and theoretical interaction energies are close to each other and higher than that of the Al cases. This could be explained, due to both HBA forming a positive charge shielded by alkyl groups, forming stronger intra H-bonding between the HBA and HBD, which also leads to larger interactions with surrounding CO_2 molecule. Furthermore, such strong H-bonding establishment in between the HBA and HBD, for all cases, but specific to Be and ChCl cases, negative deviations from ideality in mixture, can be correlated to higher CO_2 affinity establishment. Additionally, the strong H-bonding between HBA and HBD leads to very strong NADES structure and, thus, no disruption effect has been noticed when the NADES compounds were exposed to the CO_2 environment. Consequently, the characteristics of NADES has been preserved in these cases. For novel solvent design via DES or NADES, in order to confirm the leading effect of HBA for gas solubility, there is a need for a more systematic analysis, especially considering a wider range of HBA and HBD, as well as different gases.

Supplementary Materials: The following are available online at http://www.mdpi.com/2227-9717/8/12/1533/s1, Figure S1: title. CO_2 capture performance for hydrogen bond acceptors and hydrogen bond donors prior to mixing to form deep eutectic solvent at 298.15 K isotherm with (mol CO_2/mol solvent units).

Author Contributions: Conceptualization, T.A., S.A. and M.A.; methodology, S.A. and M.A.; software, S.A. and M.A.; validation, T.A., S.A. and M.A.; formal analysis, T.A., A.A., S.A. and M.A.; investigation, S.A. and M.A.; resources, A.A., S.A. and M.A.; data curation, M.A.; writing—original draft preparation, T.A., S.A. and M.A.; writing—review and editing, T.A., A.A., S.A. and M.A.; visualization, M.A.; supervision, A.A., S.A. and M.A.; project administration, M.A. All authors have read and agreed to the published version of the manuscript.

Funding: This research received no external funding.

Acknowledgments: The authors acknowledge SCAYLE (Supercomputación Castilla y León, Burgos, Spain) for providing supercomputing facilities. The statements made herein are solely the responsibility of the authors.

Conflicts of Interest: The authors declare no conflict of interest.

References

1. Macário, I.P.E.; Oliveira, H.; Menezes, A.C.; Ventura, S.P.M.; Pereira, J.L.; Gonçalves, A.M.M.; Coutinho, J.A.P.; Gonçalves, F.J.M. Cytotoxicity profiling of deep eutectic solvents to human skin cells. *Sci. Rep.* **2019**, *9*, 3932. [CrossRef]
2. Halder, A.K.; Cordeiro, M.N.D.S. Probing the Environmental Toxicity of Deep Eutectic Solvents and Their Components: An In Silico Modeling Approach. ACS Sustain. *Chem. Eng.* **2019**, *7*, 10649–10660. [CrossRef]
3. Valencia-Marquez, D.; Flores-Tlacuahuac, A.; Vasquez-Medrano, R. An optimization approach for CO_2 capture using ionic liquids. *J. Clean. Prod.* **2017**, *168*, 1652–1667. [CrossRef]
4. Bi, Y.; Hu, Z.; Lin, X.; Ahmad, N.; Xu, J.; Xu, X. Efficient CO_2 capture by a novel deep eutectic solvent through facile, one-pot synthesis with low energy consumption and feasible regeneration. *Sci. Total Environ.* **2020**, *705*, 135798. [CrossRef]
5. Leron, R.B.; Li, M.-H. Solubility of carbon dioxide in a choline chloride–ethylene glycol based deep eutectic solvent. *Thermochim. Acta* **2013**, *551*, 14–19. [CrossRef]
6. Sreedhar, I.; Nahar, T.; Venugopal, A.; Srinivas, B. Carbon capture by absorption–Path covered and ahead. Renew. *Sustain. Energy Rev.* **2017**, *76*, 1080–1107. [CrossRef]
7. Mirza, N.; Mumford, K.; Wu, Y.; Mazhar, S.; Kentish, S.; Stevens, G. Improved Eutectic Based Solvents for Capturing Carbon Dioxide (CO_2). *Energy Procedia* **2017**, *114*, 827–833. [CrossRef]
8. Hjelmaas, S.; Storheim, E.; Flø, N.E.; Thorjussen, E.S.; Morken, A.K.; Faramarzi, L.; de Cazenove, T.; Hamborg, E.S. Results from MEA Amine Plant Corrosion Processes at the CO_2 Technology Centre Mongstad. *Energy Procedia* **2017**, *114*, 1166–1178. [CrossRef]
9. Husebye, J.; Brunsvold, A.L.; Roussanaly, S.; Zhang, X. Techno Economic Evaluation of Amine based CO_2 Capture: Impact of CO_2 Concentration and Steam Supply. *Energy Procedia* **2012**, *23*, 381–390. [CrossRef]
10. Davis, J.; Rochelle, G. Thermal degradation of monoethanolamine at stripper conditions. *Energy Procedia* **2009**, *1*, 327–333. [CrossRef]
11. Luis, P. Use of monoethanolamine (MEA) for CO_2 capture in a global scenario: Consequences and alternatives. *Desalination* **2016**, *380*, 93–99. [CrossRef]
12. Abotaleb, A.; El-Naas, M.; Amhamed, A. Enhancing gas loading and reducing energy consumption in acid gasremoval systems: A simulation study based on real NGL plant data. *J. Nat. Gas Sci. Eng.* **2018**, *55*, 565–574. [CrossRef]
13. Rozyyev, V.; Thirion, D.; Ullah, R.; Lee, J.; Jung, M.; Oh, H.; Atilhan, M.; Yavuz, C.T. High-capacity methane storage in flexible alkane-linked porous aromatic network polymers. *Nat. Energy* **2019**, *4*, 604–611. [CrossRef]
14. Furukawa, H.; Cordova, K.E.; O'Keeffe, M.; Yaghi, O.M. The Chemistry and Applications of Metal-Organic Frameworks. *Science* **2013**, *341*, 1230444. [CrossRef]
15. Megías-Sayago, C.; Bingre, R.; Huang, L.; Lutzweiler, G.; Wang, Q.; Louis, B. CO_2 Adsorption Capacities in Zeolites and Layered Double Hydroxide Materials. *Front. Chem.* **2019**, *7*, 551. [CrossRef]
16. Titinchi, S.J.J. Chemically Modified Solid Adsorbents for CO_2 Capture. *Energy Procedia* **2014**, *8*, 8153–8160. [CrossRef]
17. Jiang, L.; Gonzalez-Diaz, A.; Ling-Chin, J.; Roskilly, A.P.; Smallbone, A.J. Post-combustion CO_2 capture from a natural gas combined cycle power plant using activated carbon adsorption. *Appl. Energy* **2019**, *245*, 1–15. [CrossRef]

18. Closmann, F.; Nguyen, T.; Rochelle, G.T. MDEA/Piperazine as a solvent for CO_2 capture. *Energy Procedia* **2009**, *1*, 1351–1357. [CrossRef]
19. Liu, J.; Li, X.; Zhang, Z.; Li, L.; Bi, Y.; Zhang, L. Promotion of CO_2 capture performance using piperazine (PZ) and diethylenetriamine (DETA) bi-solvent blends. Greenh. *Gases Sci. Technol.* **2019**, *9*, 349–359. [CrossRef]
20. Aghaie, M.; Rezaei, N.; Zendehboudi, S. A systematic review on CO_2 capture with ionic liquids: Current status and future prospects. *Renew. Sustain. Energy Rev.* **2018**, *96*, 502–525. [CrossRef]
21. Ramdin, M.; de Loos, T.W.; Vlugt, T.J.H. State-of-the-Art of CO_2 Capture with Ionic Liquids. *Ind. Eng. Chem. Res.* **2012**, *51*, 8149–8177. [CrossRef]
22. Chong, F.K.; Chemmangattuvalappil, N.; Foo, D.C.Y.; Atilhan, M.; Eljack, F.T. Ionic Liquid Mixture Design for Carbon Capture using Property Clustering Technique. *Chem. Eng. Trans.* **2015**, *45*, 1567–1572. [CrossRef]
23. Shiflett, M.B.; Drew, D.W.; Cantini, R.A.; Yokozeki, A. Carbon Dioxide Capture Using Ionic Liquid 1-Butyl-3-methylimidazolium Acetate. *Energy Fuels* **2010**, *24*, 5781–5789. [CrossRef]
24. García, G.; Atilhan, M.; Aparicio, S. Simultaneous CO_2 and SO_2 capture by using ionic liquids: A theoretical approach. *Phys. Chem. Chem. Phys.* **2017**, *19*, 5411–5422. [CrossRef] [PubMed]
25. Rafat, A.; Atilhan, M.; Kahraman, R. Corrosion Behavior of Carbon Steel in CO_2 Saturated Amine and Imidazolium-, Ammonium-, and Phosphonium-Based Ionic Liquid Solutions. *Ind. Eng. Chem. Res.* **2016**, *55*, 446–454. [CrossRef]
26. Bhawna; Pandey, A.; Dhingra, D.; Pandey, S. Can common liquid polymers and surfactants capture CO_2? *J. Mol. Liq.* **2019**, *277*, 594–605. [CrossRef]
27. Zulfiqar, S.; Ilyas Sarwar, M.; Mecerreyes, D. Polymeric ionic liquids for CO_2 capture and separation: Potential, progress and challenges. *Polym. Chem.* **2015**, *6*, 6435–6451. [CrossRef]
28. Kupgan, G.; Abbott, L.J.; Hart, K.E.; Colina, C.M. Modeling Amorphous Microporous Polymers for CO_2 Capture and Separations. *Chem. Rev.* **2018**, *118*, 5488–5538. [CrossRef]
29. Zhang, J.; Chai, S.-H.; Qiao, Z.-A.; Mahurin, S.M.; Chen, J.; Fang, Y.; Wan, S.; Nelson, K.; Zhang, P.; Dai, S. Porous Liquids: A Promising Class of Media for Gas Separation. *Angew. Chem.* **2015**, *127*, 946–950. [CrossRef]
30. Zhang, N.; Huang, Z.; Zhang, H.; Ma, J.; Jiang, B.; Zhang, L. Highly Efficient and Reversible CO_2 Capture by Task-Specific Deep Eutectic Solvents. *Ind. Eng. Chem. Res.* **2019**, *58*, 13321–13329. [CrossRef]
31. Trivedi, T.J.; Lee, J.H.; Lee, H.J.; Jeong, Y.K.; Choi, J.W. Deep eutectic solvents as attractive media for CO_2 capture. *Green Chem.* **2016**, *18*, 2834–2842. [CrossRef]
32. García, G.; Aparicio, S.; Ullah, R.; Atilhan, M. Deep Eutectic Solvents: Physicochemical Properties and Gas Separation Applications. *Energy Fuels* **2015**, *29*, 2616–2644. [CrossRef]
33. Ren, H.; Lian, S.; Wang, X.; Zhang, Y.; Duan, E. Exploiting the hydrophilic role of natural deep eutectic solvents for greening CO_2 capture. *J. Clean. Prod.* **2018**, *193*, 802–810. [CrossRef]
34. Mulia, K.; Putri, S.; Krisanti, E.; Nasruddin. Natural deep eutectic solvents (NADES) as green solvents for carbon dioxide capture. *AIP Conf. Proc.* **2017**, *1823*, 020022. [CrossRef]
35. Amhamed, A.; Atilhan, M.; Berdiyorov, G. Permeabilities of CO_2, H_2S and CH4 through Choline-Based Ionic Liquids: Atomistic-Scale Simulations. *Molecules* **2019**, *24*, 2014. [CrossRef]
36. Ma, C.; Sarmad, S.; Mikkola, J.-P.; Ji, X. Development of Low-Cost Deep Eutectic Solvents for CO_2 Capture. *Energy Procedia* **2017**, *142*, 3320–3325. [CrossRef]
37. Cruz, H.; Jordão, N.; Branco, L.C. Deep eutectic solvents (DESs) as low-cost and green electrolytes for electrochromic devices. *Green Chem.* **2017**, *19*, 1653–1658. [CrossRef]
38. Ullah, R.; Atilhan, M.; Anaya, B.; Khraisheh, M.; García, G.; ElKhattat, A.; Tariq, M.; Aparicio, S. A detailed study of cholinium chloride and levulinic acid deep eutectic solvent system for CO_2 capture via experimental and molecular simulation approaches. *Phys. Chem. Chem. Phys.* **2015**, *17*, 20941–20960. [CrossRef]
39. Liu, Y.; Yu, H.; Sun, Y.; Zeng, S.; Zhang, X.; Nie, Y.; Zhang, S.; Ji, X. Screening Deep Eutectic Solvents for CO_2 Capture with COSMO-RS. *Front. Chem.* **2020**, *8*, 82. [CrossRef]
40. Paiva, A.; Craveiro, R.; Aroso, I.; Martins, M.; Reis, R.L.; Duarte, A.R.C. Natural Deep Eutectic Solvents–Solvents for the 21st Century. *ACS Sustain. Chem. Eng.* **2014**, *2*, 1063–1071. [CrossRef]
41. Isaifan, R.J.; Amhamed, A. Review on Carbon Dioxide Absorption by Choline Chloride/Urea Deep Eutectic Solvents. *Adv. Chem.* **2018**, *2018*, 2675659. [CrossRef]
42. Gu, Y.; Hou, Y.; Ren, S.; Sun, Y.; Wu, W. Hydrophobic Functional Deep Eutectic Solvents Used for Efficient and Reversible Capture of CO_2. *ACS Omega* **2020**, *5*, 6809–6816. [CrossRef] [PubMed]

43. Abbott, A.P.; Boothby, D.; Capper, G.; Davies, D.L.; Rasheed, R.K. Deep Eutectic Solvents Formed between Choline Chloride and Carboxylic Acids: Versatile Alternatives to Ionic Liquids. *J. Am. Chem. Soc.* **2004**, *126*, 9142–9147. [CrossRef] [PubMed]
44. Sze, L.L.; Pandey, S.; Ravula, S.; Pandey, S.; Zhao, H.; Baker, G.A.; Baker, S.N. Ternary Deep Eutectic Solvents Tasked for Carbon Dioxide Capture. *ACS Sustain. Chem. Eng.* **2014**, *2*, 2117–2123. [CrossRef]
45. Sarmad, S.; Mikkola, J.-P.; Ji, X. Carbon Dioxide Capture with Ionic Liquids and Deep Eutectic Solvents: A New Generation of Sorbents. *ChemSusChem* **2017**, *10*, 324–352. [CrossRef]
46. García, G.; Atilhan, M.; Aparicio, S. Interfacial Properties of Deep Eutectic Solvents Regarding to CO_2 Capture. *J. Phys. Chem. C* **2015**, *119*, 21413–21425. [CrossRef]
47. Zubeir, L.F.; van Osch, D.J.G.P.; Rocha, M.A.A.; Banat, F.; Kroon, M.C. Carbon Dioxide Solubilities in Decanoic Acid-Based Hydrophobic Deep Eutectic Solvents. *J. Chem. Eng. Data* **2018**, *63*, 913–919. [CrossRef]
48. Haider, M.B.; Jha, D.; Marriyappan Sivagnanam, B.; Kumar, R. Thermodynamic and Kinetic Studies of CO_2 Capture by Glycol and Amine-Based Deep Eutectic Solvents. *J. Chem. Eng. Data* **2018**, *63*, 2671–2680. [CrossRef]
49. Altamash, T.; Amhamed, A.I.; Aparicio, S.; Atilhan, M. Combined Experimental and Theoretical Study on High Pressure Methane Solubility in Natural Deep Eutectic Solvents. *Ind. Eng. Chem. Res.* **2019**, *58*, 8097–8111. [CrossRef]
50. Altamash, T.; Nasser, M.S.; Elhamarnah, Y.; Magzoub, M.; Ullah, R.; Qiblawey, H.; Aparicio, S.; Atilhan, M. Gas solubility and rheological behavior study of betaine and alanine based natural deep eutectic solvents (NADES). *J. Mol. Liq.* **2018**, *256*, 286–295. [CrossRef]
51. Altamash, T.; Nasser, M.S.; Elhamarnah, Y.; Magzoub, M.; Ullah, R.; Anaya, B.; Aparicio, S.; Atilhan, M. Gas Solubility and Rheological Behavior of Natural Deep Eutectic Solvents (NADES) via Combined Experimental and Molecular Simulation Techniques. *ChemistrySelect* **2017**, *2*, 7278–7295. [CrossRef]
52. Ebner, A.D.; Gray, M.L.; Chisholm, N.G.; Black, Q.T.; Mumford, D.D.; Nicholson, M.A.; Ritter, J.A. Suitability of a Solid Amine Sorbent for CO_2 Capture by Pressure Swing Adsorption. *Ind. Eng. Chem. Res.* **2011**, *50*, 5634–5641. [CrossRef]
53. Karadas, F.; Yavuz, C.T.; Zulfiqar, S.; Aparicio, S.; Stucky, G.D.; Atilhan, M. CO_2 Adsorption Studies on Hydroxy Metal Carbonates M(CO_3)x(OH)y (M = Zn, Zn–Mg, Mg, Mg–Cu, Cu, Ni, and Pb) at High Pressures up to 175 bar. *Langmuir* **2011**, *27*, 10642–10647. [CrossRef] [PubMed]
54. Adams, S.; De Castro, P.; Echenique, P.; Estrada, J.; Hanwell, M.D.; Murray-Rust, P.; Sherwood, P.; Thomas, J.; Townsend, J. The Quixote project: Collaborative and Open Quantum Chemistry data management in the Internet age. *J. Cheminform.* **2011**, *3*, 38. [CrossRef] [PubMed]
55. Neese, F. The ORCA program system. *WIREs Comput. Mol. Sci.* **2012**, *2*, 73–78. [CrossRef]
56. Lee, C.; Yang, W.; Parr, R.G. Development of the Colle-Salvetti correlation-energy formula into a functional of the electron density. *Phys. Rev. B* **1988**, *37*, 785–789. [CrossRef] [PubMed]
57. Becke, A.D. Density-functional exchange-energy approximation with correct asymptotic behavior. *Phys. Rev. A* **1988**, *38*, 3098–3100. [CrossRef]
58. Grimme, S.; Antony, J.; Ehrlich, S.; Krieg, H. A consistent and accurate ab initio parametrization of density functional dispersion correction (DFT-D) for the 94 elements H-Pu. *J. Chem. Phys.* **2010**, *132*, 154104. [CrossRef]
59. Simon, S.; Duran, M.; Dannenberg, J.J. How does basis set superposition error change the potential surfaces for hydrogen-bonded dimers? *J. Chem. Phys.* **1996**, *105*, 11024–11031. [CrossRef]
60. Bader, R.F.W. *Atoms in Molecules: A Quantum Theory*; Clarendon Press: Oxford, UK, 1994.
61. Lu, T.; Chen, F. Multiwfn: A multifunctional wavefunction analyzer. *J. Comput. Chem.* **2012**, *33*, 580–592. [CrossRef]
62. Aparicio, S.; Yavuz, C.T.; Atilhan, M. Molecular Insights into Benzimidazole-Linked Polymer Interactions with Carbon Dioxide and Nitrogen. *ChemistrySelect* **2018**, *3*, 3691–3701. [CrossRef]
63. Xiao, J.; Zhao, Y.-P.; Fan, X.; Cao, J.-P.; Kang, G.-J.; Zhao, W.; Wei, X.-Y. Hydrogen bonding interactions between the organic oxygen/nitrogen monomers of lignite and water molecules: A DFT and AIM study. *Fuel Process. Technol.* **2017**, *163*, 58–64. [CrossRef]
64. Anbu, V.; Vijayalakshmi, K.A.; Karunathan, R.; Stephen, A.D.; Nidhin, P.V. Explosives properties of high energetic trinitrophenyl nitramide molecules: A DFT and AIM analysis. *Arab. J. Chem.* **2019**, *12*, 621–632. [CrossRef]

65. Johnson, E.R.; Keinan, S.; Mori-Sánchez, P.; Contreras-García, J.; Cohen, A.J.; Yang, W. Revealing Noncovalent Interactions. *J. Am. Chem. Soc.* **2010**, *132*, 6498–6506. [CrossRef] [PubMed]
66. Abranches, D.O.; Silva, L.P.; Martins, M.A.R.; Pinho, S.P.; Coutinho, J.A.P. Understanding the Formation of Deep Eutectic Solvents: Betaine as a Universal Hydrogen Bond Acceptor. *ChemSusChem* **2020**, *13*, 4916–4921. [CrossRef] [PubMed]
67. Crespo, E.A.; Costa, J.M.L.; Palma, A.M.; Soares, B.; Martín, M.C.; Segovia, J.J.; Carvalho, P.J.; Coutinho, J.A.P. Thermodynamic characterization of deep eutectic solvents at high pressures. *Fluid Phase Equilibria* **2019**, *500*, 112249. [CrossRef]
68. Wang, J.; Cheng, H.; Song, Z.; Chen, L.; Deng, L.; Qi, Z. Carbon Dioxide Solubility in Phosphonium-Based Deep Eutectic Solvents: An Experimental and Molecular Dynamics Study. *Ind. Eng. Chem. Res.* **2019**, *58*, 17514–17523. [CrossRef]
69. Altamash, T.; Atilhan, M.; Aliyan, A.; Ullah, R.; Garcia, G.; Aparicio, M. Insights into choline chloride-phenylacetic acid deep eutectic solvent for CO_2 absorption. *RSC Adv.* **2016**, *6*, 109201–109210. [CrossRef]

processes MDPI

Article

A Kraft Mill-Integrated Hydrothermal Liquefaction Process for Liquid Fuel Co-Production

Benjamin H. Y. Ong [1,2], Timothy G. Walmsley [1,*], Martin J. Atkins [1] and Michael R. W. Walmsley [1]

1 Energy Systems Integration Group, School of Engineering, University of Waikato, 3216 Hamilton, New Zealand; benjamin.ong@hslu.ch (B.H.Y.O.); martin.atkins@waikato.ac.nz (M.J.A.); walmsley@waikato.ac.nz (M.R.W.W.)

2 Competence Centre for Thermal Energy Systems and Process Engineering, University of Lucerne of Applied Sciences and Arts, 6048 Horw, Switzerland

* Correspondence: tim.walmsley@waikato.ac.nz

Received: 7 September 2020; Accepted: 25 September 2020; Published: 28 September 2020

Abstract: There is a growing awareness of the need to mitigate greenhouse gas emissions and the inevitable depletion of fossil fuel. With the market pull for the growth in sustainable and renewable alternative energy, the challenge is to develop cost-effective, large-scale renewable energy alternatives for all energy sectors, of which transport fuels are one significant area. This work presents a summary of novel methods for integrating kraft mills with a hydrothermal liquefaction process. The application of these methods has resulted in a proposed kraft mill-integrated design that produces a liquid fuel and could provide net mitigation of 64.6 kg CO_2-e/GJ, compared to conventional petrol and diesel fuels, at a minimum fuel selling price of 1.12–1.38 NZD/LGE of fuel, based on the case study. This paper concludes that a hydrothermal liquefaction process with product upgrading has promising economic potential and environmental benefits that are significantly amplified by integrating with an existing kraft mill. At the current global kraft pulp production rate, if each kraft mill transforms into a biorefinery based on hydrothermal liquefaction, the biofuel production is an estimated 290 Mt (9.9 EJ).

Keywords: hydrothermal liquefaction; black liquor; process integration; techno-economic analysis; utility system; total site heat integration

1. Introduction

The World Commission on Environment and Development [1] defined sustainable development as development that meets the needs of the present without compromising the ability of future generations to meet their own needs. With the growing awareness on the need to mitigate greenhouse gas (GHG) emissions and the inevitable depletion of fossil fuel, the world is on the journey of transitioning towards more sustainable and renewable alternatives. The need to minimise fossil fuel use and mitigate its associated GHG emissions drives the ongoing growth in sustainable and renewable alternative energy.

In the world's consumption of fossil fuel (coal, natural gas, and oil), 91% is used for energy applications. In crude oil consumption, 63% is for the global transportation sector and 16% is used to make building-block chemicals and polymers [2]. With transportation demand increasing globally, driven in part by population growth, the challenge to decrease the world's reliance on fossil fuels requires the implementation of cost-effective, large-scale, renewable energy-based transport fuel projects.

Biorefineries are the most promising route to produce biofuel and platform chemicals to support a new bio-based industry [3]. A biorefinery is an industrial facility (or network of facilities) that covers a collection of technologies to sustainably convert biomass into basic building blocks for the production of biofuels, energy, and chemicals [4]. It is analogous to current petroleum refineries. To be a feasible alternative, biorefineries must have a dependable supply of feedstock [5], which usually makes up 40 to 60% of the operating costs [6], and maximise the energy conservation between energy inputs and outputs.

Kraft pulping is a well-established process that can be converted into large-scale biorefineries, producing biofuels as a main product. Kraft mills contain critical components vital for a biorefinery [7], i.e., access to biomass feedstock and supply chains, understanding of biomass refining-type processes, and accessible residual feedstock, such as black liquor. Traditional processing in a chemical pulp mill, like a kraft mill, extracts between 40 to 60% of high-value pulp or paper products from the harvested logs while the remaining dissolved wood in the form of a liquor has relatively low economic potential. To maintain profitability, the kraft pulp sector is facing pressures to expand the range of products produced to more than just pulp, heat, and power production. A pulp mill processes a high volume of biomass feedstock and generates by-product streams like black liquor (a mixture of spent pulping chemicals and lignin), which is partially processed by the pulp production [7]. Biomass components—containing mostly hemicellulose and lignin in black liquor—supplies the energetic demand (heat and power) required by the kraft mill through combustion in a recovery boiler. The recovery boiler also plays a vital role in the inorganic chemical recovery process, which contributes to the overall economy of the kraft process. The organic component in black liquor has the potential to be transformed into bioproducts that have higher value than using black liquor as a fuel for heat and power.

Hydrothermal liquefaction (HTL) is a thermochemical process that depolymerises wet biomass into liquid fuels in a reactor operating at high temperature and pressure and sufficient time to decompose the solid natural polymeric structure to mostly liquid compounds [8]. It is a flexible conversion process due to the variability of bio-based or waste feedstock that have been successfully tested. The key advantage of why the HTL process is successful is because the feedstock of the HTL process does not have to undergo a drying process. Water in the HTL process serves as a reactant and catalyst in the subcritical region as the properties of the water change extremely. In the subcritical region, the dielectric constant of the water decreases significantly, as compared to ambient water [9]. Due to this, the solubility of hydrophobic compounds is higher than at ambient condition. In addition, the subcritical environment of the water increases the rate of acid/base-catalysed reactions due to the higher ionic product of water [10]. Liquid bio-crude is the key product of the process. With upgrading the process (like conventional fuel), this bio-crude can be transformed to the whole distillate range of petroleum-derived equivalent fuel products. When compared to gasification, pyrolysis and HTL have a simpler technical conversion of biomass to a liquid fuel [11]. However, when compared to pyrolysis oils, the lower oxygen content in HTL bio-crude makes it less corrosive and has higher heating value (30–36 MJ/kg). The higher calorific value of HTL bio-crude as compared to pyrolysis (15–22 MJ/kg) is more similar to conventional petroleum (43–46 MJ/kg) [12].

Numerous studies have been focusing on evaluating the technical and economic feasibility of the HTL process of different feedstock and algae, considering the various operating conditions of the reactors. However, techno-economic evaluation of HTL of black liquor still presents a research gap when compared to the other feedstock. Black liquor is a complex organic and inorganic mixture. The organic mixture in black liquor is mostly the remaining cellulosic fibres, lignin and hemicellulose, and the caustic inorganics that are used in the kraft process. The advantages of using black liquor in the HTL process are [13] (1) the organic component serves as feedstock to the process and (2) the inorganic component acts as a caustic catalytic solution, instead of sodium hydroxide, for the HTL process. Huet et al. [14] studied the integration of the HTL process of sulphur-free black liquor with a kraft mill, with the reactor temperature between 270 °C and 310 °C. HTL processing of black liquor produces both phenolic molecules and bio-crude. A sodium recovery of 97% was reported, which matches with kraft mill inorganics recovery technology. The sodium is recovered in the form of sodium carbonate, which is converted to caustic soda with the current available technology. Kosinkova et al. [15] conducted a study of hydrothermal liquefaction of bagasse, using co-solvents, ethanol, and black liquor. The yield of the HTL bio-crude increases as the black liquor content increases. This is because black liquor in the co-solvent contains organic residues, which provide additional reactants for conversion and the basicity supports the base-catalysed condensation reaction that leads to oil formation.

Ong et al. [16] analysed the techno-economic feasibility of HTL of radiata pine with black liquor from an existing kraft mill. The estimated minimum fuel selling price (MFSP) of this approach was 1.75 NZD/LGE of fuel (using a conversation rate of 1 NZD = 0.6 USD). Funkenbusch et al. [17] conducted a techno-economic analysis of HTL of lignin, where the lignin are extracted from BL, and determined a MFSP of 1.58 NZD/LGE. Melin et al. [18] experimentally investigated the optimum operating parameters for producing high-quality bio-oil from HTL of BL, using glycerol as the hydrogen donor and sodium hydroxide as the alkali for high heating value fuel. Lappalainen et al. [19] studied the effects of process conditions of the HTL process on the quality of bio-oil from HTL of black liquor. The process parameters studied are residence time and any additives (solvents, catalysts); these conditions are modified to optimise the quantity and quality of the bio-oil production and to minimise the production of secondary products such as biochar and gaseous products.

Prior to today, much of the research on the HTL process studies the process parameters of lab-scale batch reactors. There is, however, a shift towards commercialisation by scaling up the process to a continuous pilot-scale operation [20]. Due to the sub-critical condition required by the process, maximizing the energy efficiency of the HTL process is crucial. Okoro et al. [21] used pinch analysis to conduct heat integration for HTL of meat waste and successfully reduce the heating and cooling demands by approximately 36% and 32%, respectively. Shemfe et al. [22] applied pinch analysis to design the heat exchanger network for upgrading of bio-crude. Anastasaki et al. [20] studied the effectiveness of enhancing a custom-designed heat exchanger design through oscillating of the slurry. Magdeldin et al. [23] conducted a techno-economic assessment of a HTL process that is integrated with downstream combined heat and power (CHP) generation using waste heat and by-products of the process. The introduction of CHP in the HTL process increases the thermal efficiency of the process. Knorr et al. [24] recognized that the maximising the heat integration of the reactor design is a crucial gap. Ong et al. [25] used an iterative process integration and simulation methodology to improve the energy efficiency of integrating HTL with an existing kraft mill.

The aim of this paper is to carry out a techno-economic and carbon emissions assessment of hydrothermal liquefaction of radiata pine and black liquor to assess the techno-economic viability of the process, as compared to fossil fuel feedstock. This paper shows the benefits of integration of the HTL process with a kraft mill with a centralised utility system. The results include a thermo-economic assessment of two other options for reducing marginal fuel use in an existing kraft mill. These options are (1) black liquor evaporators with vapour recompression and (2) replacing the current aged recovery boiler with a high efficiency modern design. The study also explores the trade-off between the GHG emission cost and oil price increase on levelised profit.

2. Materials and Methods

The current study considers three scenarios and determines the cost-benefit of integrating the kraft mill with the new biorefinery technology by measuring how much the minimum fuel selling price changes for the different scenarios.

2.1. Total Site Heat Integration (TSHI)

Heat integration reduces heat demand on boilers and the consumption of fuel, e.g., natural gas, residual biomass, wood chips, and/or black liquor. The introduction of a new biorefinery process to an existing kraft mill significantly affects the site's best heat integration design as well as its overall heat and power balance. The methods used throughout all two scenarios are TSHI with the kraft mill and site heat and power utility modelling.

A site utility model of the recovery boiler and supplementary boilers, turbine, and process heat demands has been implemented in an Excel™ spreadsheet. The boilers assume a constant thermal efficiency of 75%. The turbine model incorporates the extended Willan's line approach of Medina-Flores et al. [26], where historical turbine performance data was used to define turbine model coefficients. Process heat demand from the utility system varies for each of the evaporation system options. The decrease in the low-pressure steam demand adversely impacts on turbine power generation. However, it reduces the required high-pressure steam from the marginal fuel boiler, which is fuelled partially by wood residue (50%) and natural gas (50%). The data is based on an existing kraft mill in the Central North Island of New Zealand.

2.2. Economic Assessment

The economic assessment is based on the methodology reported by Ong et al. [16]. The data used to estimate the operating costs are as presented in Table 1.

Table 1. Estimated material, utility, and carbon emission prices.

Materials	Cost
Radiata pine	95.00 NZD/t
Radiata pine (+125 km)	160.00 NZD/t
Radiata pine (+350 km)	250.00 NZD/t
Sodium hydroxide	500.00 NZD/t
Sodium sulphate	220.0 NZD/t
Hydrotreating catalyst	50.20 NZD/kg
Utilities	
Natural gas	10.00 NZD/GJ
Electricity	90.00 NZD/MWh
Cooling water	2.50 NZD/MWh
GHG emission	25.00 NZD/t

2.3. Environmental Impact

A shortcut life cycle analysis of the GHG calculation uses the method in Martinez Hernandez and Ng [27]. The GHG emission is calculated for the HTL process and its downstream benefits using the GHG factors presented in Table 2.

Table 2. Environmental impact coefficients.

Item	GHG Factor t CO_2-e/Unit	Unit
Electricity	0.085 [28]	MWh
Natural gas	0.058 [29]	GJ
Wastewater	0.188 [30]	t

The main GHG emissions considered in this study are (1) electricity, (2) natural gas, and (3) wastewater. The Emission Trading Scheme (ETS) in New Zealand excludes emissions from biofuels and covers the conventional liquid fossil fuel emissions [31]. The main products converted from the upgrading of the bio-crude process are (1) gasoline equivalent, (2) diesel equivalent, and (3) heavy fuel oil equivalent. The flue gas emission only considers the emissions from the combustion of natural gas. The combustion of off-gas is considered emission free because the source is from the biomass and this falls outside the ETS. The flue gas emission data is extracted from the HTL process simulation model.

2.4. Process Flowsheet

The flowsheet for this paper is based on a PhD work that has been carried out [32]. The thesis started out with designing a kraft mill-integrated biorefinery system. The case study applies a process synthesis technique by developing a multi-dimensional, heat and mass integration methodology [33], which combines pinch analysis, total site heat integration [34], and P-graph [35] frameworks, to select a biorefinery option. The novel method takes into account heat and mass integration with an existing kraft mill in Central North Island of New Zealand and also first-order capital costs of the options at the selection process. The result from the paper shows that hydrothermal liquefaction was the optimal choice to be integrated with the existing kraft mill.

Figure 1 shows the final design summarised in the thesis. The biomass slurry consists of radiata pine, black liquor, and water. The biomass slurry is pressurised in multi-stage pumps and reacts to produce bio-crude, aqueous phase with a fraction of organic materials (hydrocarbons, alcohol), non-condensable gaseous products, and organic solid residues (biochar). Black liquor serves as an additional organic feedstock (hemicellulose and lignin) and a catalyst (Na and K) to the process. Kraft pulp residuals, combined with virgin wood chips and sawdust, can be considered to be the primary biomass feedstock, tapping into existing processes, supply chains, and infrastructure. The upgrading process, e.g., hydro-deoxygenation of bio-crude [36], reduces the oxygen content of the bio-crude, producing hydrocarbon fuels equivalent to petroleum products (45 MJ/kg). The process model based on Aspen Plus^{TM} for the HTL process and upgrading of bio-crude is described in Ong et al. [16]. The ultimate analysis of the HTL bio-crude is presented in Table 3. However, the proposed HTL process in Ong et al. [16] does not address the black liquor inorganics recovery. Figure 1 is a process flow diagram of a new proposed HTL process.

Table 3. Ultimate analysis of HTL bio-crude, adapted from Rowlands et al. [13].

Component	Bio-Crude
Carbon	73.4
Hydrogen	6.5
Nitrogen	0.1
Sulphur	0.6
Oxygen	18.9
Ash	0.5

The inorganic chemicals and sodium and sulphur (Na/S) balance in a kraft mill are crucial for the process economics and environmental viability. The key to that is recycling the Na/S balance at a rate of approximately 97% [37]. Black liquor offers a chemical balance of sodium and sulphur (Na/S balance), which reduces the production cost of pulp and paper. Sodium salts and sulphur anions are primarily in the water and condensates of the HTL process. The concentrations of sodium and sulphur, however, are low, in terms of ppm in the aqueous phase. Evaporating the water from the aqueous phase would incur a high thermal energy cost in an already thermally intensive process. To overcome this issue, supercritical water gasification of the aqueous phase is proposed.

Supercritical water gasification (SCWG) is considered the most appropriate separation approach because:

(1) SCWG is able to treat the phenolic compounds that are in the aqueous phase,
(2) SCWG produces syngas that has lower contamination, and
(3) The alkali salts are insoluble in the SCWG processing conditions due to the change in thermophysical properties above the critical point, which is important in the inorganic recovery of the process.

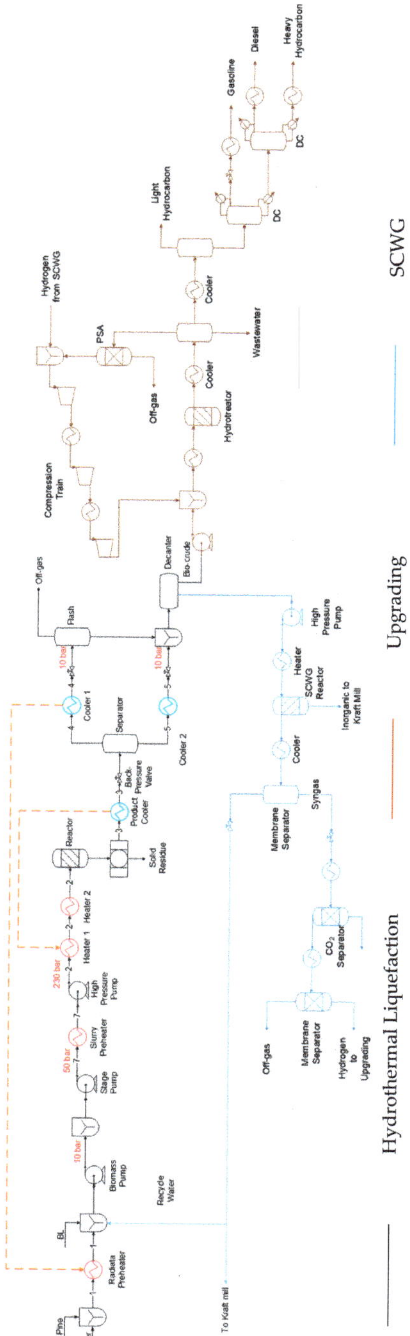

Figure 1. Process flow diagram for the kraft mill-integrated HTL process.

Additional benefits for implementing this process are the elimination of a hydrogen production plant and the reduction in wastewater treatment requirements.

The HTL, upgrading, and SCWG processes are simulated in Aspen Plus^{TM} v9.0 (AspenTech, Bedford, MA, USA, 2020). The biomass into the HTL process was based on 2000 t/day organic loading, which is converted into bio-crude through the HTL process and upgraded to produce a gasoline and diesel blend.

Hydrothermal liquefaction is an energy-intensive process that operates at high temperature (315–355 °C) and pressure (220–250 bar). With these high operating conditions, heat and exergy recovery during cooling and depressurisation of the product flow greatly affects the economic competitiveness of the process. There is still a gap focused on increasing the energy efficiency of the HTL process. In the HTL process, the process conditions affect the product characteristics, yield, and quality. However, the temperature and pressure set points of some streams can be altered without affecting the product characteristics.

Ong et al. [25] established a novel and iterative method to optimise the mass and energy flows and asset of the HTL process, using heat and mass integration simultaneously with process simulation. The iterative procedure uses a combination of existing frameworks and design tools (process simulation tools, pinch analysis, process optimisation, and heat exchanger network design), with the inclusion of the process constraints of the HTL process. The procedure includes the simultaneous need for process simulation to provide in-depth analysis of the multiple impacts of the process modification opportunities. An important section of the analysis and optimisation is the implementation of process constraints based on the best available process knowledge from literature. As a result, the stream parameters and flow sheet design are modified and improved within defined parameters that do not affect the integrity of the process and product and that respect technological limitations. Based on the possible process modification and process constraints of the HTL process that would not affect the yield and quality of the product, the optimised flowsheet of the HTL process is presented in Figure 1.

3. Results

3.1. Scenario 1: Kraft Mill with Hydrothermal Liquefaction System

Scenario 1 considers the HTL process of radiata pine and black liquor. Total site heat integration for the HTL process and kraft mill has been used to have an insight into the integration potential between the HTL process and the kraft mill. The utility model is designed to understand the reduction in power generation due to the lower production of steam from the boilers.

3.1.1. Integration of Hydrothermal Liquefaction with Kraft Mill

Figure 2 shows the utility system of the existing kraft mill. The black liquor solids (BLS) are burned in two recovery boilers, RB1 and RB2. PB1 and PB2 are the marginal fuel (MF) boiler that supplies steam to any deficit demands. The current turbine generates 30.4 MW of power, which is about half the power use of the kraft mill.

Figure 3 shows the utility system of integrating the hydrothermal liquefaction system concept with a kraft mill, using 18% of the black liquor. The three processes are internally heat integrated by exchanging mass and heat with the kraft mill, as outlined in Ong et al. [25]. The utility generated from the three processes are delivered to the kraft pulp process through the utility system. In Figure 3, it shows that the heat that is supplied from the HTL process reduces the marginal fuel of PB1 by 34.8%. Due to the lower steam production from the diversion of the black liquor solids and the decrease in the marginal fuel, the power generation reduced by 9.8 MW. The power generated is calculated using a correlated Willan's line based on the current turbine size. As a result, the inherent cost of black liquor as a feedstock includes the power generation lost by using part of the black liquor. The marginal fuel that is reduced from PB1 is sent to the HTL process as feedstock, which is cheaper due to the lower quality.

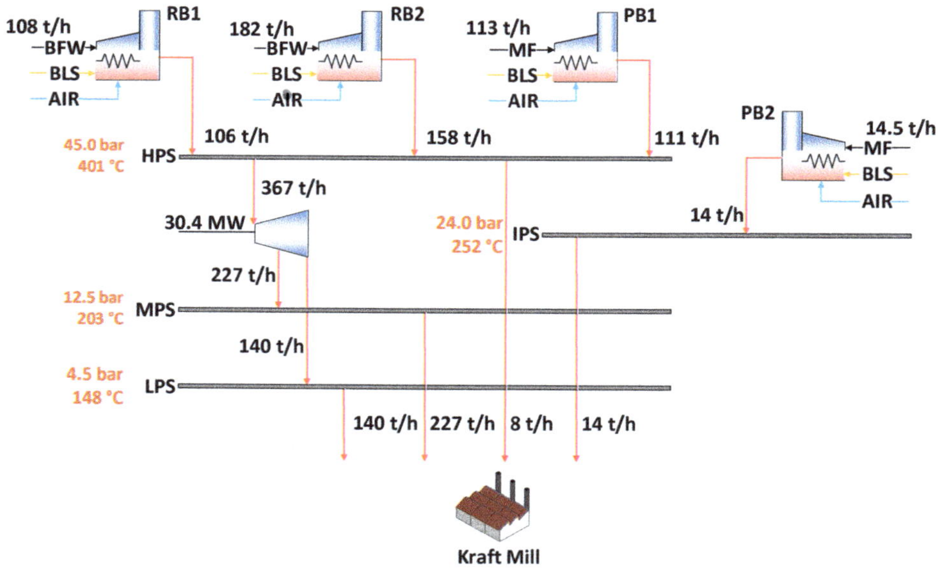

Figure 2. Simplified utility system of the existing kraft mill.

3.1.2. Economic Assessment

The economic assessment is presented in Table 4. The tax used in the study is the federal tax in New Zealand, 34% of the taxable income. The HTL process and the upgrading of the bio-crude are the two processes that contribute the highest to the capital cost, comprising 28% and 29% of the total installed cost (TIC). Efforts are needed to decrease the cost of these sub-processes. Zhu et al. [38] reported that the TIC can be reduced by approximately 10% through decreasing of the operating pressure by 37 bar and temperature of the HTL process by 20 °C. Zhu et al. [38] also investigated hydrocracking the heavy fuel oil and reported that it increases the production rate by 62.9% as compared with the case without hydrocracking. This is another important direction for further work in relation to the present study.

Table 4. Cost results for HTL system with 18% black liquor.

Installed Costs	NZD Million	Operating Cost	NZD Million/y
Biomass preparation	33.3	Variable operating cost	65.1
HTL	111.7	Feedstock	8.2
Upgrading	124.7	Natural gas	6.3
SCWG	60.0	Catalysts and chemicals	21.5
Utilities	39.8	Utilities	
Contingency	40.9	Fixed costs	40.3
Total installed cost	423.5	Revenue from by-products	33.2
Indirect costs	234.3	Capital depreciation	22.5
Total capital investment	657.8	Annualised investment	52.9
		Tax	11.8
MFSP per L of product		1.11 NZD/L	
MFSP per LGE of product		1.23 NZD/LGE	

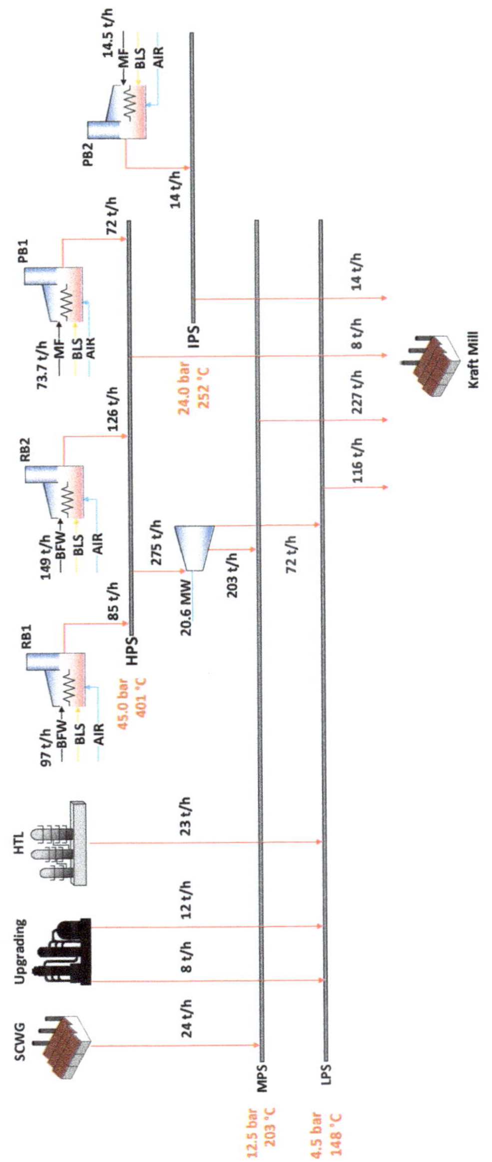

Figure 3. Simplified utility system of integrating the hydrothermal liquefaction process with the existing kraft mill.

Figure 4 shows the MFSP for redirecting a range of fractions of weak black liquor from the kraft mill to the hydrothermal liquefaction process for biofuel production. The figure shows that the MFSP decreases up to 20% of the kraft black liquor and the MFSP starts increasing again. This is due to the increase in the cost of radiata pine.

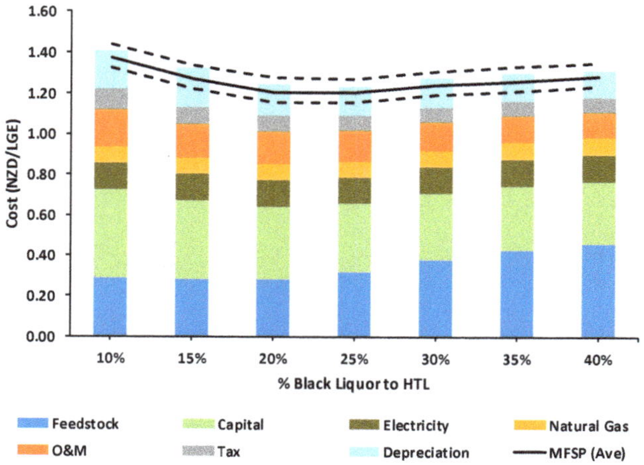

Figure 4. MFSP and the breakdown of the levelised production costs of the HTL process with different processing size.

At the current forestry activity, there is an excess feedstock supply near the kraft mill [39]. However, at 25% of black liquor, the radiata pine feedstock at that vicinity is insufficient. The cost of the feedstock for the 25% black liquor case takes into account the additional 150 km needed to travel for feedstock collection and an additional 350 km for the 30% case and higher. The cost used to calculate the additional distance travelled for the delivery of the radiata pine is costed according to Robertson [40]. Therefore, the assumption of taking 18% of the black liquor used as the basis of the calculation is near the optimum.

The main products of the HTL process are biofuels, heat utilities, and heavy fuel oil. The biofuels are costed in the MFSP and the heat utilities are delivered back to the kraft pulp process. The heavy fuel oil, in this case, is equivalent to marine fuel oil. It is a revenue stream and is priced at the same cost of fossil fuel-derived marine fuel oil. In the last year, the marine fuel oil price has fluctuated at about 250 NZD/t [41]. The selling price of the heavy fuel oil is studied at low, average, and high values: 800, 1000, and 1150 NZD/t.

Figure 5 shows the change of feedstock (radiata pine and natural gas), electricity costs, and capital cost. The range of price shows the change from −30% to +30% based on the cost used in the calculation. The capital costs were estimates, the probable accuracy of the estimate is ±30% [42]. According to Figure 5, the cost of radiata pine (i.e., feedstock) has the highest effect on the MFSP, as reported in other studies [38,43].

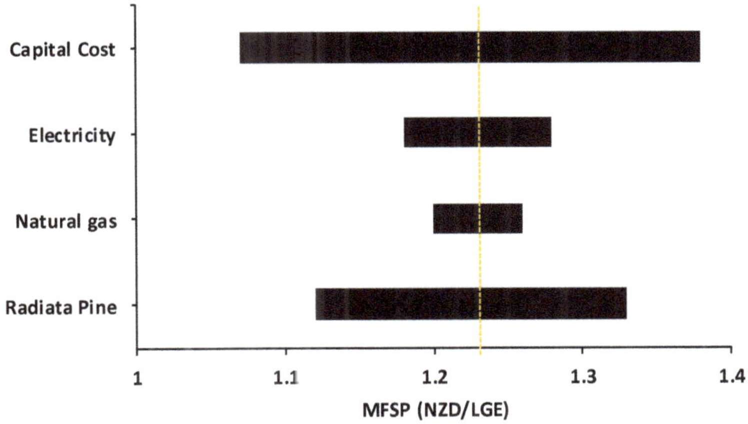

Figure 5. Sensitivity analysis of the change in materials and capital cost.

3.1.3. GHG Emission

The calculated GHG emission for the HTL process is a credit of 97.5 kt CO_2-e/y due the credits earned from sequestration of the solid residue outweighing the total emissions released by the process, 194.3 kt CO_2-e/y.

In this work, the solid residue (biochar) produced from the integrated hydrothermal liquefaction process is used for soil amendment to earn GHG credits [44]. According to He et al. [45], 77.86% of the organic compound in the biochar would break down into the soil while the remaining carbon would be released into the atmosphere as GHG emissions. The net carbon credit of the biochar is estimated at 291.8 kt CO2-e/y. The carbon credit from the by-product earns 7.3 million NZD/y with a carbon price at 25 NZD/t. Sequestering the biochar for soil while producing emissions credits has economic and environmental merit. However, there may be extra opportunities for higher-value markets for the biochar in the future.

Due to the high pressure of the HTL process, a substantial amount of electricity is required by the process. The emissions shown in Table 5 are low because 90% of the electricity in New Zealand is renewable. Table 5 compares the GHG emissions of the HTL system by comparing the emission factor for electricity in Australia. The GHG emission factor for Australia assumed is the National Electricity Market's emission factor [46]. The results in Table 5 show that the HTL production system is carbon neutral fuel, which provides a net reduction in New Zealand, due to the source of the electricity.

Table 5. Comparing GHG emissions between New Zealand and Australia.

	New Zealand	Australia
Grid emissions factor (t CO_2-e/MWh)	0.085	0.830
GHG emissions by the HTL system	kt CO_2-e/y	kt CO_2-e/y
Natural gas	160.8	160.8
Electricity	27.0	263.7
Wastewater	6.5	6.5
Carbon sequestration	291.8	291.8
Net GHG emissions from HTL system	−97.5	139.2
GHG emissions offset from fuel substitution	−344.3	−344.3
Total net GHG emissions reduction	−441.8	−205.1

A high amount of natural gas is used in the HTL process to supply the required heat demand of the kraft mill. Due to the energy-intensive HTL process, the net GHG credit is less than 100%.

One of the ways to reduce heat consumption is in the SCWG process. In the current SCWG process, the reaction process is an endothermic process. The natural gas needed by the SCWG process is used to solely heat the feed and maintain the temperature of the reactor. Studies from Castello [47] and Gutiérrez Ortiz et al. [48] suggest that the reaction process achieves an auto-thermal regime at a biomass concentration of 15 to 20 wt%. At that regime, maximum H_2 is produced, as well as methane production. Future work should include other biomasses to have a concentration that is beneficial for both hydrogen production and process energy sustainability.

Nie and Bi [44] carried out a life-cycle assessment of HTL fed with forest residues. They reported GHG emissions of 20.5 kg CO2-e/GJ. In Nie and Bi [44], the natural gas needed for producing hydrogen (as feedstock and heating) is 3.0 kg of natural gas per kg of H_2 produced. Due to the high temperature in SCWG, the required natural gas for the hydrogen production is 6.2 kg of natural gas per kg of H2 produced. However, the median value for the production of biofuel through the HTL process is 23.58 ± 4.18 kg CO2-e/GJ [44].

3.1.4. Biofuel Policy Consideration

The question to answer is under what situation would the cost of biofuel be sufficiently profitable to compensate for the high level of investment risk. New Zealand's fuel price is governed by global oil supply and demand factors. Table 6 shows the breakdown of the current petrol price in New Zealand.

Table 6. Breakdown of petrol price in New Zealand, as of February 2020 [49].

Components	NZD/LGE
Refined Fuel	0.559
Fuel excise	0.703
Goods and Services Tax (GST)	0.269
Emissions Trading Scheme (ETS)	0.062
Shipping	0.041
Importer Margin	0.450
Total	2.084

The ETS in New Zealand excludes biofuels from the emissions calculation. The overall carbon balance reported in the previous section only includes the carbon released by the HTL production system, which is calculated to be a carbon neutral system that provides a net reduction in GHG emissions. As a result, the use of biofuel would mitigate the emission of conventional fuel, which is covered by ETS. Therefore, it is assumed that because of the mitigation and that the biofuel is produced in New Zealand, the base cost for comparison with a fossil fuel should include the cost of the refined fuel, plus the ETS liability, shipping cost, and importer margin. The importer margin for the analysis is assumed as the average of 0.45 NZD/LGE. As a result, the current fuel price that the biofuel must compete against is 1.11 NZD/LGE.

Figure 6 shows the effect of GHG prices on the fuel cost. As the GHG price increases, the biofuel MFSP decreases. The increase in GHG price increases the selling price of the heavy fuel oil and revenue earned from carbon sequestration through the solid residue. The current GHG price is assumed to be 25 NZD/t.

At the present time, the GHG price needs to increase its current price to about 47 NZD/t to reach the biofuel MFSP for the current oil price. This price is well within the anticipated range of GHG prices for New Zealand in 2030 [50]. Wetterlund et al. [51] mentioned that the feasibility of the process is highly dependent on policy framework and energy market conditions.

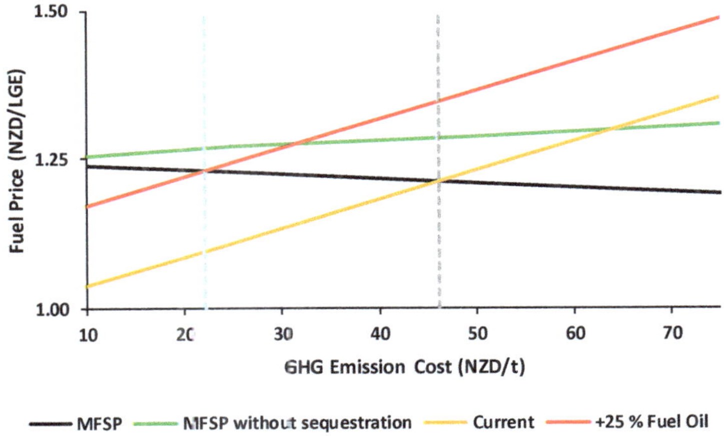

Figure 6. Effect of the carbon price on the fuel selling price.

The price of crude oil plays a significant role in the economic viability of biofuels. The crude oil price is often volatile and closely linked to political and economic climates. At the beginning of October 2018, the oil price peaked at about 142 NZD/barrel before falling sharply to around 83 NZD/barrel by the year's end. At the start of 2020, the price had slowly rebounded to 108 NZD/barrel. As a result, it is quite possible in the near future, a 25% increase in the current fuel cost is reasonable. Under this future oil price, the MFSP would be slightly exceeded, favoring implementation of the integrated HTL system.

With New Zealand on its transition to a sustainable economy, one of the ways to pull biofuels onto the market is through favorable policies [52]. In regions like Europe, the U.S., and South-East Asia, governments provide policies to grant biofuel producers tax exemptions and/or subsidies. With a similar effect, governments can also add a tax on specific classes of non-renewable energy fuels, beyond an ETS. To make the biofuel process economical, the New Zealand government could consider either to tax an extra 0.12 NZD/LGE on non-renewable fuels or give a subsidy of 0.12 NZD/LGE to the biofuel producer. An example of emerging markets is the biodiesel program in Colombia. The government implemented a combination of policies which includes the blending of fuels and tax credits for production and consumption, showing success in increasing biofuel production during the early stages of implementation [53]. Developing a comprehensive policy framework for biofuel uptake in New Zealand falls outside the scope of the current thesis but would be an interesting direction for further exploration.

3.2. Scenario 2: Mechanical Vapour Recompression of Black Liquor Evaporators

Black liquor evaporation in the recovery loop process of a kraft mill is an energy-intensive process. The energy demand of the evaporation process represents 20–30% of site-wide thermal energy demand. Usually, multi-effect evaporators (MEE) concentrate black liquor from about 18% to its firing solids (≈70% for older recovery boilers and ≈85% for modern ones). The first evaporator effect uses low-pressure steam (about 4.5 bar_{ab}) and rejects heat through a condenser (about 0.18 bar_{ab}) and cooling tower. To decrease this energy consumption, the integration of the evaporators with the kraft pulp processes is maximised. An alternate approach to energy reduction in evaporation systems is the use of vapour recompression technologies, thermal vapour recompression (TVR) and mechanical vapour recompression (MVR) [54]. MVR technology has rarely been considered for black liquor evaporators due to the availability of "free" energy gained from the recovery boiler as well as black liquor's high boiling point elevation, which reaches about 15 °C at 67 wt% solids. However, some older kraft mills burn fossil fuels in supplementary boilers but have access to renewable electricity.

As a result, integration of MVR and/or TVR technology into the MEE may be an economic opportunity to reduce fossil fuel use and emissions.

The methodology for this scenario applies TSHI to correctly integrate MVR and TVR with MEE to significantly reduce the use of fossil fuel and its associated emissions. The methodology is further presented in Walmsley et al. [55].

3.2.1. Total Site Heat Integration for Conventional 7-Effect Black Liquor Evaporators

The total site profiles (excluding the evaporator) and the site utility grand composite curve for the studied kraft mill are shown in Figure 7. The T** denotes double-shifted temperature, a notation used in Total Site Heat Integration (information available in [34]). Weak black liquor enters at the lowest pressure effect at 85.0 °C, which initially operates with a saturation temperature of 66.8 °C. For each of the following effects, black liquor increases in pressure and enters at a temperature lower than its saturation temperature. At present, the 7-effect evaporators bleed 2.6 MW of steam from the 2nd effect to the foul condensate stripper column (in place of low-pressure steam). The net SEC (specific energy consumption) for the 7-effect evaporator is 420 kJ/kg_{evap}.

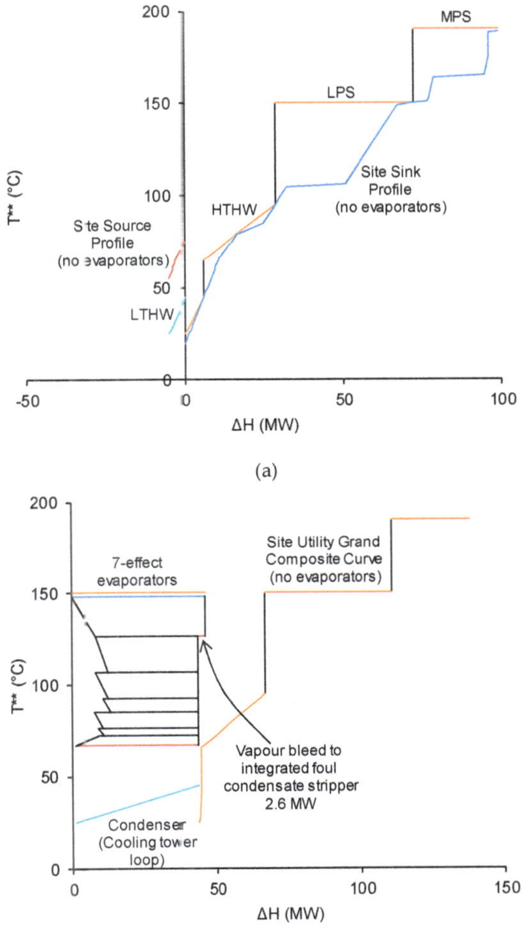

(a)

(b)

Figure 7. (**a**) Total site profiles and (**b**) site utility grand composite curve with existing integration for a conventional 7-effect black liquor evaporator.

3.2.2. Black Liquor Evaporators with Vapour Recompression

Figure 8 exhibits the TSHI of the 3-stage MVR system. There are two integration points: condensate heat recovery and a vapour bleed from stage 2 for the stripper column. Condensate from the 3-stage MVR system is available at 90 °C, which is hotter compared to 72 °C condensate from a conventional 7-effect set-up.

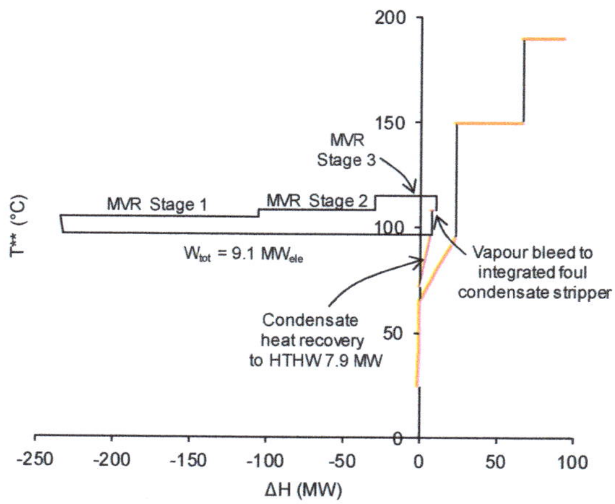

Figure 8. Total site heat integration of 3-stage mechanical vapour recompression (MVR) evaporator with the site steam and hot water utility system using the site utility grand composite curve.

A thermo-economic assessment of three options to retrofit black liquor evaporators with vapour recompression has been studied with results presented in Table 7. Table 7 shows a re-evaluated result from Walmsley et al. [55]. It is crucial to indicate that the modelling accounted for electricity use and includes the loss of cogeneration opportunity, increased heat integration, reductions in low pressure steam use, cooling tower expenses, carbon emission cost, and added maintenance due to MVR and TVR equipment.

The 2-stage MVR system attains a simple payback of 1.7 y and an internal rate of return (IRR) of 57%, which is better when compared with the 3- and 4-stage MVR systems. However, the challenge with implementing the 2-stage system is that the required saturation temperature lifts (8.8 and 9.5 °C) for the two MVR fans. These levels of saturation temperature lift push the upper design and operational limits of an MVR fan. The 3-stage MVR system requires lower temperature lifts (7.0, 3.3, and 8.0 °C) and achieves a greater levelised profit than the 2-stage MVR system. The 4-stage MVR uses less electricity than the 2- and 3-stage MVR systems but the higher capital of a fourth MVR fan offsets the energy benefit.

Figure 9 shows the MFSP of the kraft mill-integrated HTL process with the 3-stage MVR upgrade. The MFSP of the fuel is higher by 0.02 NZD/LGE, as compared to the base kraft mill due to the electricity use by the MVR process. The higher grid sourced electricity use in the HTL system increases the GHG emissions, and the production system is no longer a carbon credit production system.

Table 7. Thermo-economic assessment of retrofitting various multi-stage mechanical vapour recompression options with an existing effect evaporator. The assessment is compared to the conventional 7-effect evaporator.

	2-Stage MVR		3-Stage MVR		4-Stage MVR	
	Rate	**Benefits (NZD M/y)**	**Rate**	**Benefits (NZD M/y)**	**Rate**	**Benefits (NZD M/y)**
Electricity use						
MVR electricity use	8.3 MW	−6.20	7.5 MW	−5.60	7.1 MW	−5.32
Cogeneration reduction	10.0 MW	−7.43	9.8 MW	−7.30	9.5 MW	−7.12
Steam use						
LPS use reduction	37.9 MW		37.9 MW		37.9 MW	
Increased heat recovery	4.1 MW		3.3 MW		2.1 MW	
Steam flow reduction	74.8 t/h	17.68	73.3 t/h	17.34	71.2 t/h	16.9
Other						
Cooling tower reduction	35.1 MW	0.73	35.1 MW	0.73	35.1 MW	0.73
Carbon liability reduction	62.6 kt/y	1.56	61.4 kt/y	1.53	59.7 kt/y	1.49
Additional maintenance	1.2%	−0.14	1.2%	−0.16	1.2%	−0.18
Operation and maintenance (O&M) cost reduction		6.20		6.54		6.45
Capital cost (uninstalled)		3.49		3.91		4.32
Capital cost (installed)		10.47		11.73		12.95
Key Indicators						
Levelised profit	4.65 M NZD/y		4.83 M NZD/y		4.63 M NZD/y	
Simple payback	1.7 y		1.8 y		2.0 y	
Internal rate of return	57%		53%		47%	
SEC_{net} *	145		130		130	

* SEC_{net} is the net specific energy consumption.

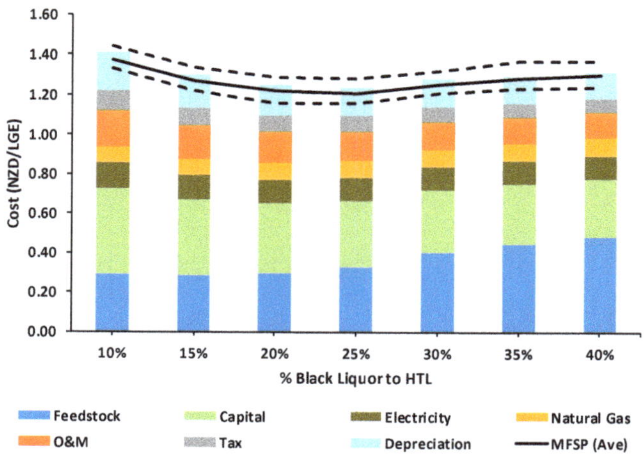

Figure 9. MFSP and the breakdown of the levelised production costs of the kraft mill-integrated HTL with MVR upgrade, with different processing size.

3.3. Scenario 3: New Modern Recovery Boiler

To increase the energy efficiency of the existing kraft mill, a new recovery boiler to replace the "old" style RB is considered. The aim of the new recovery boiler is to minimise the energy purchase and maximise power generation. This is achieved by producing higher steam parameters. The current recovery boilers in the existing kraft mills are fired with black liquor at about 67% dry solids to produce superheated steam at 400 °C at 45 bar. Over the years, recovery boiler technology has been improved and developed to fire black liquor at 72% to 85% dry solids. Environmental benefits include reductions in SO_2 and H_2S emissions with dry solids above 75% dry solids [56].

Figure 10 shows the utility system of replacing the old recovery boiler with a higher solids recovery boiler. As compared to the old recovery boiler, the new recovery boiler produces very high pressure (VHP) steam at 110 bar and 515 °C.

A mass and energy balance of the recovery boiler is carried out to evaluate the economics and operating costs using an Excel spreadsheet. The spreadsheet includes the existing supplementary boilers and turbine, process heat demands, and a new turbine. The design of the new turbine is based on satisfying the steam demand of the kraft pulp process. The new recovery boiler generates as much steam as possible and is primarily expanded in the new turbine. Any additional steam is sent to the existing turbine in the kraft mill. The splits of the expanded steam are determined by optimising maximum power generation in both turbines.

Comparing Figure 10 with Figure 2, PB2 was eliminated and the heat supplied by PB1 is reduced by half. The low-pressure steam demands increase slightly due to the higher evaporation demand needed to concentrate the black liquor solids. The difference in the performance of the boilers is presented in Table 8.

Table 8. Performance data.

	Old Recovery Boiler	New Recovery Boiler	New Recovery Boiler + HTL
Steam produced (t/h)	362	352	288.7
Power generation (MW)	30.5	54.1	41.7
Wood residue used (t/h)	72	8.4	5.4

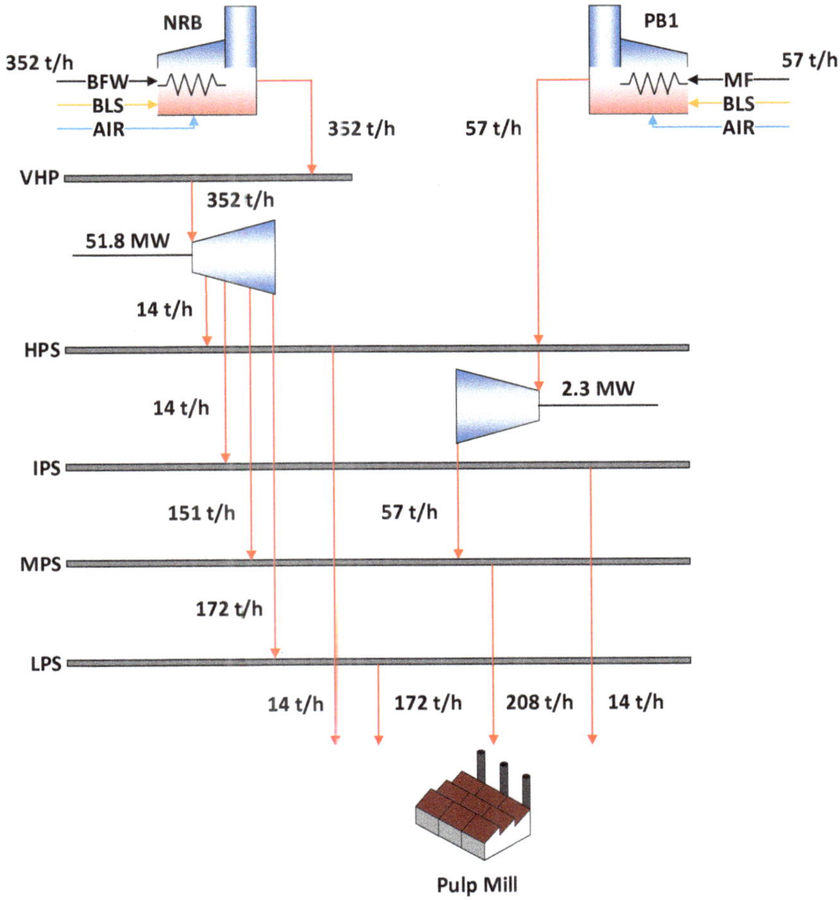

Figure 10. Utility system for kraft mill with a new recovery boiler.

Table 8 also compares integrating HTL into the new recovery boiler case, taking 18% of the black liquor solids for biofuel production, as shown in Figure 1. The wood residue used as marginal fuel in PB1 reduces from 8.4 to 5.4 t/h. However, the power generation decreases with the reduction in the marginal fuel used.

Figure 11 shows the effect of increasing the black liquor solids in the HTL process on the power generation and marginal fuel used. The marginal fuel decrease is due to the lower steam demand from the kraft process, which is supplied by the HTL process. When about 52% of the black liquor is diverted to the hydrothermal liquefaction process, the kraft mill will be self-sufficient, in terms of energy demand. The 72 t/h of wood residue would be used as low-cost feedstock in the HTL process.

Figure 12 shows the MFSP of the kraft mill-integrated HTL process with a new recovery boiler. The new recovery boiler generates more electricity through the expansion of the VHP steam to lower quality steam as compared to the current recovery boiler. The decrease in black liquor flow has a higher impact on the decrease in electricity generation. Therefore, higher electricity is needed to replace the losses, which increase both the cost of electricity and the GHG emission cost.

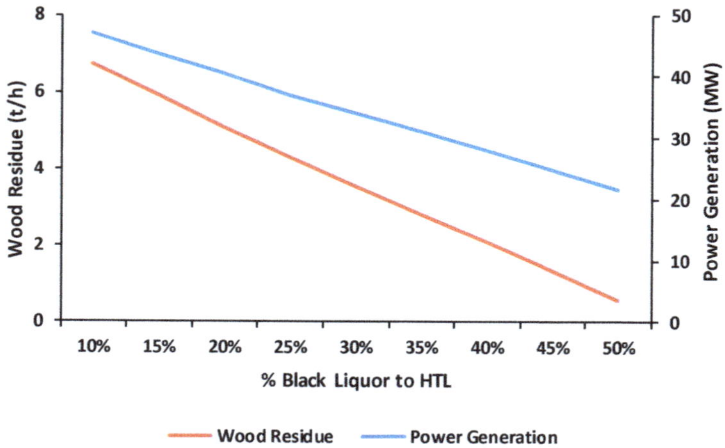

Figure 11. Effect of black liquor solids in HTL.

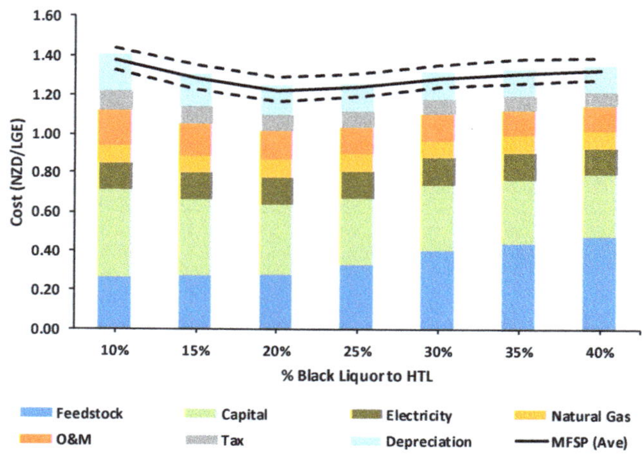

Figure 12. MFSP and the breakdown of the levelised production costs of the kraft mill-integrated HTL with a new recovery boiler, with different processing size.

4. Conclusions and Directions of Future Work

Hydrothermal liquefaction is a promising biorefinery technology that could be integrated with existing kraft mills. The comprehensive flowsheet of the hydrothermal liquefaction, upgrading of bio-crude, and auxiliary processes analysed in this study was developed in a PhD thesis. The economic evaluation of the integration of the HTL system design has been undertaken to establish key price points that would indicate full-scale implementation can compete with conventional fuels. The net reduction in GHG emissions of the system is 441.8 kt CO_2-e/y due to the substitution of conventional petrol and diesel fuels with the biofuel and sequestration of the biochar product. Vapour recompression technologies and a new high solids recovery boiler are considered for the integration with the HTL process. Vapour recompression can be economically integrated into a multi-effect evaporator at kraft mills with older recovery boiler technology, causing a step reduction in steam use. Since vapour recompression acts as an open cycle heat pump, the benefit gained from reducing carbon emissions in supplementary fossil fuel boilers is magnified. A new high solids recovery boiler produces very high-pressure steam that is expanded to generate electricity. The high solids recovery boiler eliminates the use of the natural gas boiler and increases the power generated by 76.8%. Integrating the HTL process in these scenarios increases the MFSP due to the higher electricity consumption of the process.

Future work should focus on improving the economics of the process by reducing the temperature and pressure of the process. The current study of the HTL process is fixed at 355 °C and 230 bar due to the availability of data at this condition. Experimental testing on the effect of temperature and pressure on the bio-crude yield and quality should be tested, as the processing conditions play significant roles in the outcome. Looping of a co-solvent in the hydrothermal liquefaction process has been proven to reduce the energy requirement of the HTL process.

Author Contributions: This paper is based on the work carried out by B.H.Y.O. in his Ph.D., under the supervision of M.J.A., T.G.W., and M.R.W.W. Writing—original draft preparation, B.H.Y.O.; writing—support, T.G.W.; writing—review and editing, T.G.W. and M.J A.; supervision, T.G.W. All authors have read and agreed to the published version of the manuscript.

Funding: This research was funded by the New Zealand Ministry of Business, Innovation and Employment (MBIE), through project "Catalysing Investment in New Zealand Wood-Energy Industrial Symbiosis Opportunities", grant number CONT-37659-EMTR-FRI.

Conflicts of Interest: The authors declare no conflict of interest.

References

1. Brundland, G.H. World Commission on Environment and Development. In *Our Common Future*; Oxford University Press: Oxford, NY, USA, 1987.
2. Statista. *Oil Demand Distribution by Sector Worldwide 2016*; Statista: Hamburg, Germany, 2019.
3. Fernando, S.; Adhikari, S.; Chandrapal, C.; Murali, N. Biorefineries: Current Status, Challenges, and Future Direction. *Energy Fuels* **2006**, *20*, 1727–1737.
4. Morais, A.R.C.; Bogel-Lukasik, R. Green chemistry and the biorefinery concept. *Sustain. Chem. Process.* **2013**, *1*, 18. [CrossRef]
5. Stephen, J.D.; Sokhansanj, S.; Bi, X.; Sowlati, T.; Kloeck, T.; Townley-Smith, L.; Stumborg, M.A. The impact of agricultural residue yield range on the delivered cost to a biorefinery in the Peace River region of Alberta, Canada. *Biosyst. Eng.* **2010**, *105*, 298–305. [CrossRef]
6. Ghatak, H.R. Biorefineries from the perspective of sustainability: Feedstocks, products, and processes. *Renew. Sustain. Energy Rev.* **2011**, *15*, 4042–4052.
7. Pettersson, K.; Mahmoudkhani, M.; von Schenk, A. Opportunities for biorefineries in the pulping industry. In *Systems Perspectives on Biorefineries*; Chalmers University of Technology: Göteborg, Sweden, 2013.
8. Elliot, D.C.; Biller, P.; Ross, A.B.; Schmidt, A.J.; Jones, S.B. Hydrothermal Liquefaction of Biomass: Developments from Batch to Continuous Process. *Bioresour. Technol.* **2015**, *178*, 147–156. [CrossRef] [PubMed]
9. Uematsu, M.; Frank, E.U. Static Dielectric Constant of Water and Steam. *J. Phys. Chem. Ref. Data* **1980**, *9*, 1291–1306.
10. Toor, S.S.; Rosendahl, L.; Rudolf, A. Hydrothermal liquefaction of biomass: A review of subcritical water technologies. *Energy* **2011**, *36*, 2328–2342. [CrossRef]
11. Xu, C.; Lad, N. Production of Heavy Oils with High Caloric Values by Direct Liquefaction of Woody Biomass in Sub/Near-Critical Water. *Energy Fuels* **2008**, *22*, 635–642. [CrossRef]
12. Demirbas, A. Competitive Liquid Biofuels from Biomass. *Appl. Energy* **2011**, *88*, 17–28. [CrossRef]
13. Rowlands, W.N.; Humphreys, L.J.; Thew, R.W.C.; Spankie, J.A.; Uloth, V.C.; Watson, P.A.; Pudlas, M.W. Integrated Kraft Pulp Mill and Thermochemical Conversion System. 2017. Available online: https://patentscope.wipo.int/search/en/detail.jsf?docId=WO2016058098 (accessed on 24 February 2019).
14. Huet, M.; Roubaud, A.; Chirat, C.; Lachenal, D. Hydrothermal treatment of black liquor for energy and phenolic platform molecules recovery in a pulp mill. *Biomass Bioenergy* **2016**, *89*, 105–112. [CrossRef]
15. Kosinkova, J.; Ramirez, J.A.; Nguyen, J.; Ristovski, Z.; Brown, R.; Lin, C.S.K.; Rainey, T.J. In the field: Hydrothermal liquefaction of bagasse using ethanol and black liquor as solvents. *Biofuels Bioprod. Biorefin.* **2015**, *9*, 630–638. [CrossRef]
16. Ong, B.H.Y.; Walmsley, T.G.; Atkins, M.J.; Walmsley, M.R.W. Hydrothermal liquefaction of Radiata Pine with Kraft black liquor for integrated biofuel production. *J. Clean. Prod.* **2018**, *199*, 737–750. [CrossRef]

17. Funkenbusch, L.T.; Mullins, M.E.; Vamling, L.; Belkhieri, T.; Srettiwat, N.; Winjobi, O.; Shonnard, D.R.; Rogers, T.N. Technoeconomic assessment of hydrothermal liquefaction oil from lignin with catalytic upgrading for renewable fuel and chemical production. *Wires Energy Environ.* **2018**, *8*, 1–12. [CrossRef]
18. Melin, K.; Välimäki, A.; Oasmaa, A.; Lehtonen, J. The Effect of Hydrothermal Liquefaction of Black Liquor in Bio-Oil Quality. In Proceedings of the 27th European Biomass Conference and Exhibition, Lisbon, Portugal, 27–30 May 2019; pp. 1144–1145.
19. Lappalainen, J.; Baudouin, D.; Hornung, U.; Schuler, J.; Melin, K.; Bjelić, S.; Vogel, F.; Konttinen, J.; Joronen, T. Sub- and Supercritical Water Liquefaction of Kraft Lignin and Black Liquor Derived Lignin. *Energies* **2020**, *13*, 3309. [CrossRef]
20. Anastasakis, K.; Biller, P.; Madsen, R.; Glasius, M.; Johannsen, I. Continuous Hydrothermal Liquefaction of Biomass in a Novel Pilot Plant with Heat Recovery and Hydraulic Oscillation. *Energies* **2018**, *11*, 2695. [CrossRef]
21. Okoro, O.V.; Sun, Z.; Birch, J. Techno-Economic Assessment of a Scaled-Up Meat Waste Biorefinery System: A Simulation Study. *Materiel* **2019**, *12*, 1030. [CrossRef]
22. Shemfe, M.B.; Fidalgo, B.; Gu, S. Heat integration for bio-oil hydroprocessing coupled with aqueous phase steam reforming. *Chem. Eng. Res. Des.* **2016**, *107*, 73–80. [CrossRef]
23. Magdeldin, M.; Kohl, T.; Järvinen, M. Techno-economic assessment of the by-products contribution from non-catalytic hydrothermal liquefaction of lignocellulose residues. *Energy* **2017**, *137*, 679–695. [CrossRef]
24. Knorr, D.; Lukas, J.; Schoen, P.; Harris Group Inc. *Production of Advanced Biofuels via Liquefaction: Hydrothermal Liquefaction Reactor Design*; National Renewable Energy Lab.(NREL): Atlanta, GA, USA, 2013.
25. Ong, B.H.Y.; Walmsley, T.G.; Atkins, M.J.; Varbanov, P.S.; Walmsley, M.R.W. A heat- and mass-integrated design of hydrothermal liquefaction process co-located with a Kraft pulp mill. *Energy* **2019**, *189*, 116235. [CrossRef]
26. Medina-Flores, J.M.; Picón-Núñez, M. Modelling the power production of single and multiple extraction steam turbines. *Chem. Eng. Sci.* **2010**, *65*, 2811–2820. [CrossRef]
27. Martinez Hernandez, E.; Ng, K.S. Design of biorefinery systems for conversion of corn stover into biofuels using a biorefinery engineering framework. *Clean Technol. Environ. Policy* **2017**, *20*, 1–14. [CrossRef]
28. Ministry of Business, Innovation and Employment. *Electricity Demand and Generation Scenariost*; Ministry of Business, Innovation and Employment: Wellington, New Zealand, 2018.
29. Ministry for the Environment. *Guidance for Voluntary, Corporate Greenhouse Gas. Reporting: Data and Methods for the 2007 Calendar Year*; Ministry for the Environment: Wellington, New Zealand, 2008.
30. Ministry for the Environment. *New Zealand's Greenhouse Gas. Inventory*; Ministry for the Environment: Wellington, New Zealand, 2017.
31. Environmental Protection Authority. Liquid Fossil Fuels 2019. Available online: https://www.epa.govt.nz/industry-areas/emissions-trading-scheme/industries-in-the-emissions-trading-scheme/liquid-fossil-fuels/?accordion-anchor=75 (accessed on 2 June 2020).
32. Ong, B.H.Y. *A Kraft Mill-Integrated Biorefinery Design for New Zealand*; University of Waikato: Hamilton, New Zealand, 2019.
33. Ong, B.H.Y.; Walmsley, T.G.; Atkins, M.J.; Walmsley, M.R.W. Total site mass, heat and power integration using process integration and process graph. *J. Clean. Prod.* **2017**, *167*, 32–43. [CrossRef]
34. Klemes, J.J. *Handbook of Process. Integration (PI): Minimisation of Energy and Water Use, Waste and Emissions*; Woodhead Publishing in Energy, Woodhead Pub: Cambridge, UK, 2013; ISBN 9780857095930.
35. Friedler, F.; Tarjan, K.; Huang, Y.W.; Fan, L.T. Graph-Theoretic Approach to Process Synthesis: Axioms and Theorems. *Chem. Eng. Sci.* **1992**, *47*, 1973–1988. [CrossRef]
36. Elliot, D.C. Historical Developments in Hydroprocessing Bio-Oils. *Energy Fuels* **2007**, *21*, 1792–1815. [CrossRef]
37. Tran, H.; Vakkilainnen, E.K. *The Kraft Chemical Recovery Process*; TAPPIE Press: St. Petersburg, FL, USA, 2012.
38. Zhu, Y.; Biddy, M.J.; Jones, S.B.; Elliot, D.C.; Schmidt, A.J. Techno-economic analysis of liquid fuel production from woody biomass via hydrothermal liquefaction ({HTL}) and upgrading. *Appl. Energy* **2014**, *129*, 384–394. [CrossRef]
39. Alcaraz, S.; Hall, P. *Mapping of Primary Processing Heat Demand and Forestry Resources to Allow Identification of Industrial Symbiosis Opportunities at A Regional Level Scion Report for MBIE*; Scion: Rotorua, New Zealand, 2018.

40. Robertson, K. *Estimating Regional Supply and Delivered Cost of Forest and Wood Processing Biomass Available for Bioenergy*; University of Canterbury: Christchurch, New Zealand, 2006.
41. Ship & Bunker. Singapore Bunker Prices. Ship Bunker 2019. Available online: https://shipandbunker.com/prices/apac/sea/sg-sin-singapore (accessed on 2 June 2020).
42. Gerrard, A.M. *Guide to Capital Cost Estimating*, 4th ed.; Institution of Chemical Engineers (IChemE): Warwickshire, UK, 2000.
43. Magdeldin, M.; Kohl, T.; Järvinen, M. Techno-economic Assessment of Integrated Hydrothermal Liquefaction and Combined Heat and Power Production from Lignocellulose Residues. *J. Sustain. Dev. Energy Water Environ. Syst.* **2018**, *6*, 89–113. [CrossRef]
44. Nie, Y.; Bi, X. Life-cycle assessment of transportation biofuels from hydrothermal liquefaction of forest residues in British Columbia. *Biotechnol. Biofuels* **2018**, *11*, 23. [CrossRef]
45. He, Y.; Zhou, X.; Jiang, L.; Li, M.; Du, Z.; Zhou, G.; Shao, J.; Wang, X.; Xu, Z.; Bai, S.H.; et al. Effects of biochar application on soil greenhouse gas fluxes: A meta-analysis. *Gcb Bioenergy* **2017**, *9*, 743–755. [CrossRef]
46. Department of the Environment and Energy. *National Greenhouse Accounts Factors*; Department of the Environment and Energy: Canberra, Australia, 2017.
47. Castello, D. *Supercritical Water Gasification of Biomass*; University of Trento: Trento, Italy, 2013.
48. Gutiérrez Ortiz, F.J.; Ollero, P.; Serrera, A.; Galera, S. An energy and exergy analysis of the supercritical water reforming of glycerol for power production. *Int. J. Hydrog. Energy* **2012**, *37*, 209–226. [CrossRef]
49. Motoring, A.A. How Petrol Prices are Calculated. Available online: https://www.aa.co.nz/cars/owning-a-car/fuel-prices-and-types/how-petrol-prices-are-calculated/ (accessed on 1 June 2020).
50. New Zealand Productivity Commission. *Low—Emissions Economy*; New Zealand Productivity Commission: Wellington, New Zealand, 2018.
51. Wetterlund, E.; Pettersson, K.; Harvey, S. Systems analysis of integrating biomass gasification with pulp and paper production—Effects on economic performance, CO_2 emissions and energy use. *Energy* **2011**, *36*, 932–941. [CrossRef]
52. Suckling, I.D.; De Miguel Mercader, F.; Monge, J.J.; Wakelin, S.J.; Hall, P.W.; Bennett, P.J. *New Zealand Biofuels Roadmap Summary Report: Growing A Biofuelled New Zealand*; Scion: Rotorua, New Zealand, 2018.
53. Martins, F.; Gay, J.C. Biofuels: From Boom to Bust? 2017. Available online: https://www.bain.com/insights/biofuels-from-boom-to-bust/ (accessed on 1 June 2020).
54. Sharan, P.; Bandyopadhyay, S. Integration of thermo-vapor compressor with multiple-effect evaporator. *Appl. Energy* **2016**, *184*, 560–573. [CrossRef]
55. Walmsley, T.G.; Atkins, M.J.; Ong, B.H.Y.; Klemeš, J.J.; Walmsley, M.R.W. Total site heat integration of multi-effect evaporators with vapour recompression for older kraft mills. *Chem. Eng. Trans.* **2017**, *61*, 265–270.
56. Hyöty, P.A.; Ojala, S.T. Super Combustion of Black Liquor. In Proceedings of the International Recovery Boiler Conferen Cetampere Hall, Tampere, Finland, 11 June 2014; pp. 49–53.

processes

MDPI

Article

Thermochemical Conversion of Napier Grass for Production of Renewable Syngas

Mohamad Syazarudin Md Said [1,*], Wan Azlina Wan Abdul Karim Ghani [1], Tan Hong Boon [1], Siti Aslina Hussain [1] and Denny Kok Sum Ng [2]

1 Sustainable Process Engineering Research Centre (SPERC), Department of Chemical and Environmental Engineering, Faculty of Engineering, Universiti Putra Malaysia, Serdang 43400, Selangor Darul Ehsan, Malaysia; wanazlina@upm.edu.my (W.A.W.A.K.G.); hboonn90@gmail.com (T.H.B.); aslina@upm.edu.my (S.A.H.)

2 School of Engineering and Physical Sciences, Heriot-Watt University Malaysia, No 1, Jalan Venna P5/2, Precinct 5, Putrajaya 62200, Malaysia; denny.ng@hw.ac.uk

* Correspondence: syazarudin@upm.edu.my

Received: 12 July 2019; Accepted: 3 September 2019; Published: 5 October 2019

Abstract: Fuel resource diversification is a global effort to deviate from non-renewable fossil fuels. Biomass has been identified as an alternative solid biofuel source due to its desirable properties and carbon neutrality. As reported in the literature, biomass can positively contribute towards combating climate change while providing alleviation for energy security issue. As part of efforts to diversify biomass resources, this work intends to explore the potential of Napier grass, one type of energy crop, for the production of renewable syngas via gasification. This energy crop is originally from Africa, which is highly productive with low cost (40 tonnes per year per hectare). Limited studies were conducted to analyze the potential of such an energy crop as a fuel source, which is the subject of this work. In order to analyze the full potential of such energy crop, the physical and chemical characteristics of this biomass was first analyzed. To determine the productivity of syngas from this biomass, fluidized bed gasifier was used in this work. The effects of gasification process parameters (i.e., equivalence ratio and temperature) on product yield and producer gas compositions were examined. Besides, the effects of equivalence ratio towards higher heating value of syngas and carbon conversion efficiency were analyzed. Based on the ultimate analysis results, the molecular formula of Napier gas was $CH_{1.56}O_{0.81}N_{0.0043}$. Meanwhile, the higher heating value of such biomass was determined as 16.73 MJ/kg, which was comparable to other biomasses. It is noted that in this work, the volatile matter was determined as 85.52% and this promoted gasification process remarkably. The dynamics of the reactions involved were observed as a significant variation in product yield and biogas components were recorded at varying equivalence ratio and gasifier operating temperature.

Keywords: energy crop; solid biofuel; renewable energy; Napier grass; gasification

1. Introduction

The increasing scarcity of conventional fossil fuels has led to diversification of energy resources. In addition, the combustion process of fossil fuels for electricity generation emits greenhouse gasses and criteria pollutants, which are harmful to both living organisms and the planet. The global carbon emission has been increasing at an alarming rate. Average annual global carbon dioxide emission from burning of fossil fuels was 3.1 GtC per year in the 1960s. Recently, the rate has recorded an increment higher than threefold where 9.4 GtC per year was emitted during 2008–2017 [1]. Combustion of finite non-renewable fossil fuels for energy production in various sectors such as transportation and industrial activities has been reported to be the main perpetrator to this worrying situation. This dire situation prompts for cooperative and collective effort at a global scale as manifested by the Kyoto Protocol and Paris Agreement. Many countries around the world are phasing out and rendering non-renewable fossil fuels as an obsolete option for energy production.

Reducing our reliance on finite fossil fuels and exploration of potential renewable resources for energy generation have become a focus at the global scale, and Malaysia is not left behind in this worldwide trend. The initiative has gained support at the governmental level as evidenced by the introduction of the Five-Fuel Diversification Policy. Under this policy, renewable energy is included as the fifth fuel in the supply mix where utilization of abundant biomass is one of the strategies being encouraged [2,3]. According to the Malaysia Energy Commission, 100,721 ktoe of energy was supplied in 2015 where 95.5% of the energy was generated from non-renewable resources, mainly natural gas (61.7%) and crude oil (32.2%). Biomass, on the other hand, contributed a small fraction of 0.2% to the total energy supply for that year [4]. Biomass is derived from living organisms through the photosynthesis process where solar energy is converted into carbohydrates. A wide range of biomass is available that entails significant variation in their properties, characteristics and chemical compositions. In general, major constituents of biomass consist of oxygen, carbon and hydrogen. The use of biomass for energy production is considered carbon neutral due to carbon fixation process during photosynthesis [5].

Conversion processes (physical, biological, thermochemical, etc.) of the ample biomass produce renewable syngas, which provides alleviation for both energy security and global warming issues. Additionally, various types of value-added products can be produced from the conversion processes. For this purpose, a range of thermochemical conversion processes is available such as pyrolysis, gasification, liquefaction and direct combustion. The main difference between thermochemical technologies is the availability of oxygen during the process. In some applications, more than one thermochemical conversion process is combined to enhance the quality of producer gas as conducted by Alipour Moghadam, et al. [6] where both pyrolysis and air-steam gasification processes are integrated together. Four possible biomass thermochemical conversion routes for renewable energy production have been discussed and compared by Mohammed, Salmiaton, Wan Azlina, Mohammad Amran, Fakhru'l–Razi and Taufiq–Yap [5].

Thermochemical conversion of biomass produces syngas with half energy density of natural gas. The reactions involved during biomass conversion process are summarized in Table 1 [7].

Table 1. Chain of reactions involved in the biomass thermochemical process.

Eq.	Reaction	Chemical Equation	Enthalpy (Negative Value Indicates Exothermicity)
1	Combustion (complete)	$C + O_2 \rightarrow CO_2$	−283 MJ/kmol
2	Combustion (incomplete)	$C + 1/2O_2 \rightarrow CO$	−111 MJ/kmol
3	Boudouard	$C + CO_2 \leftrightarrow 2CO$	+172 MJ/kmol
4	Water-gas	$C + H_2O \leftrightarrow CO + H_2$	+131 MJ/kmol
5	Methanation	$C + 2H_2 \leftrightarrow CH_4$	−75 MJ/kmol
6	Water-gas shift	$CO + H_2O \leftrightarrow CO_2 + H_2$	−41 MJ/kmol
7	Steam-methane reforming	$CH_4 + H_2O \leftrightarrow CO + 3H_2$	+206 MJ/kmol
8	Dry reforming	$CH_4 + CO_2 \leftrightarrow 2H_2 + 2CO$	+260 MJ/kmol
9	Methanation	$C + 2H_2O \leftrightarrow CH_4 + CO_2$	+103 MJ/kmol

In Malaysia, many work related to the thermochemical conversion of biomass has been concentrated on palm oil derived biomass due to its abundancy and wide availability [8,9]. In order to broaden the range of biomass utilized for renewable energy generation, which directly supports the fuel diversification policy of Malaysia, new potential renewable energy resources are being explored. Napier grass (NG) has gained considerable attention in recent years due to its desirable characteristics as potential renewable fuel. This energy crop of African origin is highly productive with low establishment cost [10]. The annual yield is 40 tonnes per hectare with multiple harvest frequency. There is limited information on the potential of producing green energy from Napier grass reported in the literature where the works have been concentrated on using the pyrolysis conversion process [11–13].

Fluidized bed gasification has been reported to be a versatile technology for biomass conversion. Intensive mixing in the bed enhances heat and mass transfer that leads to a high reaction rate [14]. Abdoulmoumine, et al. [15] reported that operation parameters have a major influence on the kinetics of reactions involved, which directly affect yield and the quality of producer gas. To our knowledge, the potential of generating renewable fuel from gasification of Napier grass has never been conducted. It is the aim of this study to evaluate the feasibility of syngas production from Napier grass via the bench-scale gasifier system at varying operating conditions.

2. Materials and Methods

2.1. Sample Preparation

Mature Napier grass was sourced from Crops for the Future Research Centre (CFFRC), Semenyih, Selangor, Malaysia. The biomass was dried in an oven at 105 °C according to BS EN12048 standard prior to size reduction by using the Retsch rotor beater mill. The sample size was reduced to 0.2 and 2 mm and kept in air-tight plastic bags for further analysis.

2.2. Proximate Analysis of Mature Napier Grass

Proximate analysis was conducted on the shredded form of Napier grass by using a thermogravimetric analyzer (TGA; TGA/SDTA851, Mettler Toledo, Columbus, OH, USA) to determine fixed carbon, volatile matter, moisture and ash contents in Napier grass. The details of the experimental procedure can be found elsewhere [16].

2.3. Ultimate Analysis of Mature Napier Grass

An ultimate analysis was conducted to determine elemental composition of mature Napier grass by using the CHNS/O analyzer (model LECO CHN628 and 628S, St. Joseph, MI, USA) according to the ASTM D-5291 standard method.

2.4. Measurement of the Higher Heating Value of Napier Grass

The gross calorific value of mature Napier grass was measured by using the Parr 6100 oxygen bomb calorimeter (Moline, IL, USA) according to BS EN 14918.

2.5. Gasification of Napier Grass for Syngas Production

The gasification of the shredded Napier grass was conducted in a fluidized bed gasifier. The reactor was cylindrical with 370 mm high and 54 mm wide, made of stainless steel. The schematic of the experimental rig is shown in Figure 1.

The procedure began with charging the reactor with 20 g of sand as the bed material to obtain good temperature distribution, to stabilize the fluidization and to prevent coking inside the reactor. Air stream and biomass feedstock were introduced from the bottom and top of the reactor respectively as the bed temperature achieved the steady state condition. The experiment was carried out at five different temperatures between 650 °C and 850 °C at 50 °C temperature increment and three different equivalence ratio (ER; 0.2, 0.3 and 0.4).

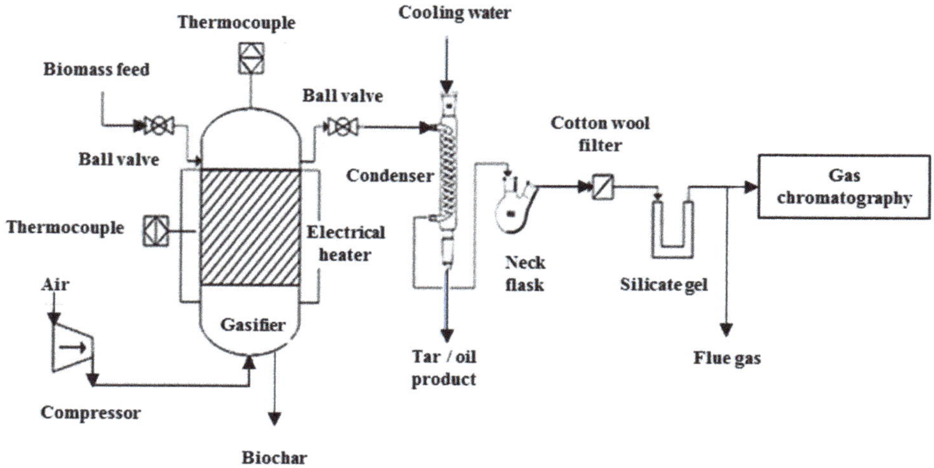

Figure 1. Process flow diagram of lab scale gasifier setup.

2.5.1. Gas Analysis

Syngas produced from gasification process were analyzed using gas chromatography (GC) (model Agilent Technologies 6890N, Mundelein, IL, USA) equipped with a thermal conductivity detector (TCD).

2.5.2. Condensed Liquid Characterization

A visually brown liquid was produced from gasification of Napier grass. Prior to characterization, the brown liquid was decanted and diluted with dichloromethane solvent (99.8%) at a volume ratio of 1:1. The mixed solvent and brown liquid product were then centrifuged at 4000 rpm for 5 min by using the Hettich EBA 21 Centrifuge (Tuttlingen, Germany) to separate the organic phase from aqueous phase and char traces. The upper layer of the solution (bio-oil and solvent) was extracted and analyzed using gas chromatography mass spectrometry (GC–MS model QP2010 Plus SHIMADZU, Japan) equipped with a Zebron ZB-5MS capillary column (30 m long, 0.25 mm inner diameter and 0.25 μm thick). The injection and detector temperatures were set at 250 °C and 200 °C respectively. The flow rate of the carrier gas, He, was 1.0 mL/min.

2.5.3. Analysis of Inorganic Compounds in Ash

The inorganic compounds in the Napier grass ash were analyzed by using an energy dispersive X-ray fluorescence spectrometer (model SHIMADZU EDX-720, Japan).

3. Results and Discussion

3.1. Proximate and Ultimate Analysis of Napier Grass Feedstock

The results of the ultimate and proximate analysis of NG and other biomasses reported in the literature are presented in Table 2. TGA was conducted at three different heating rates (5, 10 and 20 °C/min) and almost similar results were produced. Therefore, the result acquired at 10 °C/min was considered for further discussion. As shown in Figure 2, NG in air atmosphere recorded a higher moisture content (8.78%) compared to the N_2 atmosphere (7.73%). The same pattern was observed for volatile matter where a high content was recorded in air (81.49%) compared to nitrogen (62.70%). This is due to the presence of oxygen in the air that promotes biomass devolatilization. In contrast, lower fixed carbon content was recorded in air atmosphere (3.54%) compared to the N_2 atmosphere (9.69%). Ash content was three times higher under N_2 as compared to air atmospheres. According to

Ptasinski [17] substantial variation in characteristics of biomass might have a direct effect on syngas composition. In addition, due to the environmental conditions and harvesting techniques, the biomass composition may vary remarkably [18].

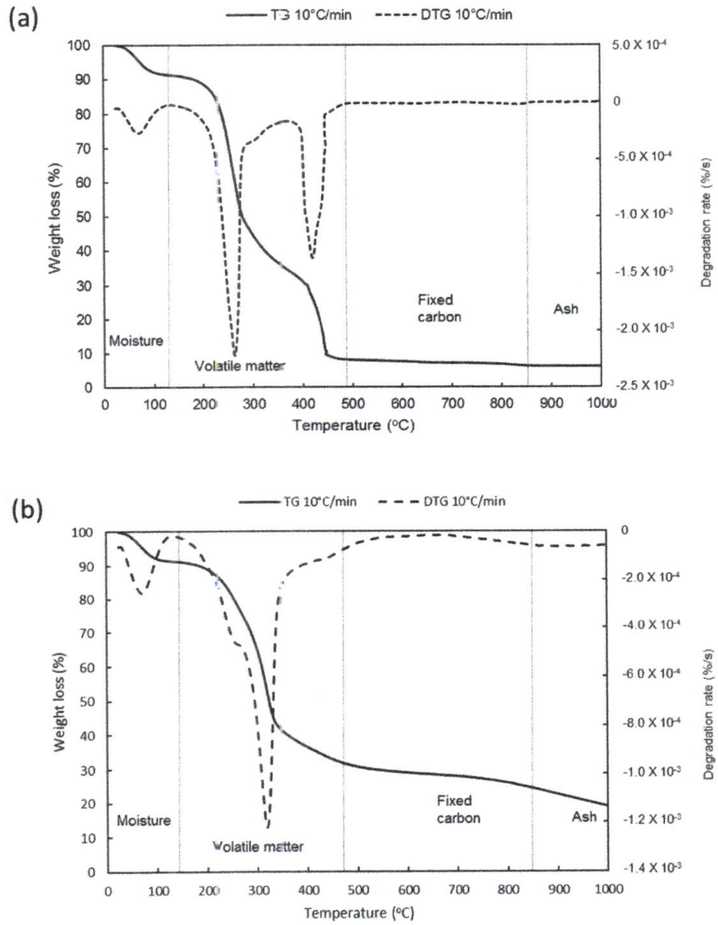

Figure 2. Proximate analysis of Napier grass in (**a**) air and (**b**) nitrogen at a constant flowrate of 25 mL/min.

NG contains a high fraction of volatile matter (>80%) and moisture (30%) but low ash content (<7%). The high content of volatile matter in the Napier grass is desirable since this feature enhances the gasification process and reduces the amount of char produced [7]. High moisture content (>40 wt%) of biomass tends to degrade the gasification performance as more energy is required in the drying process [19]. Besides, moderate fixed carbon compared to other biomasses indicates shorter burning time is required as the solid-gas combustion reactions are slower than gas–gas reactions [7].

Ultimate analysis of NG found that NG contained 48.52% oxygen, 45.10% carbon, 5.94% hydrogen, 0.45% nitrogen and 0% sulphur. These values were comparable to NG, *Miscanthus* and coconut shell reported in literature. However, in comparison to all biomasses except for the NG in literature, NG had higher oxygen but lower carbon content, indicating a lower calorific value. Higher oxygen composition (carbon–oxygen bonds) would have a significant negative effect on the heating value of biomass [20]. NG is considered as an environmentally friendly feedstock as it contains a low amount of nitrogen and no sulphur, which may improve fuel quality and reduce toxic gas emission. The heating value of NG (16.73 MJ/kg) was comparable to coconut shell but slightly lower than other biomasses in literature because of its high level of moisture and oxygen. These desirable characteristics indicate that NG is a good solid biofuel candidate for energy generation via gasification.

Table 2. Comparison of Napier grass (NG) feedstock characteristics with various types of biomass in literature.

Property	Napier Grass (Current Study)	Napier Grass [13]	Cardoon [21]	*Miscanthus* [21]	Rubber–Wood–Sawdust [22]	Palm Kernel Shell [23]	Coconut Shell [23]
Moisture (wt%)	30.07	9.43	12.00	6.48	-	7.96	4.89
			Proximate analysis (wt%, dry basis)				
Volatile matter	85.52	72.58	76.02	78.36	51.39	72.47	30.62
Fixed carbon	8.17	8.35	9.19	14.90	14.29	18.56	26.41
Ash content	6.31	9.68	14.80	6.74	22.67	8.97	42.98
			Ultimate analysis (wt%, dry basis)				
C	45.10	42.40	56.01	46.97	53.40	51.63	45.24
H	5.94	5.96	6.46	5.57	6.70	5.52	5.04
N	0.45	1.71	0.99	1.37	3.10	1.89	1.46
S	0.00	0.09	0.22	0.28	0.00	0.05	0.06
O (by difference)	48.52	45.32	36.10	45.82	36.80	40.91	48.2
Calorific value (MJ/kg)	16.73	-	17.33	18.73	18.30	22.97	16.07

3.2. Inorganic Compounds in Napier Grass Ash

Table 3 compares inorganic compounds in ash of NG and other biomasses reported in literature. More than half of inorganic compounds detected in NG ash were potassium (54.39%), with some iron (15.53%), silicon (9.81%), chlorine (8.84%), calcium (8.20%) and sulphur (2.30%). The ash composition of NG in literature is rich in silicon (43.00%), potassium (30.5%), magnesium (9.90%) and phosphorus (7.20%). The differences of NG ash properties in the current study as compared to the literature were due to the fact that the trees were dependent on growth conditions and other environmental factors. The ash composition has a negative effect on the gasification performance reactor when subjected to high temperature combustion [24].

According to Obernberger and Thek [25], a high potassium content in ash causes agglomeration at high temperature, which may bring severe damage to the gasifier system. Arvelakis, et al. [26] reported that potassium content in ash could react with bed material such as silica sand and break Si–O–Si bonds to form silicates that deposit on the reactor wall and on the bed particle surface, causing agglomeration. According to Arvelakis, Gehrmann, Beckmann and Koukios [26], high proportion of SiO_2 found in the ash of *Miscanthus* (62.21%) and agricultural residues (89.57%) would cause severe agglomeration when the potassium reacts with SiO_2.

Table 3. Inorganic contents in ash of Napier grass and other types of biomass.

Ash Basis (wt%)	Napier Grass (Current Study)	Olive Tree Residue [27]	EFB [28]	Napier Grass [12]	Cardoon [21]	*Miscanthus* [21]	Agricultural Residue [29]
K_2O	54.39	9.26	44.00	30.5	24.91	14.00	1.65
Fe_2O_3	15.53	1.38	3.00	1.4	1.77	2.63	2.95
SiO_2	9.81	11.84	27.00	43.0	8.34	62.21	89.57
Cl	8.84	-	5.30	-	-	-	1.30
CaO	8.20	54.82	8.00	1.9	38.33	8.32	0.77
SO_3	2.03	-	2.70	-	-	-	-
MnO	0.44	0.10	0.11	-	-	-	-
Rb_2O	0.37	-	0.12	-	-	-	-
Br	0.14	-	0.018	-	-	-	-
CuO	0.10	-	0.039	-	-	-	-
ZnO	0.10	-	0.092	0.08	-	-	-
As_2O_3	0.66	-	-	-	-	-	-
Al_2O_3	-	2.60	0.97	<0.1	3.50	5.47	1.32
MgO	-	4.36	4.80	9.9	5.74	3.16	0.76
Na_2O	-	0.16	0.55	<0.01	13.08	0.53	1.15
TiO_2	-	0.35	0.08	0.03	0.10	0.32	7.56
P_2O_5	-	3.40	3.60	7.2	4.23	3.37	1.04
NiO	-	-	0.01	-	-	-	-
SrO	-	-	0.03	0.03	-	-	-
BaO	-	-	-	0.08	-	-	-

Mohammed, Salmiaton, WanAzlina and Mohamad Amran [16] stated that CaO acts as a CO_2 adsorbent where its presence might accelerate the secondary reaction and therefore improve hydrogen content in the syngas. In addition, Mahishi and Goswami [30] supported that the presence of CaO has a significant effect on hydrogen production during the gasification process at an elevated temperature. In relation to iron content, NG in the current study contains a high amount of iron, which was 15.53%. According to Lahijani and Zainal [28], magnesium, iron and calcium are good agents for reducing agglomeration. Low silica content (9.81%) was found in Napier grass in this current study and this amount was comparable to olive residues (11.84%) and cardoon (8.34%). Mohammed, Salmiaton, Wan Azlina and Mohamad Amran [16] mentioned that the combination of low silica content with reasonable amount of MgO, Fe_2O and Al_2O_3 is effective in reducing agglomeration. Furthermore, NG in the current study contained a significant amount of Cl (8.84%), which could react with potassium to form potassium chloride and subsequently promote potassium devolatilization [26].

3.3. Components of Bio-Liquid

Gasification of biomass generates bio-liquid as one of the by-products. The liquid is visually dark brown, usually comprised of water, oxygenated hydrocarbons and other hydrocarbons. Bio-liquid generated from pyrolysis and gasification can be used for fuel in direct combustion or as chemical products after further treatment processes. Table 4 presents the fraction of individual compounds detected in the bio-liquid produced from gasification of Napier grass. The liquid contains a mixture of hydrocarbon, oxygenated and nitrogenated compounds. Phosphonic acid, (p-hydroxyphenyl) was found to be the major constituent (31.94%), followed by pentane, 2,2-dimethyl (16.10%). The oxygenated compounds in bio-liquid were corrosive in nature with low pH value. Bio-liquid with high amount of oxygenated compounds should undergo further treatment such as hydrothermal processing to produce biofuels or value-added chemicals [13].

Table 4. Chemical compounds detected in bio-liquid obtained from gasification of Napier grass at 850 °C and equivalence ratio (ER) of 0.2.

Compound Name	Area (%)	Formula
Pentane, 2,2-dimethyl- (Al)	16.10	C_7H_{16}
Dimethoxydimethylsilane (E)	4.01	$C_4H_{12}O_2Si$
Pentane, 3,3-dimethyl- (Al)	2.36	C_7H_{16}
Cyclohexane (Al)	7.12	C_6H_{12}
Pentane, 2,3-dimethyl- (Al)	1.33	C_7H_{16}
Hexane, 1-chloro- (Al)	3.63	$C_6H_{13}Cl$
Hexane, 3-methyl- (Al)	6.09	C_7H_{16}
Pyridine (N)	2.79	C_5H_5N
Pyrrole (N)	1.57	C_4H_5N
2,2-Dimethoxybutane (K)	1.41	$C_6H_{14}O_2$
Phosphonic acid, (p-hydroxyphenyl)- (A)	31.94	$C_6H_7O_4P$
Phenol, 2-methyl- (P)	4.01	C_7H_8O
Phenol, 3-methyl- (P)	9.49	C_7H_8O
Total	91.85	

A = carboxylic acid; Al = alkane; E = ester; K = ketal; N = nitrogenated; P = phenols.

3.4. *Effects of ER on the Product Yield and Composition of Producer Gas*

Equivalence ratio (ER) is defined as the ratio of the amount of air supplied during the thermochemical process to the amount of air required for stoichiometric combustion of the fuel (Equation (1)). The concept of ER is applicable in gasification when air or oxygen is injected for partial combustion of the biomass feedstock. In comparison to the combustion process, which requires excess air and ER to be greater than 1 for complete fuel combustion, the range of ER for gasification is usually limited to a value below 0.4.

$$\text{ER} = (\text{Actual weight air/weight of dry biomass})/(\text{Stoichiometric air/biomass ratio}). \quad (1)$$

The formula for stoichiometric combustion with oxygen is:

$$CH_{1.56}O_{0.81} + 0.985O_2 \rightarrow 0.78H_2O + CO_2, \quad (2)$$

where:

$$AC_{stq} = (M_{O2} + 3.76M_{N2})/100 \times (C/M_c + H/2M_{H2} + S/M_S - O/M_{O2}) = 5.12. \quad (3)$$

Taking the information below into consideration.

$$\text{Density of air, } \rho = m/V. \quad (4)$$

$$1.18 = 0.032/V. \quad (5)$$

$$V = 0.027 \text{ m}^3 = 27 \textit{ liters}. \quad (6)$$

$$\text{Superficial velocity} = V/A. \quad (7)$$

$$\text{Diameter of reactor, D} = 0.054 \text{ m}. \quad (8)$$

$$\text{Surface area of reactor, A} = \pi \times (D/2)^2 = 2.29 \times 10^{-3} \text{ m}^2. \quad (9)$$

$$\text{Superficial velocity} = 3.0 \text{ L/min} \times 1 \text{ m}^3/1000 \text{ L} \times 1/2.29 \times 10^{-3} \text{ m}^2 \times 1 \text{ min}/60 \text{ s} = 0.0218 \text{ m/s}. \quad (10)$$

In this study, the ER was manipulated by varying the air flowrate while the biomass feeding rate remained unchanged. The effects of ER on product yield are shown in Figure 3. Bio-gas and bio-liquid yields recorded an upward trend with increasing ER while the bio-char demonstrated the opposite.

The ER is an indicator of the quantity of oxygen supplied to the reactor and gasification temperature under autothermal operation [29]. Higher ER leads to higher gasification temperature, accelerates oxidation reactions and leads to enhanced product quality. Conversely, lower ER limits the amount of oxygen available for gasification reactions and therefore is not a favorable condition [31].

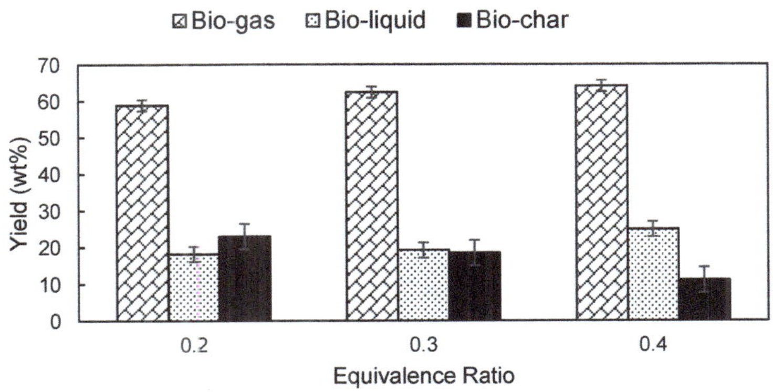

Figure 3. Effect of ER on the NG gasification product yield.

Figure 4 shows the gas composition at different ER. As ER increased from 0.20 to 0.40, H_2 and CH_4 contents in the producer gas decreased from 11.54 mol% to 8.29 mol% and 9.85 mol% to 7.09 mol% respectively. The effect of ER on gas composition is attributed to the oxidation reactions. Higher ER implies that more air (oxygen) is injected into the reactor, which in turn promotes combustion of CH_4 with O_2 while the CH_4 formation by methanation reaction is inhibited at high temperature [32]. Therefore, the volume fraction of CH_4 decreases as ER increases. Furthermore, Kuo, et al. [33] also reported that the fraction of H_2, CO and CH_4 decreases as ER increases. As the amount of oxygen supplied decreases with decreasing ER, the carbon converts to CO instead of CO_2 through oxidation and partial combustion reactions [30]. As the amount of CO increased, more CH_4 and H_2 are formed through methanation and water-gas shift reactions. Hence, higher concentration of CH_4 and H_2 are detected in producer gas at lower ER.

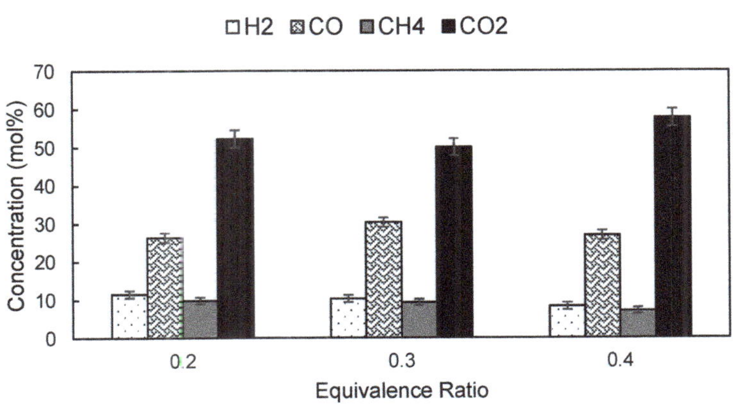

Figure 4. Effect of ER on the composition of producer gas produced from the gasification of Napier grass at a temperature of 850 °C.

At ER of 0.20, the concentration of CO and CO_2 were 26.36 mol% and 52.26 mol%, respectively. As ER increased to 0.30, the concentration of CO increased to 30.32 mol% and decreased to 26.94 mol%

as ER was further increased to 0.40. The observation is in good agreement with finding reported by Ghassemi and Shahsavan–Markadeh [34]. Meanwhile, the concentration of CO_2 decreased slightly to 50.04 mol% as ER increased to 0.3, and increased to 57.68 mol% as ER was further increased to 0.40. The phenomena could be deciphered by reversible water-gas shift and dry reforming reactions. Excess air would promote oxidation of bio-char and other combustible species and consequently leads to elevated CO_2 production.

3.5. Effects of Temperature on Product Yield and Quality of Producer Gas

In air gasification of biomass, gasification temperature is one of the crucial operating parameters, which is usually manipulated to investigate thermodynamic behavior of the reactions. Production of syngas consists of multi-step chemical reactions where temperature has significant impacts on the kinetics of the reactions involved. The influence of temperature on syngas production from the gasification of Napier grass was investigated at five different reactor temperatures (650 °C, 700 °C, 750 °C, 800 °C and 850 °C) while ER was fixed at 0.25.

The experimental results are shown in Figures 5 and 6. As observed in Figure 5, the bio-gas yield increased from 56.92 wt% to 67.56 wt% while the bio-char yield decreased from 27.40 wt% to 17.88 wt% with the rise of temperature, showing a divergent trend. The yield of bio-liquid showed an upward trend with increasing temperature and recorded a peak of 20.12 wt% at 800 °C. The yield subsequently declined as the temperature was further increased. A high operating temperature provides a conducive condition and supplies sufficient thermal energy for Boudouard, water-gas and methanation reactions, consuming more solid carbon to produce combustible gases. Furthermore, as the temperature increased from 650 °C to 800 °C, more H_2 was produced and reacted with O_2 to form water, and thus increased the yield of bio-liquid. As the temperature further increased to 850 °C, secondary reactions such as tar-cracking consume water and therefore reduce the yield of bio-liquid.

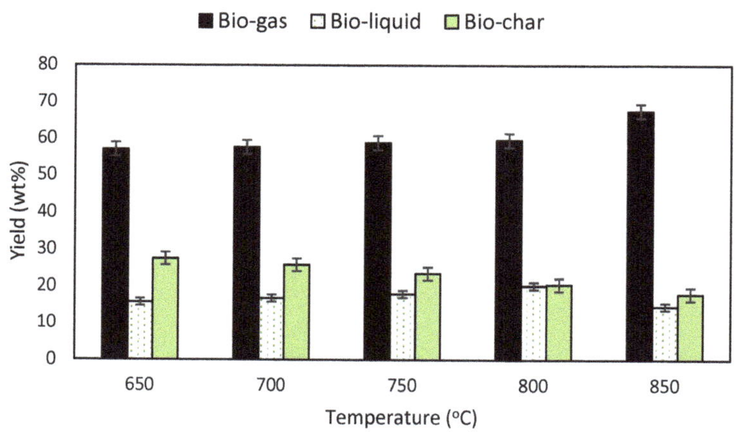

Figure 5. Effects of temperature on Napier grass gasification yield.

Figure 6 illustrates the effects of temperature on the composition of H_2, CO, CH_4 and CO_2 in the gases produced from gasification of NG at ER of 0.25. H_2 gas concentration increased remarkably from 7.31 mol% to 11.47 mol% as the operating temperature increased from 650 °C to 800 °C and dropped slightly to 10.27 mol% at 850 °C. The high operating temperature provides favorable conditions for endothermic reversible steam methane reforming, water-gas and dry reforming reactions. The production of hydrogen is enhanced as the heat absorbing reactions shift the equilibrium to the right [28].

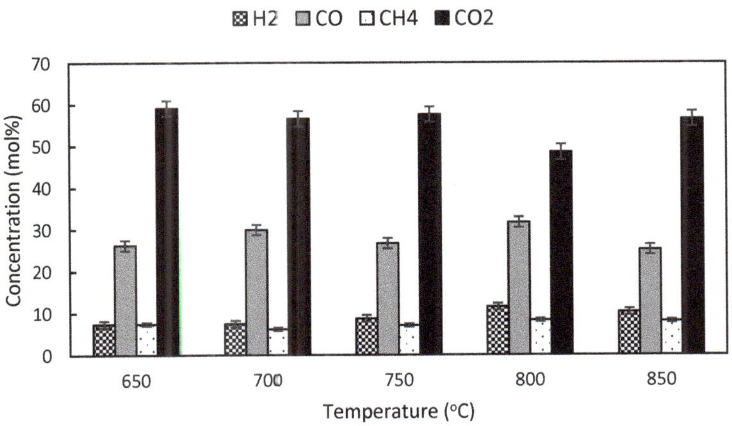

Figure 6. Effects of temperature on gas composition at ER of 0.25.

No distinctive trend can be observed for CO and CO_2 production. CO concentration increased from 26.23 mol% to 29.93 mol% as the temperature was increased from 650 °C to 700 °C. Subsequently the CO content decreased to 26.59 mol% at 750 °C and dropped further to 25.16 mol% at 850 °C. Within the gasifier operating temperature range of 650–850 °C, the recorded CO_2 concentration was within the range of 48.48–59.07 mol%. The pattern in CO_2 production at varying operating temperatures was opposite to CO. The high bed temperature stimulated complete combustion while deterring incomplete combustion processes, and accelerated solid carbon burning to produce CO_2 instead of CO.

3.6. Effects of ER on the Higher Heating Value (HHV) of Syngas and Carbon Conversion Efficiency (CCE)

The calculation of the gross calorific value or higher heating value (HHV) of Napier grass can be performed by using Equation (11) [35]. Carbon conversion efficiency (CCE) is one of the key indicators of gasification performance, which provides the information pertaining to the degree of reaction completion. CCE can be determined from carbon element content in biomass feedstock and syngas composition, and can be calculated by using Equation (12) [36].

$$\mathrm{HHV\ (MJ/Nm^3)} = (\mathrm{CO\%} \times 3018 + \mathrm{H_2\%} \times 3052 + \mathrm{CH_4\%} \times 9500)(0.01 \times 4.1868), \tag{11}$$

where CO%, H_2% and CH_4% are molar fraction of syngas components.

$$\mathrm{CCE\ (\%)} = (1 - M_a/M_0) \times 100, \tag{12}$$

where M_a and M_0 are final and initial total mass of biomass respectively.

The use of air as a gasifying agent is cheap and widely practiced. However, the presence of abundant nitrogen in air dilutes the concentration of syngas and consequently reduces the syngas heating value. The values of producer gas HHV and biomass CCE at varying ER are summarized in Table 5. The calculation of HHV is performed for both with and without taking nitrogen content in the air into consideration. With the presence of nitrogen, the highest HHV was found at ER of 0.20, and the value decreased as ER was further increased. According to Lv, et al. [37], ER is more than just a measurement of oxygen supply. ER represents the border-line between combustion and gasification reactions in the gasification system. Jayathilake and Rudra [38] reported that lower ER values resulted in higher CH_4 and H_2 concentration. The higher amount of air supplied at higher ER promotes combustion of H_2 and CH_4 components in the syngas. In addition, CH_4 formation by methanation reaction is retarded at higher temperature that comes with higher ER. Consequently, the reduction of H_2 and CH_4 concentration in syngas at higher ER will directly reduce the HHV of the producer gas. The findings are in good agreement with research done by Sheth and Babu [39]. Syngas HHV under N_2 free condition increased slightly from 8.72 MJ/m^3 to 8.78 MJ/m^3 as ER increased from 0.20 to 0.30, and decreased to 7.28 MJ/m^3 with a further increase in the ER to 0.4. On the other hand, as shown in Table 5, higher CCE is achieved at higher ER. The increased amount of air supplied during gasification improves the combustion process and contact with solid carbon, thus enhances the carbon conversion rate.

The effects of varying gasification temperature on producer gas HHV and biomass CCE are presented in Table 6. The HHV appeared to be increasing with increasing temperature. A peak was recorded at a gasification temperature of 800 °C where HHV subsequently decreased with a further temperature rise. Gasification temperature had a significant effect on syngas composition, which directly influenced the HHV of producer gas. As reported by Wu, et al. [40], elevated CO_2 concentration from high temperature biomass combustion process dilutes the concentration of H_2 and CH_4, results in a lower calorific value. As depicted in Table 6, CCE demonstrates a positive correlation with gasification temperature. The highest CCE is recorded to be 82.12% at 850 °C. The findings are in good agreement with work reported by Lv, Xiong, Chang, Wu, Chen and Zhu [37]. As shown in Tables 5 and 6, the dilution of the syngas by nitrogen can degrade the HHV of the syngas by at least twofold. The syngas produced from gasification thermochemical conversion of biomass can be used to produce heat and electricity in the combined heat and power (CHP) system, internal combustion engines or other applications.

Table 5. Effects of ER on higher heating value (HHV) and carbon conversion efficiency (CCE) at a gasification temperature of 850 °C.

ER	0.20	0.30	0.40
HHV (MJ/kg)	3.37	2.68	1.99
HHV (N_2 free) (MJ/kg)	8.72	8.78	7.28
CCE (%)	77.04	81.00	89.08

Table 6. Effects of temperature on HHV and CCE.

Temperature (°C)	650	700	750	800	850
HHV (MJ/kg)	2.42	2.29	2.70	3.37	3.07
HHV (N_2 free) (MJ/kg)	7.19	7.43	7.81	8.95	8.51
CCE (%)	72.60	74.56	76.32	77.16	82.12

4. Conclusions

Napier grass energy crop demonstrates a good potential as a renewable solid biofuel. Its calorific value of 16.73 MJ/kg is comparable to other biomasses reported in literature. The high content of volatile matter in Napier grass is highly desirable as this feature promotes the gasification process. Thermochemical gasification of Napier grass produces syngas and value-added by-products such as bio-char and bio-liquid. The presence of a high amount of potassium in Napier grass ash might impose a further problem for the long-term operation. The dynamics of the reactions involved were observed as a significant variation in product yield and biogas components were recorded at varying ER and gasifier operating temperatures. There was a positive correlation between ER and gasification temperature with CCE. Enhancement of combustion process at elevated temperature and air supply produced CO_2 that degraded the syngas quality and resulted in low HHV. The highest HHV was recorded at a gasification temperature of 800 °C with ER of 0.2 and thus these conditions were determined as the optimum operational conditions for bench-scale gasification of Napier grass. The findings from this study were encompassed within the limited explored range and further research (scale-up) will be carried out as future development.

Author Contributions: Conceptualization, T.H.B. and W.A.W.A.K.G.; methodology, W.A.W.A.K.G. and D.K.S.N.; formal analysis, T.H.B.; investigation, T.H.B.; resources, D.K.S.N.; writing—original draft preparation, M.S.M.S. and T.H.B.; writing—review and editing, M.S.M.S. and W.A.W.A.K.G.; supervision, W.A.W.A.K.G. and S.A.H.; funding acquisition, W.A.W.A.K.G.

Funding: This research was funded by Ministry of Education Malaysia, grant number LRGS/2013/UKM/KPT.

Acknowledgments: Feedstocks provided by Crops for the Future Research Centre (CFFRC) are gratefully acknowledged.

Conflicts of Interest: The authors declare no conflict of interest.

References

1. Le Quéré, C.; Andrew, R.M.; Friedlingstein, P.; Sitch, S.; Hauck, J.; Pongratz, J.; Pickers, P.A.; Korsbakken, J.I.; Peters, G.P.; Canadell, J.G.; et al. Global Carbon Budget 2018. *Earth Syst. Sci. Data* **2018**, *10*, 2141–2194. [CrossRef]
2. Bujang, A.S.; Bern, C.J.; Brumm, T.J. Summary of energy demand and renewable energy policies in Malaysia. *Renew. Sustain. Energy Rev.* **2016**, *53*, 1459–1467. [CrossRef]
3. Rebitanim, N.Z.; Wan Ab Karim Ghani, W.A.; Rebitanim, N.A.; Amran Mohd Salleh, M. Potential applications of wastes from energy generation particularly biochar in Malaysia. *Renew. Sustain. Energy Rev.* **2013**, *21*, 694–702. [CrossRef]
4. Energy Commission. *2017 Malaysia Energy Statistics Handbook*; Suruhanjaya Tenaga (Energy Commission): Putrajaya, Malaysia, 2017.
5. Mohammed, M.A.A.; Salmiaton, A.; Wan Azlina, W.A.K.G.; Mohammad Amran, M.S.; Fakhru'l-Razi, A.; Taufiq-Yap, Y.H. Hydrogen rich gas from oil palm biomass as a potential source of renewable energy in Malaysia. *Renew. Sustain. Energy Rev.* **2011**, *15*, 1258–1270. [CrossRef]
6. Alipour Moghadam, R.; Yusup, S.; Azlina, W.; Nehzati, S.; Tavasoli, A. Investigation on syngas production via biomass conversion through the integration of pyrolysis and air–steam gasification processes. *Energy Convers Manag.* **2014**, *87*, 670–675. [CrossRef]
7. Mallick, D.; Mahanta, P.; Moholkar, V.S. Co-gasification of coal and biomass blends: Chemistry and engineering. *Fuel* **2017**, *204*, 106–128. [CrossRef]
8. Ismail, W.M.S.W.; Mohd Thaim, T.; Abdul Rasid, R. Biomass gasification of oil palm fronds (OPF) and Koompassia malaccensis (Kempas) in an entrained flow gasifier: A performance study. *Biomass Bioenergy* **2019**, *124*, 83–87. [CrossRef]
9. Chan, Y.H.; Quitain, A.T.; Yusup, S.; Uemura, Y.; Sasaki, M.; Kida, T. Liquefaction of palm kernel shell in sub- and supercritical water for bio-oil production. *J. Energy Inst.* **2018**, *91*, 721–732. [CrossRef]
10. Hlavsová, A.; Corsaro, A.; Raclavská, H.; Juchelková, D.; Škrobánková, H.; Frydrych, J. Syngas Production from Pyrolysis of Nine Composts Obtained from Nonhybrid and Hybrid Perennial Grasses. *Sci. World J.* **2014**, *2014*, 11. [CrossRef]
11. Suntivarakorn, R.; Treedet, W.; Singbua, P.; Teeramaetawat, N. Fast pyrolysis from Napier grass for pyrolysis oil production by using circulating Fluidized Bed Reactor: Improvement of pyrolysis system and production cost. *Energy Rep.* **2018**, *4*, 565–575. [CrossRef]
12. Strezov, V.; Evans, T.J.; Hayman, C. Thermal conversion of elephant grass (Pennisetum Purpureum Schum) to bio-gas, bio-oil and charcoal. *Bioresour. Technol.* **2008**, *99*, 8394–8399. [CrossRef] [PubMed]
13. Lee, M.-K.; Tsai, W.-T.; Tsai, Y.-L.; Lin, S.-H. Pyrolysis of napier grass in an induction-heating reactor. *J. Anal. Appl. Pyrol.* **2010**, *88*, 110–116. [CrossRef]
14. Farzad, S.; Mandegari, M.A.; Görgens, J.F. A critical review on biomass gasification, co-gasification, and their environmental assessments. *Biofuel Res. J.* **2016**, *3*, 483–495. [CrossRef]
15. Abdoulmoumine, N.; Kulkarni, A.; Adhikari, S. Effects of temperature and equivalence ratio on mass balance and energy analysis in loblolly pine oxygen gasification. *Energy Sci. Eng.* **2016**, *4*, 256–268. [CrossRef]
16. Mohammed, M.A.A.; Salmiaton, A.; Wan Azlina, W.A.K.G.; Mohamad Amran, M.S. Gasification of oil palm empty fruit bunches: A characterization and kinetic study. *Bioresour. Technol.* **2012**, *110*, 628–636. [CrossRef] [PubMed]
17. Ptasinski, K.J. Thermodynamic efficiency of biomass gasification and biofuels conversion. *Biofuel. Bioprod. Biorefin.* **2008**, *2*, 239–253. [CrossRef]

18. Monti, A.; Di Virgilio, N.; Venturi, G. Mineral composition and ash content of six major energy crops. *Biomass Bioenergy* **2008**, *32*, 216–223. [CrossRef]
19. Hosseini, M.; Dincer, I.; Rosen, M.A. Steam and air fed biomass gasification: Comparisons based on energy and exergy. *Int. J. Hydrog. Energy* **2012**, *37*, 16446–16452. [CrossRef]
20. Sher, F.; Pans, M.A.; Sun, C.; Snape, C.; Liu, H. Oxy-fuel combustion study of biomass fuels in a 20 kWth fluidized bed combustor. *Fuel* **2018**, *215*, 778–786. [CrossRef]
21. Karampinis, E.; Vamvuka, D.; Sfakiotakis, S.; Grammelis, P.; Itskos, G.; Kakaras, E. Comparative study of combustion properties of five energy crops and Greek lignite. *Energy Fuels* **2012**, *26*, 869–878. [CrossRef]
22. Ghani, W.A.W.A.K.; Mohd, A.; da Silva, G.; Bachmann, R.T.; Taufiq-Yap, Y.H.; Rashid, U.; Ala'a, H. Biochar production from waste rubber-wood-sawdust and its potential use in C sequestration: Chemical and physical characterization. *Ind. Crop. Prod.* **2013**, *44*, 18–24. [CrossRef]
23. Wan Ab Karim Ghani, W.; Moghadam, R.A.; Salleh, M.; Alias, A. Air gasification of agricultural waste in a fluidized bed gasifier: Hydrogen production performance. *Energies* **2009**, *2*, 258–268. [CrossRef]
24. González-Vázquez, M.P.; García, R.; Gil, M.V.; Pevida, C.; Rubiera, F. Unconventional biomass fuels for steam gasification: Kinetic analysis and effect of ash composition on reactivity. *Energy* **2018**, *155*, 426–437. [CrossRef]
25. Obernberger, I.; Thek, G. Physical characterisation and chemical composition of densified biomass fuels with regard to their combustion behaviour. *Biomass Bioenergy* **2004**, *27*, 653–669. [CrossRef]
26. Arvelakis, S.; Gehrmann, H.; Beckmann, M.; Koukios, E.G. Preliminary results on the ash behavior of peach stones during fluidized bed gasification: Evaluation of fractionation and leaching as pre-treatments. *Biomass Bioenergy* **2005**, *28*, 331–338. [CrossRef]
27. Cuenca, J.; Rodríguez, J.; Martín-Morales, M.; Sánchez-Roldán, Z.; Zamorano, M. Effects of olive residue biomass fly ash as filler in self-compacting concrete. *Constr. Build. Mater.* **2013**, *40*, 702–709. [CrossRef]
28. Lahijani, P.; Zainal, Z.A. Gasification of palm empty fruit bunch in a bubbling fluidized bed: A performance and agglomeration study. *Bioresour. Technol.* **2011**, *102*, 2068–2076. [CrossRef] [PubMed]
29. Ghani, W.; Alias, A.; Savory, R.; Cliffe, K. Co-combustion of agricultural residues with coal in a fluidised bed combustor. *Waste Manag.* **2009**, *29*, 767–773. [CrossRef]
30. Mahishi, M.R.; Goswami, D.Y. An experimental study of hydrogen production by gasification of biomass in the presence of a CO2 sorbent. *Int. J. Hydrog. Energy* **2007**, *32*, 2803–2808. [CrossRef]
31. Miccio, F.; Moersch, O.; Spliethoff, H.; Hein, K.R.G. Generation and conversion of carbonaceous fine particles during bubbling fluidised bed gasification of a biomass fuel. *Fuel* **1999**, *78*, 1473–1481. [CrossRef]
32. Guo, F.; Dong, Y.; Zhang, T.; Dong, L.; Guo, C.; Rao, Z. Experimental Study on Herb Residue Gasification in an Air-Blown Circulating Fluidized Bed Gasifier. *Ind. Eng. Chem. Res.* **2014**, *53*, 13264–13273. [CrossRef]
33. Kuo, J.-H.; Lin, C.-L.; Wey, M.-Y. Effect of agglomeration/defluidization on hydrogen generation during fluidized bed air gasification of modified biomass. *Int. J. Hydrog. Energy* **2012**, *37*, 1409–1417. [CrossRef]
34. Ghassemi, H.; Shahsavan-Markadeh, R. Effects of various operational parameters on biomass gasification process; a modified equilibrium model. *Energy Convers. Manag.* **2014**, *79*, 18–24. [CrossRef]
35. Xiao, R.; Zhang, M.; Jin, B.; Huang, Y.; Zhou, H. High-temperature air/steam-blown gasification of coal in a pressurized spout-fluid bed. *Energy Fuels* **2006**, *20*, 715–720. [CrossRef]
36. Hernández, J.J.; Aranda-Almansa, G.; Bula, A. Gasification of biomass wastes in an entrained flow gasifier: Effect of the particle size and the residence time. *Fuel Process. Technol.* **2010**, *91*, 681–692. [CrossRef]
37. Lv, P.M.; Xiong, Z.H.; Chang, J.; Wu, C.Z.; Chen, Y.; Zhu, J.X. An experimental study on biomass air–steam gasification in a fluidized bed. *Bioresour. Technol.* **2004**, *95*, 95–101. [CrossRef] [PubMed]
38. Jayathilake, R.; Rudra, S. Numerical and Experimental Investigation of Equivalence Ratio (ER) and Feedstock Particle Size on Birchwood Gasification. *Energies* **2017**, *10*, 1232. [CrossRef]
39. Sheth, P.N.; Babu, B.V. Experimental studies on producer gas generation from wood waste in a downdraft biomass gasifier. *Bioresour. Technol.* **2009**, *100*, 3127–3133. [CrossRef]
40. Wu, C.Z.; Yin, X.L.; Ma, L.L.; Zhou, Z.Q.; Chen, H.P. Operational characteristics of a 1.2-MW biomass gasification and power generation plant. *Biotechnol. Adv.* **2009**, *27*, 588–592. [CrossRef]

MDPI
St. Alban-Anlage 66
4052 Basel
Switzerland
Tel. +41 61 683 77 34
Fax +41 61 302 89 18
www.mdpi.com

Processes Editorial Office
E-mail: processes@mdpi.com
www.mdpi.com/journal/processes

www.ingramcontent.com/pod-product-compliance
Lightning Source LLC
Chambersburg PA
CBHW061518210926
51850CB00008B/1962
9783036577357